Sustainable Aviation

Series Editors

T. Hikmet Karakoc

Eskisehir Technical University, Faculty of Aeronautics and Astronautics, Eskisehir, Turkiye

Information Technology Research and Application Center, Istanbul Ticaret University, Istanbul, Turkiye

C. Ozgur Colpan, Department of Mechanical Engineering, Dokuz Eylül University, Buca, Izmir, Türkiye

Alper Dalkiran, School of Aviation, Süleyman Demirel University, Isparta, Türkiye

The Sustainable Aviation book series focuses on sustainability in aviation, considering all aspects of the field. The books are developed in partnership with the International Sustainable Aviation Research Society (SARES). They include contributed volumes comprising select contributions to international symposiums and conferences, monographs, and professional books focused on all aspects of sustainable aviation. The series aims at publishing state-of-the-art research and development in areas including, but not limited to:

- Green and renewable energy resources and aviation technologies
- Aircraft engine, control systems, production, storage, efficiency, and planning
- Exploring the potential of integrating renewables within airports
- Sustainable infrastructure development under a changing climate
- Training and awareness facilities with aviation sector and social levels
- Teaching and professional development in renewable energy technologies and sustainability

* * *

T. Hikmet Karakoc • Nadir Yilmaz •
Alper Dalkiran • Ali Haydar Ercan
Editors

New Achievements in Unmanned Systems

International Symposium on Unmanned Systems and the Defense Industry 2021

Springer

Editors
T. Hikmet Karakoc
Eskisehir Technical University
Faculty of Aeronautics and Astronautics
Eskisehir, Turkiye

Information Technology Research
and Application Center
Istanbul Ticaret University
Istanbul, Turkiye

Nadir Yilmaz
Mechanical Engineering
Howard University
Washington, DC, USA

Alper Dalkiran
School of Aviation
Süleyman Demirel University
Keciborlu, Isparta, Türkiye

Ali Haydar Ercan
Porsuk Vocational School
Eskisehir Technical University
Eskisehir, Eskisehir, Türkiye

ISSN 2730-7778 ISSN 2730-7786 (electronic)
Sustainable Aviation
ISBN 978-3-031-29932-2 ISBN 978-3-031-29933-9 (eBook)
https://doi.org/10.1007/978-3-031-29933-9

This Springer imprint is published by the registered company Springer Nature Switzerland AG
The registered company address is: Gewerbestrasse 11, 6330 Cham, Switzerland

Preface

In both military and civilian application fields, unmanned systems have qualified exceptional levels of growth. Fixed-wing aircraft, rotorcraft aircraft, and vertical takeoff and landing aircraft are employed for surveillance, exploration and rescue, fire detection, agricultural imaging, traffic monitoring, and other unnamed applications.

Understanding and discussing recent advances in air defense is essential, so it is good that unmanned systems have been used in recent wars. Also, it is vital to look at how these systems change over time. People sometimes mix up the ideas of unmanned systems and autonomy. Autonomy and automated expressions are not particularly beneficial for classifying advanced equipment, many of which perform at least some operations automatically. Autonomy can encompass several distinct machine functions, and the labels are not particularly useful in this regard. In the context of military hardware and software, the idea of autonomy can be broken down into several different categories. In this way, unmanned systems can be put into three groups: fully autonomous, fully autonomous with human supervision, and semi-autonomous.

In contrast, unmanned systems are gaining more capabilities as technology advances in the modern world. In addition, they have varied in their characteristics. A systematic and efficient methodology is required to create new UAV futures to focus on the optimal aim for the time constraints under consideration. Complexity in selecting alternative development objectives is ideally suited to analytical methodologies. Conferences and symposia are the finest venues for identifying fresh approaches to topics such as unmanned systems.

International Symposium on Unmanned Systems and Defense Industry (ISUDEF '21), an international and multi-disciplinary symposium, was held online on October 26–28, 2021, to address current issues in the field of aerospace and defense, including such topics as Autonomous Vehicles, Unmanned Aircraft Technologies, MRO, Avionics, and Radar Systems and Air Defense. We kindly invite academics, scientists, engineers, practitioners, policymakers, and students to attend ISUDEF'21 to share knowledge, demonstrate new technologies and breakthroughs, and debate the future direction, strategies, and goals in aviation sustainability. This conference featured keynote presentations by invited speakers and general papers in oral and poster sessions.

We would like to thank Springer's editorial team for their support towards the preparation of this book and the chapter authors and reviewers for their outstanding efforts.

On the other hand, we would like to give special thanks to the SARES Editorial office members for gathering these chapters, who are the heroes behind the veil of the stage. Hursit Degirmenci and Kemal Keles played a significant role in sharing the load and managing the chapters with Betul Ozcan, Betul Kacar, and Dilara Kılıc Patatur. Their efforts in the long run for symposium author communications, following the standards, are necessary to creating a proceedings book.

Editorial Assistants
Kemal Keleş
Eskisehir Technical University
Betül Özcan
Eskisehir Technical University
Betül Kaçar
Eskisehir Technical University
Dilara Kılıç Patatur
Eskisehir Technical University

Eskisehir, Turkiye — T. Hikmet Karakoc
Washington, DC, USA — Nadir Yilmaz
Keciborlu, Turkiye — Alper Dalkiran
Eskisehir, Turkiye — Ali Haydar Ercan

Contents

A Short Review on New Development Achievements and Market Opportunities in Unmanned Systems

Selcuk Ekici, Alper Dalkiran, and T. Hikmet Karakoc

Nomenclature

AFT	Autonomous Flying Taxi
AI	Artificial Intelligence
GCS	Ground Control Systems
HTS	Home Travel Survey
LiDAR	Light Detection and Ranging
MOOS	Mission-oriented Operating System
UAM	Urban Air Mobility
UAS	Unmanned Air System
UAVs	Unmanned Air Vehicles
UGV	Unmanned Ground Vehicles
UMVs	Unmanned Maritime Vehicles
WSN	Wireless Sensor Network

S. Ekici
Department of Aviation, Iğdır University, Iğdır, Turkiye

A. Dalkiran (✉)
School of Aviation, Suleyman Demirel University, Keciborlu, Turkiye
e-mail: alperdalkiran@sdu.edu.tr

T. H. Karakoc
Eskisehir Technical University, Faculty of Aeronautics and Astronautics, Eskisehir, Turkiye

Information Technology Research and Application Center, Istanbul Ticaret University, Istanbul, Turkiye
e-mail: hkarakoc@eskisehir.edu.tr

T. H. Karakoc et al. (eds.), *New Achievements in Unmanned Systems*, Sustainable Aviation, https://doi.org/10.1007/978-3-031-29933-9_1

1 Introduction

Unmanned systems are key robotics technologies that, in the broadest sense, do not involve human components unless they have missions attached to them, maybe operated remotely or autonomously, and carry out specified duties (Zhang et al., 2017). Exceptions to this rule include systems that require missions to them. The use of robotics technology is expanding into a wider variety of fields daily (Wu & Zhang, 2020). In recent years, there has been considerable interest in the investigation because of the successful uses of unmanned systems in hazardous and remote locations such as battlefields, outer space, the depths of the ocean, and other places like that (Wang & Liu, 2012). Deep space probes (Re et al., 2009), spacecraft, unmanned air vehicles (UAV) (Ji et al., 2019), unmanned ground vehicles (UGV) (Bonadies & Gadsden, 2019), unmanned ships (Eriksen & Lützen, 2022), unmanned submarines (Tocón et al., 2022), unmanned weapons (National Research Council, 2002), unmanned robotics (Ding et al., 2019), and unmanned sensors (Balestrieri et al., 2021) are all examples of unmanned systems. As already mentioned, the development of unmanned systems extends to many fields. In the military field, more and more studies are being done on unmanned aerial vehicle systems daily (Noorman, 2014). The utilization of unmanned aerial vehicles (UAVs) in military confrontation has recently increased, and emerging trends in military robotics have triggered debates about the future of war and the ethics of using autonomous technologies on the battlefield (Arkin, 2010).

On the other hand, being capable of moral thinking, autonomous robots would act more morally than human troops under the terrible circumstances of combat since they would not be affected by fear, exhaustion, rage, frustration, or sentiments of vengeance (Noorman, 2014). This has subsequently led to consistent attempts being made to improve the level of complexity of the platforms, their payloads, armaments, communications, and Ground Control Systems (GCS), all of which are jointly referred to as unmanned aerial systems (UAS). Other countries are noticing the phenomenon, and the market for military UAVs and research activities throughout the world has stepped up (Mohsan et al., 2022).

2 Market Opportunities and Developments

There is a growing tendency among developers to simplify and create control algorithms that allow unmanned ground vehicles to independently select a task, an operating condition, and a route. To be effective on today's battlefields, unmanned ground vehicles need to function both independently and as part of coordinated teams (Mathiassen et al., 2021). As a result, they'll need to be able to exchange information and function at least somewhat independently of one another if the telerobotic link should ever be severed. When threatened, these autonomous vehicles will need to enter a self-protection mode, using AI software and collaborative links

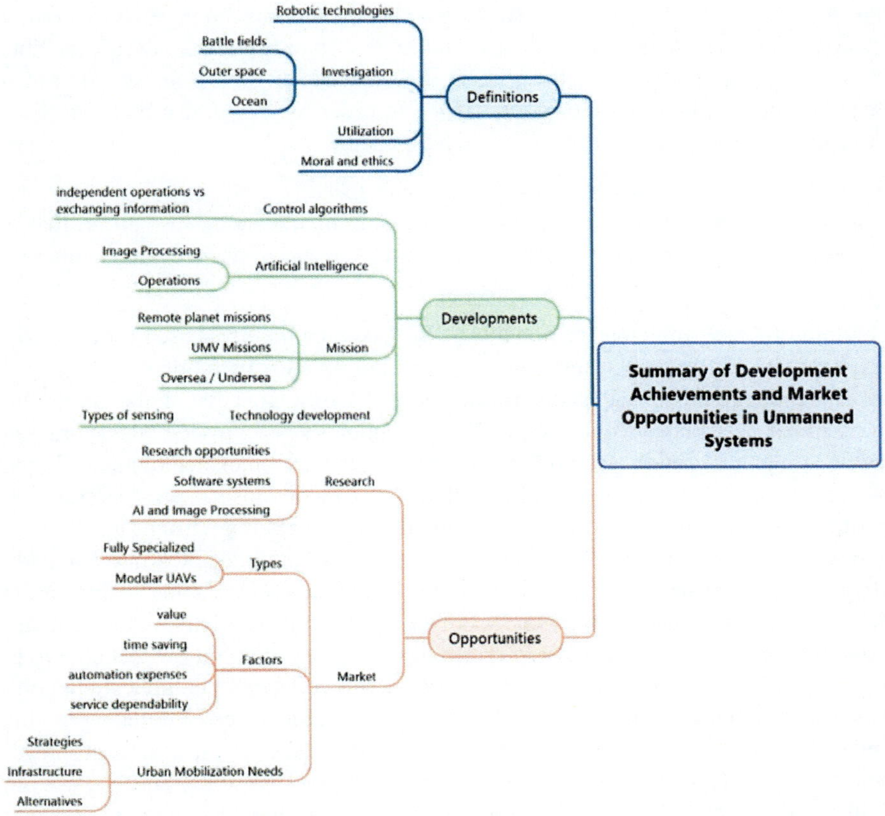

Fig. 1 Summary of development achievements and market opportunities in unmanned systems

with other robots on the battlefield to ensure their survival and the success of the mission (Blokhin et al., 2015). The connection between the unmanned ground systems is very important because unmanned ground vehicles such as mine clearing, firefighter, and rescue robots have wider coverage areas and have to use well-established connections (Aldabbas et al., 2016). Moreover, several recent studies have shown how useful Wireless Sensor Network (WSN) technology can be for massive remote monitoring, including space exploration and military surveillance. The information produced by sensor nodes must be processed and made available to users in a timely way (Aldabbas et al., 2016). The summary of development achievements and market opportunities have been grouped under three main groups the definitions, the developments, and the market in Fig. 1 to demonstrate the unmanned systems.

Upcoming space exploration missions will need unmanned robots that can access more difficult research objectives and travel farther per day than the existing Mars rovers (Gonzalez & Iagnemma, 2018). Along with a number of other enhancements, accurate slippage estimates and compensating mechanisms will play a crucial part in the realization of a navigation system that is both safer and more effective (Ding et al., 2011). The maximization of the scientific information gathered during the time that the rover is operational is the primary objective of any robotic mission designed for planetary exploration (Gonzalez, 2014). In order to accomplish that objective, two primary design concerns need to be addressed:

- To keep the rover moving at the highest safe speed
- To ensure that the rover is physically capable of traversing the terrain that is required to achieve the scientific aim that has been desired (Iagnemma & Dubowsky, 2011)

Both of these problems become more difficult when the rover used for planetary exploration must move through softer soils and/or up and down hills.

Helicopters, ships, and submarines are vital to the success of the navy, but Unmanned Maritime Vehicles (UMVs) technology is proving its worth in augmenting these forces (Abreu et al., 2016). Due to their quick deployment, simple scalability, and high reconfigurability—all promise reductions in operating time and cost,—MVs are rapidly showing their potential for upgrading current naval capabilities (Costanzi et al., 2020). When numerous UMVs are integrated into a quickly deployable and scalable system, it reduces the danger to workers while keeping them in the loop about important decisions. Due to the proliferation of non-standard solutions for the mission control interface of the UMVs, it might be challenging to operate UMVs as part of an organic structure when conducting complicated operational experiments (Carrera et al., 2016). Furthermore, the undersea domain presents major communication obstacles, such as multipath arrival structures, channel spread, and low data exchange rates. These challenges may be overcome, but they are not without difficulty (Doniec et al., 2010). The mission-oriented operating system, often known as MOOS, makes it possible for surveys conducted for commercial and scientific objectives to effectively use the technology provided by Unmanned Marine Vehicles (Barrera et al., 2021). MOOS makes use of cognitive analyses that are derived from machine learning and deep learning algorithms, and it enables the execution of missions of this kind using a supervisory control methodology (Dai et al., 2012).

Recent years have seen tremendous development in unmanned aircraft systems thanks to several advances in the fields of electronics, optics, computer science, and energy storage. The advancement of UAS technology on aircraft systems may be attributed in part to the enhancement of technologies such as Global Navigation Satellite Systems and Light Detection and Ranging (LiDAR) (González-Jorge et al., 2017). Remote sensing applications on unmanned aircraft systems are becoming increasingly popular day by day. Some of the important UAV remote-sensing applications can be categorized as follows (Mohsan et al., 2022):

- Climate change monitoring
- Forest inventory
- Wildfire fighting
- Water quality monitoring

- Precision agriculture spraying
- Topographic mapping
- Precision agriculture spraying
- Search and rescue
- Real-time monitoring of road traffic
- Security systems

These applications can be used in many different ways to develop technology, making it possible to develop systems. On the other hand, according to the findings of the many studies conducted on UAVs, the delegation of responsibilities to automated systems does not only replace human pilots. Armed unmanned aerial vehicles (UAVs) can automate various functions raditionally carried out by human actors (Fayyad et al., 2020). These duties include flight control and navigation between waypoints. The human pilot is responsible for keeping an eye on the autopilot to make sure it's doing its job at several stages of the flight. Even yet, armed UAVs still require ground-based support from a pilot, sensor operator, and mission intelligence coordinator (Ouma et al., 2011). In the last 10 years, there has been an increase in the construction and production of UAVs capable of planetary exploration (Pergola & Cipolla, 2016). UAVs can map a vast portion of the planet's surface and collect information from intelligent ecosystems. In addition, they have a higher resolution than previous satellites and orbiters (Petritoli & Leccese, 2021). Due to the fact that UAVs are remote-controlled spacecraft, they may have enough station time (Sharma et al., 2022).

Research opportunities and concerns vary among unmanned systems. We can see that the areas where there is the least potential for UAV-specific substantial innovation also have the highest need for study on a systemic basis (Chahl, 2015). Systems that defy categorization into traditional UAS categories are likely to provide the greatest return on investment. This clearly shows that more research and development needs to be done on unmanned aerial vehicle systems (Borghesan et al., 2022). In this context, research on software systems such as artificial intelligence and image processing on unmanned aerial vehicle systems development should be undertaken.

All those research efforts let market to grow in two main directions for aforementioned missions and civil purposes:

- Fully specialized solutions to make UAVs or other unmanned systems to be developed for specific target because of highest security concerns.
- Demountable and remountable modular UAV systems or other unmanned systems to be used for multiple solutions in an affordable and low-cost advantage (Tripolitsiotis et al., 2017).

The other side of the market is with the urban mobility problems (Al Haddad et al., 2020) grouped under Urban Air Mobility (UAM). The factors of mobility need that all had a significant amount of influence can be listed as:

- The value of time savings,
- The perception of expenses associated with automation.
- Service dependability.

3 Conclusions

Nevertheless, the market still has an X factor that represents an "unsure" behavior comparable to that of late adopters and non-adopters. The HTS (home travel survey) in Zurich is used to satisfy the need for urban mobilization by the chains of daily activities through statistical matching (Balac et al., 2019). This survey covers the entire nation. The findings indicate that only a very small percentage of consumers will have the option to use mobilization in the distant future. After key trade-offs between vehicle attributes, pricing structure, and processing times at improved infrastructure, such as the availability of take-off and landing stations, the demand for a UAM service in an urban region may be modified. The situation in larger cities with a high population density, extremely congested, and typified by low levels of public transit service can also be a game changer.

However, the UAM is a novel candidate to add a solution of patterns and travel time for travel inside the cities. The research is to solve the inter-traffic problems between the other modes of travel and air traffic problems by AI and image processing software solutions. The confluence of technology and new business models made available by the digitization and electrification of propulsion is making it feasible to study UAM as a new means for people to move about inside urban areas. This is making it possible to investigate UAM as a potential new way for people to commute. The transport modes preferences, most notably the adoption of Autonomous Flying Taxi (AFT) and UAM, by estimating the potential influence of service attributes that may affect people's choices among given transport alternatives and identifying the characteristics of the potential user groups with a higher propensity to accept (Fu et al., 2019) AFT and UAM services. The transport modes preferences and the adoption of AFT and UAM can be accomplished by determining the potential influence of service attributes which may affect people's choices among given transport alternatives.

References

Abreu, P., Antonelli, G., Arrichiello, F., Caffaz, A., Caiti, A., Casalino, G., et al. (2016). Widely scalable mobile underwater sonar technology: An overview of the H2020 WiMUST project. *Marine Technology Society Journal, 50*(4), 42–53. https://doi.org/10.4031/MTSJ.50.4.3

Al Haddad, C., Chaniotakis, E., Straubinger, A., Plötner, K., & Antoniou, C. (2020). Factors affecting the adoption and use of urban air mobility. *Transportation Research Part A: Policy and Practice, 132*, 696–712.

Aldabbas, O., Abuarqoub, A., Hammoudeh, M., Raza, U., & Bounceur, A. (2016). Unmanned ground vehicle for data collection in wireless sensor networks: Mobility-aware sink selection. *The Open Automation and Control Systems Journal, 8*(1), 35–46. https://doi.org/10.2174/1874444301608010035

Arkin, R. C. (2010). The case for ethical autonomy in unmanned systems. *Journal of Military Ethics, 9*(4), 332–341. https://doi.org/10.1080/15027570.2010.536402

Balac, M., Rothfeld, R. L., & Hörl, S. (2019). The prospects of on-demand urban air mobility in Zurich, Switzerland. In *2019 IEEE intelligent transportation systems conference (ITSC)* (pp. 906–913). IEEE.

Balestrieri, E., Daponte, P., de Vito, L., & Lamonaca, F. (2021). Sensors and measurements for unmanned systems: An overview. *Sensors (Basel, Switzerland), 21*(4). https://doi.org/10.3390/s21041518

Barrera, C., Padron Armas, I., Luis, F., Llinas, O., & Marichal, N. (2021). Trends and challenges in unmanned surface vehicles (USV): From survey to shipping. *TransNav, 15*(1), 135–142. https://doi.org/10.12716/1001.15.01.13

Blokhin, A., Koshurina, A., Krasheninnikov, M., & Dorofeev, R. (2015). The analytical review of the condition of heavy class military and dual-purpose unmanned ground vehicle. *MATEC Web of Conferences, 26*, 4002. https://doi.org/10.1051/matecconf/20152604002

Bonadies, S., & Gadsden, S. A. (2019). An overview of autonomous crop row navigation strategies for unmanned ground vehicles. *Engineering in Agriculture, Environment and Food, 12*(1), 24–31. https://doi.org/10.1016/j.eaef.2018.09.001

Borghesan, F., Zagorowska, M., & Mercangöz, M. (2022). Unmanned and autonomous systems: Future of automation in process and energy industries. *IFAC-PapersOnLine, 55*(7), 875–882. https://doi.org/10.1016/j.ifacol.2022.07.555

Carrera, A., Tremori, A., Caamaño, P., Been, R., Crespo Pereira, D., & Bruzzone, A. G. (2016). HLA interoperability for ROS-based autonomous systems. In J. Hodicky (Ed.), *Modelling and simulation for autonomous systems* (pp. 128–138). Springer.

Chahl, J. (2015). Unmanned aerial systems (UAS) research opportunities. *Aerospace, 2*(2), 189–202. https://doi.org/10.3390/aerospace2020189

Costanzi, R., Fenucci, D., Manzari, V., Micheli, M., Morlando, L., Terracciano, D., et al. (2020). Interoperability among unmanned maritime vehicles: Review and first in-field experimentation. *Frontiers in Robotics and AI, 7*, 91. https://doi.org/10.3389/frobt.2020.00091

Dai, S.-L., Wang, C., & Luo, F. (2012). Identification and learning control of ocean surface ship using neural networks. *IEEE Transactions on Industrial Informatics, 8*(4), 801–810. https://doi.org/10.1109/TII.2012.2205584

Ding, L., Gao, H., Deng, Z., Nagatani, K., & Yoshida, K. (2011). Experimental study and analysis on driving wheels' performance for planetary exploration rovers moving in deformable soil. *Journal of Terramechanics, 48*(1), 27–45. https://doi.org/10.1016/j.jterra.2010.08.001

Ding, X., Guo, P., Xu, K., & Yu, Y. (2019). A review of aerial manipulation of small-scale rotorcraft unmanned robotic systems. *Chinese Journal of Aeronautics, 32*(1), 200–214. https://doi.org/10.1016/j.cja.2018.05.012

Doniec, M., Detweiler, C., Vasilescu, I., & Rus, D. (2010). Using optical communication for remote underwater robot operation. In *2010 IEEE/RSJ international conference on intelligent robots and systems; 18.10.2010–22.10.2010*. IEEE.

Eriksen, S., & Lützen, M. (2022). The impact of redundancy on reliability in machinery systems on unmanned ships. *The WMU Journal of Maritime Affairs, 21*(2), 161–177. https://doi.org/10.1007/s13437-021-00259-7

Fayyad, J., Jaradat, M. A., Gruyer, D., & Najjaran, H. (2020). Deep learning sensor fusion for autonomous vehicle perception and localization: A review. *Sensors (Basel, Switzerland), 20*(15). https://doi.org/10.3390/s20154220

Fu, M., Rothfeld, R., & Antoniou, C. (2019). Exploring preferences for transportation modes in an urban air mobility environment: Munich case study. *Transportation Research Record, 2673*(10), 427–442.

Gonzalez, R. (2014). *Autonomous tracked robots in planar off-road conditions: Modelling, localization, and motion control* (1st ed.). Springer.

Gonzalez, R., & Iagnemma, K. (2018). Slippage estimation and compensation for planetary exploration rovers. State of the art and future challenges. *Journal of Field Robotics, 35*(4), 564–577. https://doi.org/10.1002/rob.21761

González-Jorge, H., Martínez-Sánchez, J., Bueno, M., & Arias aP. (2017). Unmanned aerial systems for civil applications: A review. *Drones, 1*(1), 2. https://doi.org/10.3390/drones1010002

Iagnemma, K., & Dubowsky, S. (2011). *Mobile robots in rough terrain: Estimation, motion planning, and control with application to planetary rovers*. Springer.

Ji, Z., Qin, J., Cheng, K., Liu, H., Zhang, S., & Dong, P. (2019). Thermodynamic analysis of a solid oxide fuel cell jet hybrid engine for long-endurance unmanned air vehicles. *Energy Conversion and Management, 183*, 50–64. https://doi.org/10.1016/j.enconman.2018.12.076

Mathiassen, K., Schneider, F. E., Bounker, P., Tiderko, A., de Cubber, G., Baksaas, M., et al. (2021). Demonstrating interoperability between unmanned ground systems and command and control systems. *IJIDSS, 6*(2), 100. https://doi.org/10.1504/IJIDSS.2021.115236

Mohsan, S. A. H., Khan, M. A., Noor, F., Ullah, I., & Alsharif, M. H. (2022). Towards the unmanned aerial vehicles (UAVs): A comprehensive review. *Drones, 6*(6), 147. https://doi.org/10.3390/drones6060147

National Research Council. (2002). *Technology development for army unmanned ground vehicles*. National Academies Press.

Noorman, M. (2014). Responsibility practices and unmanned military technologies. *Science and Engineering Ethics, 20*(3), 809–826. https://doi.org/10.1007/s11948-013-9484-x

Ouma, J. A., Chappelle, W. L., & Salinas, A. (2011). *Facets of Occupational Burnout among U.S. Air Force Active Duty and National Guard/Reserve MQ-1 Predator and MQ-9 Reaper Operators*.

Pergola, P., & Cipolla, V. (2016). Mission architecture for Mars exploration based on small satellites and planetary drones. *IJIUS, 4*(3), 142–162. https://doi.org/10.1108/IJIUS-12-2015-0014

Petritoli, E., & Leccese, F. (2021). Unmanned autogyro for Mars exploration: A preliminary study. *Drones, 5*(2), 53. https://doi.org/10.3390/drones5020053

Re, E., Di Cintio, A., Busca, G., Giunta, D., & Sanchez, M. (2009). Novel time synchronization techniques for deep space probes. In *2009 IEEE international frequency control symposium joint with the 22nd European frequency and time forum; 20.04.2009–24.04.2009*. IEEE.

Sharma, M., Gupta, A., Gupta, S. K., Alsamhi, S. H., & Shvetsov, A. V. (2022). Survey on unmanned aerial vehicle for Mars exploration: Deployment use case. *Drones, 6*(1), 4. https://doi.org/10.3390/drones6010004

Tocón, A., Vásquez, C., & Vinces, L. (2022). Design of a hydrodynamic profile for an unmanned underwater device using numerical simulation. In Y. Iano, O. Saotome, G. L. Kemper Vásquez, C. Cotrim Pezzuto, R. Arthur, & G. Gomes de Oliveira (Eds.), *Proceedings of the 7th Brazilian technology symposium (BTSym'21)* (pp. 488–496). Springer.

Tripolitsiotis, A., Prokas, N., Kyritsis, S., Dollas, A., Papaefstathiou, I., & Partsinevelos, P. (2017). Dronesourcing: A modular, expandable multi-sensor UAV platform for combined, real-time environmental monitoring. *International Journal of Remote Sensing, 38*(8–10), 2757–2770.

Wang, Y., & Liu, J. (2012). Evaluation methods for the autonomy of unmanned systems. *Chinese Science Bulletin, 57*(26), 3409–3418. https://doi.org/10.1007/s11434-012-5183-2

Wu, C., & Zhang, T. (2020). Intelligent unmanned systems: Important achievements and applications of new generation artificial intelligence. *Frontiers of Information Technology & Electronic Engineering, 21*(5), 649–651. https://doi.org/10.1631/FITEE.2030000

Zhang, T., Li, Q., Zhang, C.-s., Liang, H.-w., Li, P., Wang, T.-m., et al. (2017). Current trends in the development of intelligent unmanned autonomous systems. *Frontiers of Information Technology & Electronic Engineering, 18*(1), 68–85. https://doi.org/10.1631/FITEE.1601650

Inference of Civil Infrastructure Vibrations Using Unmanned Aerial Vehicles

Abeer Jazzar and Utku Kale

1 Introduction

The structures' sustainability and durability are structural safety and health monitoring primary goals. The structures weak points are identified to take pre/post cautions for hazardous events such as earthquakes. Structural Health Monitoring (SHM) is an inspection method of the structures' dynamic behaviour in a rapid, remote, and automated way. Current procedures of analytical and experimental techniques for structure performance evaluation are based on a good foundation these techniques may involve some uncertainties due to hardware and human error. In addition to that, sensors-based SHM have high hardware, installation, and maintenance costs. It also requires precise installation which is time-consuming and costly.

Therefore, new SHM methods should be time- and cost-efficient. As acknowledged from previous research, recording the dynamic responses of the structure to understand its dynamics using sensors may provide more accurate results (Feng et al., 2015). Keeping dynamic response records provides a better model analysis of the structure and necessary arrangements can be made when noticing abnormal changes in the dynamic response.

However, the high cost of sensors installation and locations that are hard to access for maintenance are some of the main problems of this method. Previous studies demonstrated that in terms of vibration frequency, using a smartphone accelerometer provides 1% error for the new smartphone generation and 5% for the old generation smartphone. However, there are many problems with smartphones such as noise filtering, workable phone to structure couplings (Feng et al., 2015). Also, placing smartphones on structures like dams and bridges can be hard and dangerous for the

A. Jazzar (✉) · U. Kale
Budapest University of Technology and Economics, Budapest, Hungary
e-mail: abeer.jazzar@edu.bme.hu; ukale@vrht.bme.hu

T. H. Karakoc et al. (eds.), *New Achievements in Unmanned Systems*, Sustainable Aviation, https://doi.org/10.1007/978-3-031-29933-9_2

inspector to access. Therefore, using a UAV – also known as a drone – provides mobility and accessibility to hard-to-reach and inaccessible places, plus UAV has lower cost and doesn't require installation which makes using it a convenient solution (Ozer & Feng, 2019).

In the early stages of the study, many tests were conducted using smartphones to create a database to validate the vibration data obtained with the drone's help. After collecting data using a drone, data analysis was performed and finite element model and time history analysis were conducted to compare and validate the possibility of collecting vibration data using UAV sensors.

2 Method

This chapter aims to understand the structural behaviour under different conditions by using vibration data and test the possibility of using UAVs to obtain vibration data. To monitor the structural safety, durability, and degree of deformation, an accessible steel structure testbed bridge with concrete shell (roof and floor made of concrete) was chosen for the study analysis. Vibration data were collected for the bridge using smartphone sensors which provided a solid database for later comparison. A series of ambient vibration tests and dynamic loading experiments were conducted.

A finite element model (FEM) was created to implement the model analysis. Mode superposition method, a linear dynamic-response process that describes displacement patterns by measuring and overlaying free-vibration mode shapes. Mode superposition explains how a structure naturally displaces. Vibration data obtained using sensors were analysed using the finite element model. The natural frequencies were found by applying different load combinations. The natural frequency was then compared with the frequency obtained from the MATLAB test data analysis. A MATLAB modal was developed to analyse the dynamic response of the structure. To determine the frequency of the structure, an application called 'Seismometer' was used. This application stores data in the x-y-z-axis and time domain. After taking the measurements, data was transferred as a CSV file to MATLAB. Data analysis was performed using MATLAB software to observe the peak acceleration values of the structure. After converting acceleration measurements from the time domain to the graphs of Power Spectral Density (PSD) values versus frequency, the peak value was determined. PSD demonstrate how the structure vibrates under certain frequencies.

2.1 SHM System Using UAVs

This subsection introduces an SHM system concept using UAVs vibration sensor data measuring and updating. The following framework shows the system concept in which drones would have a schedule and structural point assigned to be monitored.

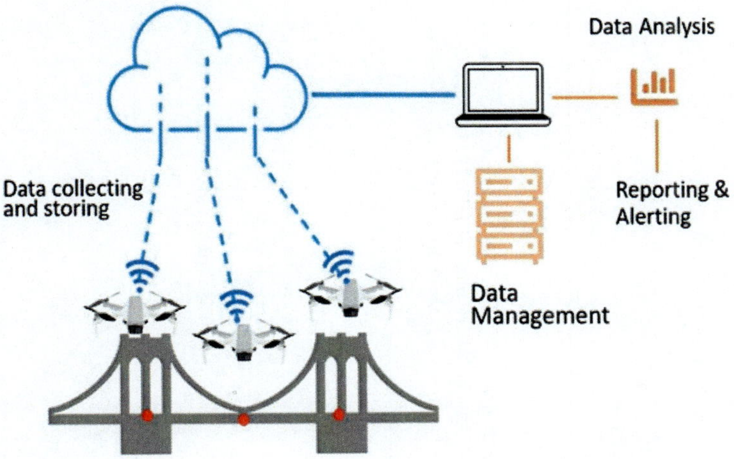

Fig. 1 SHM system concept using UAVs

So, without the need for inspectors to go to the field and remote control the drone, the drone will land on the predefined point and send the measuring data to the clouds which would create a database for each structure. This data is then transmitted to the SHM centre for further data analysis and reporting (Fig. 1).

2.2 *Test Design*

The most deformable parts of the bridge are the mid-span and the one-fourth of it. To take the data, UAV and smartphone were used. Taking the data from the structures has three main parts:

- Accurately landing the drone on the critical points of the structure
- Taking the data from the drone and analysing, filtering the drone mechanical system noise
- Validating the accuracy and reliability of the data

There were two options for obtaining the vibration data using the drone:

1. Using UAV's embedded accelerometer
2. Using the UAV as a carrier to a smartphone (Table 1)

Table 1 Summary list of experiments performed using smartphones and drone

	Sensor location		Applied force	Activity	Note
	(1/2 span)	(1/4 span)			
Smartphone	✓	✓	Two pedestrians, approximate weight is 1.0 KN	Pedestrians walking randomly, marching, and jumping	MPU6500 Accelerometer sensor, resolution 0.0011971008 m/s2, maximum range ± 19.6133 m/s2
	✓	✓	10+ pedestrians, approximate weight is 5.0 KN		
	✓	–	Ambient vibration (9 pedestrians passed during the test)	Empty bridge	
Drone	✓	–	4 pedestrians approximately 2.5 kN	Ambient, walking, jumping	Using iPhone 10 Accelerometer sensor (seismometer mobile application) by placing on the top of the drone
	✓	–	Two pedestrians, approximate weight is 1.0 KN		The drone landed on the midspan; the phone was attached to the bottom of the drone.

3 Results and Discussion

In the attached figures, time histories and PSDs of acceleration measurements are presented.

Based on an overview of the dataset, possible modal frequencies are 11.8, 12.8, 14.8, 33.7, and 38.8 Hz. Among these, 11.8 Hz possibly corresponds to the first mode. The second mode may not be visible looking at the dataset since the sensors were located at the mid span. It is important to note that vertical and torsional modes cannot be distinguished at this stage. Currently, these results are speculative, and it is essential to have tested with different sensor configurations and more advanced analysis to come up with precise estimations (Figs. 2 and 3).

In the previous tests, the system was the bridge itself. And from the bridge, the input spectrum was taken by placing the smartphone directly on the structure and the smartphones' embedded accelerometer gives the output spectra. In both of the (smartphone/drone) configurations, the mechanical system of the drone had to be

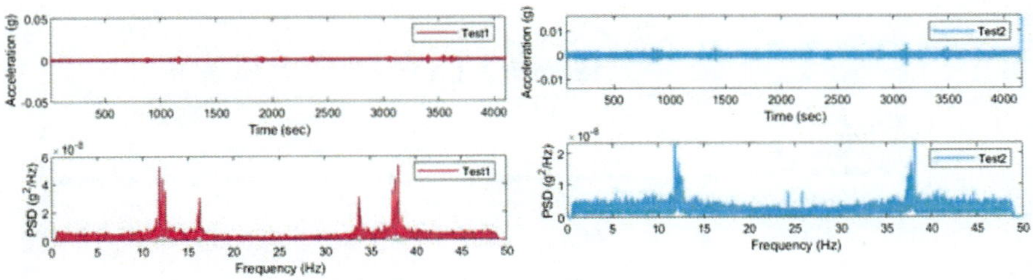

Fig. 2 Test 1 (smartphone sensor only), test 2 (smartphone attached to UAV)

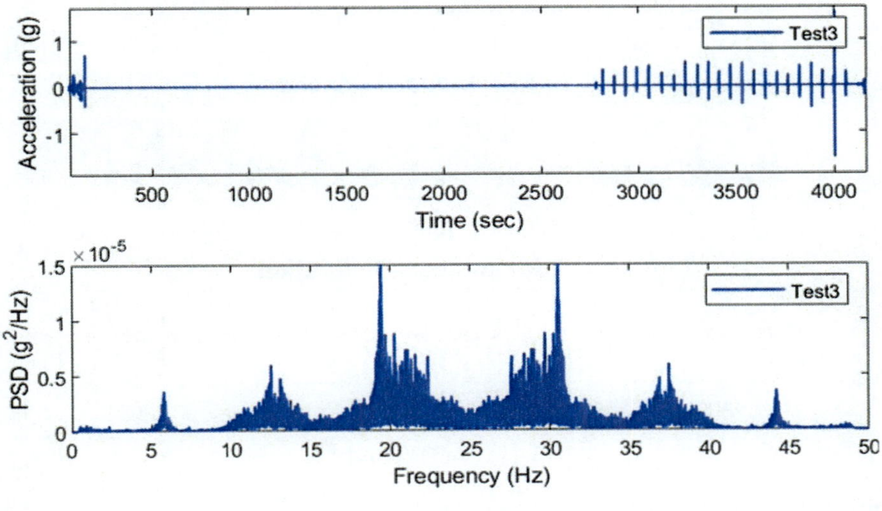

Fig. 3 Test 3 sensor data including the UAV mechanical system vibration

considered. For the configuration (smartphone/drone) the bridge was the input, the smartphones' sensor was the output, and the drone was the transfer system.

The tests were done to understand the mechanical system effects on the data. Processing vibration signals through the systems in series can be interpreted in terms of spectral changes in the frequency domain. According to this framework, smartphone sensor signals were a combination of the structural and the drone features. Eliminating drone features from the sensor data will result in structural features. Drone sensor features can be investigated by being measured while in a rigid zone.

The smartphone-based vibration measurements results are qualitatively compared with the data obtained from smartphone attached to the drone and SAP2000 modal frequencies. Looking at the results, a similar peak-to-peak order in the frequency domain is found. Results from time-domain analysis also validate the results obtained from the frequency domain. To cancel out the drone mechanical system effect following a transfer function development and conversion procedure, the system spectra can be used to eliminate drone-induced vibration content and produce pure vibration spectra (Fig. 4).

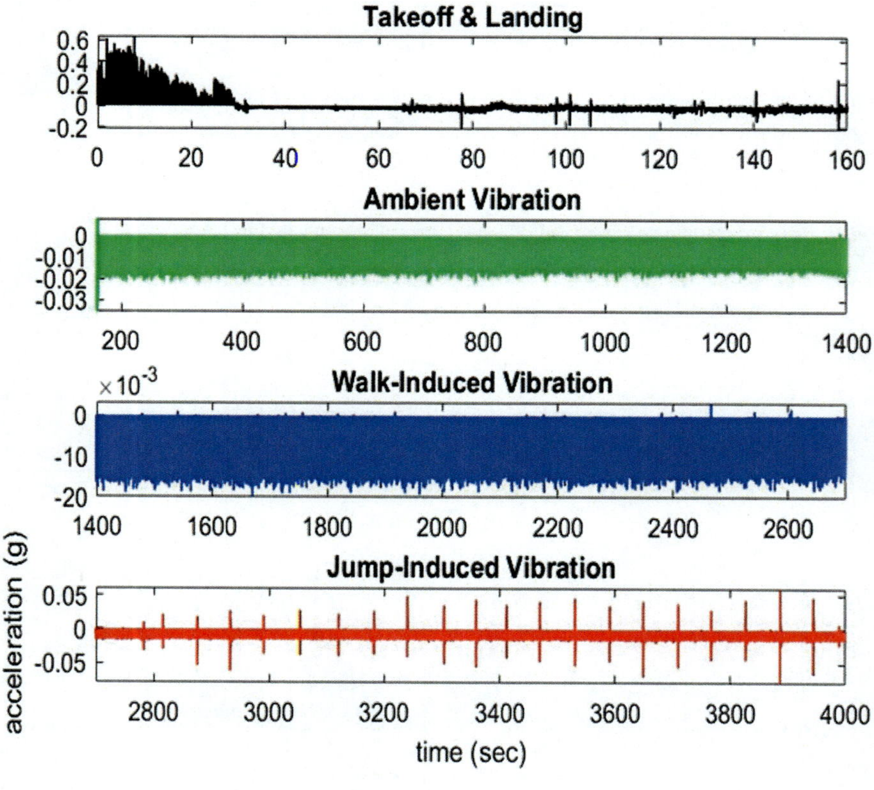

Fig. 4 Time history, power spectral density of each activity

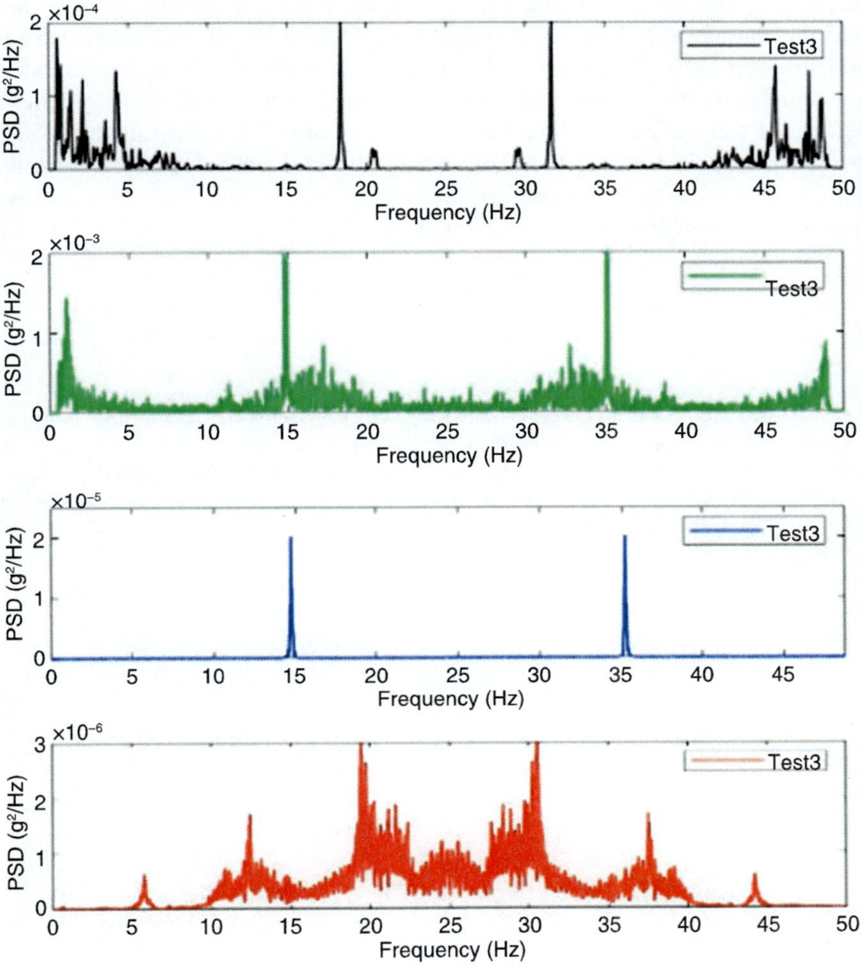

Fig. 4 (continued)

4 Conclusion

This study examined the potential of using vibration data obtained with the help of a UAV to identify the structural damage and reliability. It also illustrates a possible automated SHM system using UAVs, this would allow for a bigger database for high-risk structures mainly bridges. The vibrations measured by the smartphone on the top of the drone needs to be filtered from the drones' mechanical system which works as a transfer function. Thus, this study addresses this challenge by removing the UAV mechanical effect from the data to extract the bridge structural vibration and subsequently the structural modal parameter. Also, this study shows the possibility of finding precise, low cost, workable methods for sensor-based SHM. Two major realistic pedestrian mobility scenarios are taken into consideration such as walking and jumping. In real life, pedestrian activity is uncertain, hard to estimate, and time-dependent. Nevertheless, the two cases demonstrated throughout this study are foundations of a novel methodology moving from typical SHM systems to automated and suitable for future smart cities. Further advancements in this field are prospective in terms of how UAVs and advanced technologies are combined with structural health monitoring and eventually help in evaluating structural features in a sustainable and self-governing way.

References

Feng, M., et al. (2015). Citizen sensors for SHM: Use of accelerometer data from smartphones. *Sensors (Switzerland), 15*(2), 2980–2998. https://doi.org/10.3390/s150202980

Ozer, E., & Feng, M. Q. (2019). Structural reliability estimation with participatory sensing and mobile cyber-physical structural health monitoring systems. *Applied Sciences (Switzerland), 9*(14). https://doi.org/10.3390/app9142840

Electrical System Design for Very Light Aircraft

Merve Aluc and Guven Komurgoz

Nomenclature

VLA	Very light aircraft
LSA	Light sport aircraft
ULA	Ultralight aircraft
ELA	Electric load analysis
EASA	European Aviation Safety Agency
CS-VLA	Certification scheme for very light aircraft
SAE-AS	Society of Automotive Engineers
TAI	Turkish aerospace industry
ITU	Istanbul technical university
AH	Amper-Hour
ECU	Engine control unit

1 Introduction

Airplanes, which have become increasingly common since the first invention, were first used in the military arena during the First and Second World War. Along with the developing technology, many areas are being used today, especially transportation. Airplanes used in areas such as commercial and military area, hobby and training flights, and cargo transportation are classified according to various characteristics. These classification criteria depend on the number of wings, number of motors and types, the purpose of use, and weight. According to the purposes of use:

M. Aluc (✉) · G. Komurgoz
Department of Electrical Engineering, Istanbul Technical University, Istanbul, Turkey
e-mail: aluc@itu.edu.tr; komurgoz@itu.edu.tr

T. H. Karakoc et al. (eds.), *New Achievements in Unmanned Systems*, Sustainable Aviation, https://doi.org/10.1007/978-3-031-29933-9_3

passenger aircraft, transport aircraft, reconnaissance aircraft, training aircraft, etc. ("Aviation Writer", 2018). According to their weight: Vla (Very Light Aircraft), Lsa (Light Sport Aircraft), Ula (Ultra-Light Aircraft). These aircraft in the light aircraft category are much lower in weight than other types of aircraft. The maximum take-off weight of very light aircraft is between 600 and 750 kg. The maximum take-off weight of light sport aircraft is between 450 and 600 kg. The maximum take-off weight of ultralight aircraft is 300–450 kg.

The design of electrical systems for airplanes is usually based on the type of aircraft, the power it needs, the power source, and other features. The aircraft electrical power system provides electrical power to various other systems in the air (avionics, flight control, propulsion, fuel, lighting, etc.) and allows them to operate (Koç, n.d.).

In this study, the electrical system was designed for a single-motor, very light aircraft. The system architectures to be created, designs, electrical equipment to be selected according to the design, in-plane layout of these equipment, cables and connectors to be selected, lights were selected. The EASA CS-VLA (European Aviation Safety Agency Certification Scheme for Very Light Aircraft) Standards (EASA CS-VLA, n.d.) were applied. First, a motor type was selected for a light aircraft (which will meet its power needs). A generator was chosen to meet the electrical power requirement of the selected aircraft. According to the output power of this generator, the electrical equipment was selected and possible connections with each other will be made to create a suitable system architecture. Then an electrical load analysis was done according to the current and power values of all the electrical equipment in the aircraft considering the flight phases to be determined. At the end of the electrical load analysis (ELA), the duration of the battery resistance was calculated and a suitable battery was selected for the analysis result. According to SAE-AS 50881 Cable Standards (SAE-AS, n.d.), cables of appropriate diameter, length, and number was selected between the equipment. Depending on the cable selected, the appropriate number and type of connectors was selected considering the layouts in the aircraft at the same time. The selection of lights to be found at the wing tips was done using similar aircraft searches and light catalogs. Later, drawings were made with drawing programs. Then, the in-flight equipment layouts (in coordination with other systems) can easily be done.

2 Electrical Systems

Aircraft electrical systems are divided into two subsystems according to the equipment used. These are electric power generation system and electric power distribution system (Koç, n.d.).

The electric power generation system is also divided into a primary power generation system and a secondary power generation system.

The primary power generation system is equipment that converts mechanical energy into electrical energy. According to the working logic, the generators are

Table 1 Electric power generator system

Primary power generation system	Alternator (generator)
Secondary power generation system	Battery

Table 2 Electric power distribution system

Primary power distribution system	Busbars
Secondary power distribution system	Protection circuits/circuit breakers

integrated into the engine of the aircraft. That is, the generator generates electric energy for the aircraft by driving the motor. Also, generators in aircraft systems generally have protection and control elements. It has protection elements such as reverse current protection and high/low voltage protection and control elements such as voltage regulation. The generator is the primary source of power for the aircraft. In other words, aircraft electrical systems can be fed by the generator as long as the engine is running. Here, the generator will convert the mechanical power of the motor to the electric power and transfer it to the system by rectifying and correcting it with the regulator rectifier. This will feed both avionics equipment and other electrical systems. There is a generator switch for the generator to be controlled by the pilot. This switch is usually located alongside the battery switch on the instrument panel located on the cockpit.

The electric power distribution system is divided into two parts as the primary power distribution system and the secondary power distribution system (protection circuits/circuit breakers) (Koç, n.d.).

Primary power distribution systems include the distribution of electrical power to the busbar and the elements that provide protection against a failure that may occur in important areas (power supplies or buses).

Secondary power distribution systems include the protection circuits in the transmission of electrical power to loads. These are basically circuit breakers (Koç, n.d.) (Tables 1 and 2).

2.1 *Electrical Equipment*

The electrical power requirements and the equipment of the electrical system vary according to the situation and size of the aircraft.

Electrical system equipment that must be found on a basic aircraft:

- Battery
- Alternator/generator
- Voltage regulator
- Master–switch (alternator/generator switch)
- Fuses/circuit breakers
- Circuit breaker switches

- Power distribution bars
- Ammeter
- Related electrical cables
- External power receptable
- Starter motor
- Starter (ignition) switch

3 Electrical Load Analysis

Electric load analysis must be done for the selection of battery of light aircraft. As a result of the calculations made in ELA, the battery capacity to be used in the aircraft is obtained. Electric load analysis (ELA) is the sum of the electrical charges in the electrical system at certain flight phases of the aircraft. ELA shows the listing of the respective power requirements of each electrical equipment (Civil Aviation Safety Authority, n.d.). This is done according to various flight phases or operation conditions of the aircraft. These operation conditions are:

- *G1:* Ground operation and loading: The time it takes to perform the ground tests before the aircraft engine runs.
- *G2:* Engine start: The time the aircraft engine started to work.
- *G3:* Taxi: Condition from the first movement of the aircraft to the start of its take-off and from its descent to the engine shut-off.
- *G4:* Take off & climb: Condition starting with the take-off and ending with the cruising.
- *G5:* Cruise: Condition during the flight of the plane.
- *G6:* Landing: Condition of transition from cruise to taxi.
- *G7:* Generators failure (emergency): Condition that occurs following a loss of all normal electrical generating power sources (generators) or another (Civil Aviation Safety Authority, n.d.).

ELA is performed according to the normal operating conditions of the equipment (5 s, 5 min, and continuous status). Range of equipment operating times:

5-s analysis: This analysis includes all loads that last longer than 0.3 s.
5-min analysis: This analysis includes all loads that last longer than 5 s.
Continuous Analysis: This analysis includes all loads that last more than 5 min ("Standard Guide," n.d.).

ELA is based on normal operating conditions (5 s, 5 min, and continuous status analysis) and flight phases, depending on the various systems of the aircraft and their equipment.

ELA also shows which equipment is used in which phase and how much current is passed through the equipment. ELA should be maintained for the life of the aircraft as it exhibits the electrical balance between equipment and normal operating conditions. Table 3 shows some parts of the load analysis made in this study. For

Table 3 Electrical load analysis

	Name	Operating duration	Nominal current (in ampere)			Operating conditions						
			5 s	5 min	Cont.	G1	G2	G3	G4	G5	G6	G7
Air conditioning system	Fan	Random	0,3	0,3	0,3		*	*	*	*	*	
Flight control system	Flap linear actuator	5 min	7	7	7	*	*	*	*	*	*	*
	Elevator trim servo	Con.	0.150	0.150	0.151	*	*	*	*	*	*	

example, in Table 3, the fan in the air conditioning system draws 0.3 amperes at all phases except G1 and G7 phases and in three cases (5 , 5 min, and continuous) under normal operating conditions. The current drawn by the fan in phases G2, G3, G4, G5, and G6 is indicated by the symbol (*) in Table 3.

4 Battery Analysis

In very light aircrafts, the battery has to supply the aircraft electrical system for about 30 min in the event of generator failures. This is an emergency for the aircrafts. And this emergency is shown in the ELA to determine the capacity and the duration of the battery to be used in an emergency. It is seen that the capacity of the battery is 20 AH and the endurance of the battery is 30,705 min when the battery analysis is made in Table 4.

4.1 Battery Duration Calculation Emergency

In very light aircraft, an emergency occurs for the aircraft when the generators fail. And in this emergency, the battery supplies the aircraft's electricity needs. The G7 phase, in the ELA represents the desired amount of current in the battery during emergency occurrences in the generator failure. These 5 min in the G7 phase and the current demand in the continuous state are used to calculate the battery resistance time. In the G7 phase, the battery endurance time is calculated with the current demands in the continuous state and in the 5 min. The symbols representing the inputs used in the battery calculations are given below.

A: 5 min Current demand in case of G7, generators failure (Amp)
B: Continuous current demand in case of G7, generators failure (Amp)
C: Battery capacity (AH)
D: Available battery capacity (derated with %72) (AH)
E: Battery endurance (min)

For the battery endurance time account, the appropriate battery capacity is first calculated. The appropriate battery capacity account, represented by symbol D, is

Table 4 Battery analysis

Battery analysis		
5 min current demand in case of G7, generators failure:	28,138	Amps
Continuous current demand in case of G7, generators failure:	28,138	Amps
Battery capacity:	20	AH
Available battery capacity (derated with %72):	14,4	AH
Battery endurance:	30,70580709	Minutes

based on 72% of the normal capacity of the battery. For this calculator, the normal battery capacity is multiplied by its 72% capacity and is rated at 14.4 AH. Then, the battery endurance time account will be made according to this 14.4 AH. A suitable battery capacity value is entered to find the battery resistance time approximately 30 min. This value is given as 20 AH as shown in Table 4 and is indicated by the C symbol. To find the battery resistance time approximately 30 min, first multiply the D symbol by 60 min. Then multiply by the symbol A (5 min Current demand in case of G7, generators failure (Amp)) for 5 min. And these two values obtained are subtracted from each other. The result is divided by symbol B (continuous current demand in case of G7, generators failure [Amp]) and added to it over 5 min. As a result, battery resistance time is 30,705 min.

5 Electrical System Architectures

System architecture was created by taking into consideration all the systems working with the electric power in the aircraft and electrical load analysis. There are two permanent magnet generators integrated into the motor, selected for very light aircraft. One of these generators is the generator A, and the other is the generator B. The generator A supplies the motor's electrical equipment and the generator B supplies the electrical equipment inside the aircraft. Figure 1 shows the electrical system architecture of Generator B. There are two buses in the system. One of them is the main bus and the other is the essential bus. Primary priority equipment is connected to main bus such as lights, displays, etc. Secondary priority equipment is connected to essential bus such as avionic equipment, flow pump, etc. The battery is connected to the essential bar via the battery contactor, the battery switch, and the battery circuit breaker. At the same time, an ammeter shunt is connected to indicate the charge and discharge status of the battery. The external power system has also been added to the architecture. The external power system is used as a backup battery on the ground. When the plane is on the ground, it is used to start the engine and to energize the equipment. External power system consists of external power plug, external power relay, and external power switch.

6 Results and Discussion

The battery has to supply the aircraft electrical system for about 30 min in the event of generator failures, in the light aircrafts. In Table 4, it is seen that the results of the battery analysis meet this situation. A similar analysis was applied for generators. With the help of this study, the situation of meeting the total electrical load demand of the generation capacity of the generators has been tested. In Table 5, the status of generators according to current demands, 5-s analysis, and 6 different phases has been examined, and it has been seen that they meet the demands.

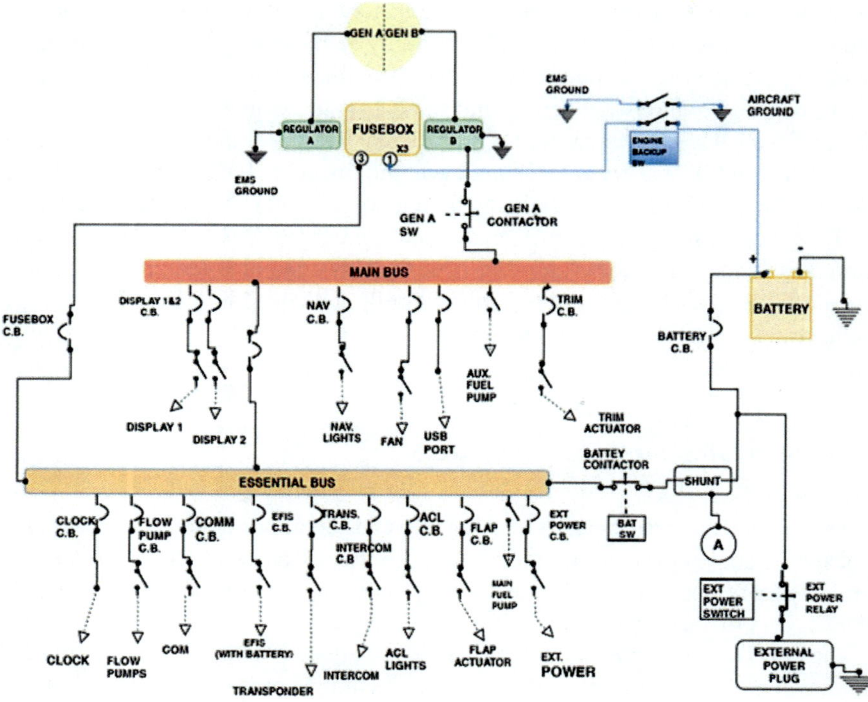

Fig. 1 Generator B electrical system architecture

Table 5 Total generator 5-s analysis

	G1	G2	G3	G4	G5	G6
Total current demand (amps.)	35,60	21,84	39,82	47,82	37,82	37,82
Current capacity (amps.)	92	92	92	92	92	92
Growth capacity (%)	64,57	76,26	56,72	48,02	58,89	58,89

7 Conclusion

In this study, the electrical system design of the very light aircraft has been made. This aircraft is designed as a single engine and training aircraft. Literature study was done during electrical system design and then electrical equipment was determined. Although the dimensions and quantities of these equipment vary according to the aircraft used, they are very similar in nature. These equipment (power supply elements and circuit protection and distribution elements) works in coordination to form the aircraft electrical system. Then, electric load analysis (ELA) was performed with the system architecture created. As a result of ELA, battery endurance time and battery capacity are determined. The aircraft electrical system contains many electrical equipment.

References

https://aviationwriter.wordpress.com/. Accessed 24 Jan 2018.

Koç, B. (n.d.). Uçak Elektrik Güç Sistemleri Teknolojileri ve Eğilimleri. Türk Havacılık ve Uzay Sanayii A.Ş. TUSAŞ, Ankara.

EASA CS-VLA (European Aviation Safety Agency Certification Scheme for Very Light Aircraft) Standards. (n.d.). European Aviation Safety Agency. https://www.easa.europa.eu/sites/default/files/dfu/decision_ED_2003_18_RM.pdf. Accessed 15 Apr 2018.

SAE-AS 50881 Cable Standards. (n.d.). https://www.sae.org/standards/content/as50881/. Accessed 13 Apr 2018.

Aircraft Electrical Load Analysis and Power Source Capacity. (n.d.). ADVISORY CIRCULAR AC 21-38 v2.0, Australian Government, Civil Aviation Safety Authority.

Standard Guide for Aircraft Electrical Load and Power Source Capacity Analysis1. (n.d.).

A Test-Bed for Attitude Determination and Control System of Nanosatellite

Aykut Kutlu, Demet Cilden-Guler, and Chingiz Hajiyev

1 Introduction

Attitude Determination and Control System (ADCS) algorithms can be simulated using a computer software but they need to be validated in a suitable test setup. The ground verification approaches are presented in many studies for the designed algorithms to be tested prior to launch of the spacecraft (Al-Majed & Alsuwaidan, 2009; Cardoso da Silva et al., 2016; Modenini et al., 2020; Schwartz et al., 2012; Tavakoli et al., 2017). The ADCS is one of the most challenging subsystems to test in necessary conditions, e.g. disturbance-free rotations. In this study, additional adjusting algorithms for hardware and software in the loop are also developed because of having laboratory conditions instead of space environment. The most commonly used simulator platform for spacecraft rotational dynamics is based on air-bearings (Schwartz et al., 2012). This study has also employed a testbed over an air-bearing table with low friction motion with unconstrained rotation over one and constrained motion in tilt angles.

The air-bearing platform is utilized for advancing the nanosatellite model by allowing a nearly torque-free rotational motion. The main torque disturbances are caused from the environment, platform, static and dynamic unbalance torques, and

A. Kutlu
Esen System Integration Ltd., Ankara, Turkey
e-mail: aykut.kutlu@esensi.com.tr

D. Cilden-Guler (✉)
Istanbul Technical University, Istanbul, Turkey
e-mail: cilden@itu.edu.tr

C. Hajiyev
Istanbul Technical University, Istanbul, Turkey
e-mail: cingiz@itu.edu.tr

T. H. Karakoc et al. (eds.), *New Achievements in Unmanned Systems*, Sustainable Aviation, https://doi.org/10.1007/978-3-031-29933-9_4

torques due to vibrations and electromagnetic interaction (Modenini et al., 2020). These torques can be eliminated in design stage or by active systems.

In the next section, parameters of the equipment in the test setup is given. Then, the structural design of the testbed is presented. The physical properties of the designed and manufactured system, the equipment that will act as sensors and actuators (reaction wheels), and the closed-loop control system outputs designed for the intended attitude maneuvers are given.

2 Design and Assembly of the Attitude Determination Testbed

In this section, the design and assembly of the test setup and the system architecture are given. Closed-loop control loops and Speedgoats hardware are used as the Real-Time Central Processing Unit Computer, which run all algorithms and manage the entire system. With this computer, which is suitable for rapid prototyping, the algorithms and models prepared in the Matlab Simulink environment are directly translated into C language via xPC Target and run on this computer in real time.

3-axis Inertial Measurement Unit (IMU) is used in order to make the attitude-related measurements of the system. This set of sensors is 3-axis angular accelerometer, 3-axis accelerometer, and a 3-axis magnetometer. Considering the sensitivity expected from the system and considering the lead times and costs, a MEMS-based sensor set is used. The equipment assembled on the platform is listed in Table 1. The solid model of the system within the scope of the attitude determination and control platform and the conceptual layout of the equipment are given in Fig. 1. The total mass of the octagonal test table is 22.6 kg with 70 cm outer circular diameter and 0.8 cm thickness.

The assembly of the equipment on the test table is seen in Fig. 2. The sensors/IMU, reaction wheels, and modem in the current picture are the equipment used in the main system. The computer is placed in the middle of the test table. There is also

Table 1 Test table equipment list

Equipment list	
Real-time computer	
(Central processing unit)	1 unit
Power supply	1 set
Battery	1 block
Sensor set (IMU)	1 unit
Reaction wheels and drivers	3 unit
RF modem	1 unit
Cabling	1 set
Fine balance masses (1 unit in each 3-axis)	3 set
Coarse balance masses (Z axis direction)	4-unit 1 set
Test table	1 unit

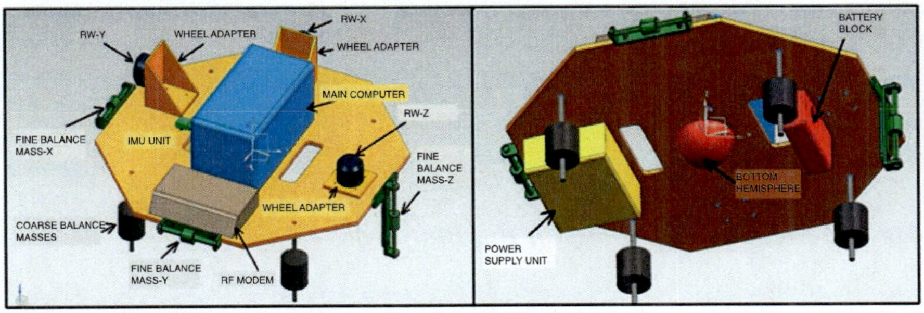

Fig. 1 Conceptual layout of the equipment placed on the platform

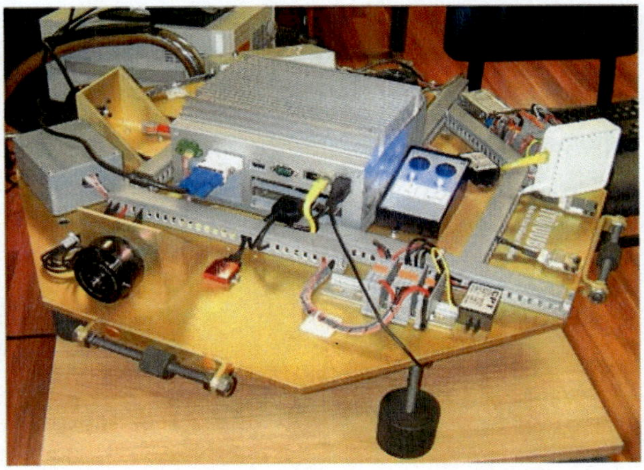

Fig. 2 Manufactured test table assembly

a battery block at the bottom surface of the table, and it is placed in the power distribution box.

3 Attitude Control System Software

The attitude control system model is created in Matlab/Simulink environment. The model that is created using Matlab Real Time XpC Target tool being translated into C language and used on the air-bearing main control computer.

The Matlab/Simulink-based Attitude Control System model consists of the modules listed as:

- Air-bearing table with dynamic and kinematic mathematical model
- Attitude control algorithm (simple PID control)
- Reaction wheel mathematical model (according to the parameters of the U7 motors)
- Attitude error calculator

The purpose of this control model is to demonstrate the maneuverability of the wheels to be used with the test table.

4 Results and Discussion

4.1 Maneuver Capability of the Platform

The analysis results of a maneuver performed on the 30-degree Roll and Pitch axes are shown in Figs. 3 and 4. The maximum speed of the wheels is determined as 9240 RPM, and the limit value is entered in the simulation environment. Control coefficients are adjusted for a 60-s maneuver so as to avoid long-term saturation of the wheels as much as possible. The initial speeds of the wheels are taken at 500 RPM. As a result, it is seen that the system goes to a steady state within 60 s and the maximum speed value for steady state on all three wheels is less than 1000 RPM. It is observed that the wheels remain within the specified torque and angular momentum capacity limits. Thus, they remained away from the risk of saturation for the next maneuver.

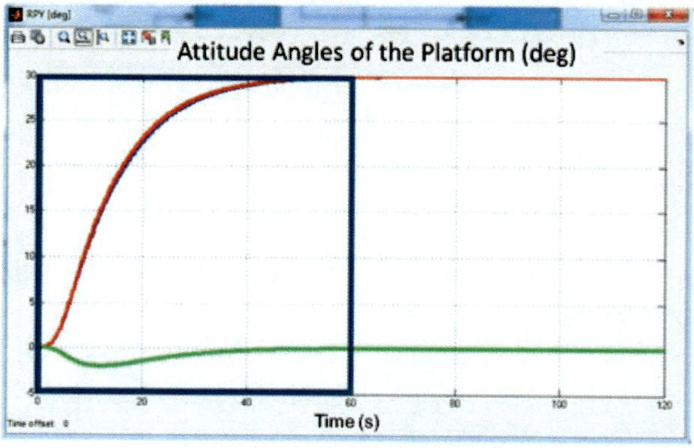

Fig. 3 Maneuver: air-bearing attitude angle values

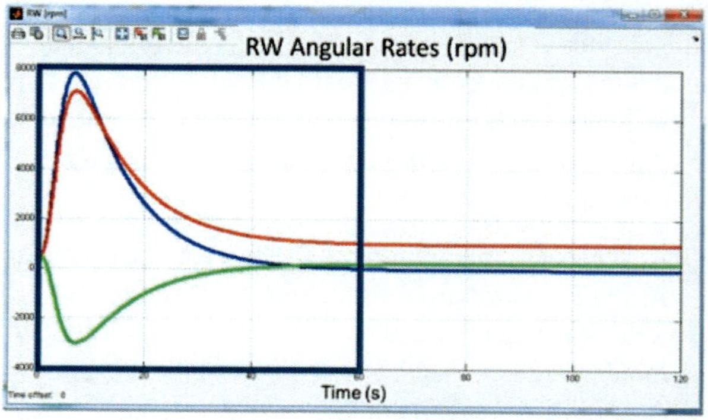

Fig. 4 Maneuver: reaction wheel angular velocity values

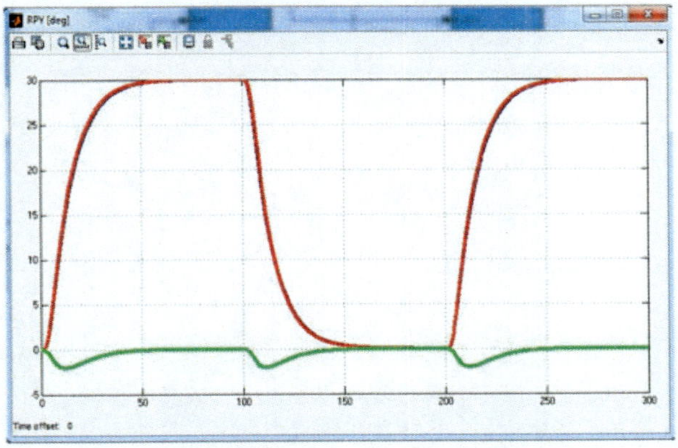

Fig. 5 Successive maneuver: air-bearing attitude angle values

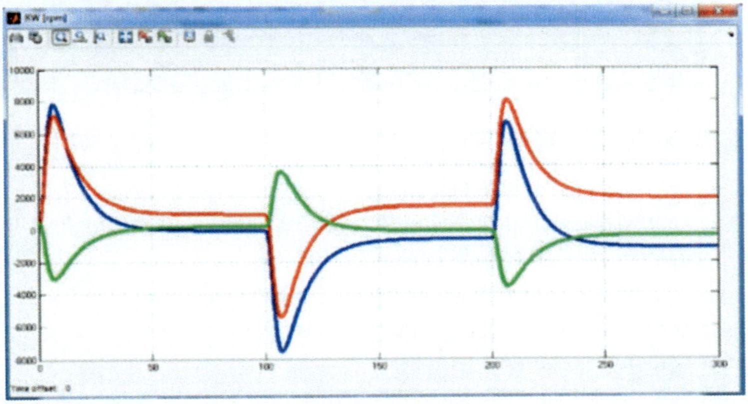

Fig. 6 Successive maneuver: reaction wheel angular velocity values

The same analysis is performed for successive maneuvers, and the saturation status of the wheels is checked. It is seen in the Figs. 5 and 6 that the wheels do not undergo saturation in these successive maneuvers as well.

4.2 Testing the Attitude Estimation and Magnetometer Calibration Algorithms

The attitude estimation algorithm is designed to obtain the angular information of the test table in three axes. This designed algorithm estimates Roll, Pitch, and Yaw angles by using angular velocity measurements in three axes and accelerometer measurements in three-axis. Angular velocity data are linear measurements that feed the estimation algorithm. Attitude information can be obtained from the accelerometer data. The angular motion in the horizontal plane is calculated by comparing the accelerometer data read from the sensor with the gravity vector.

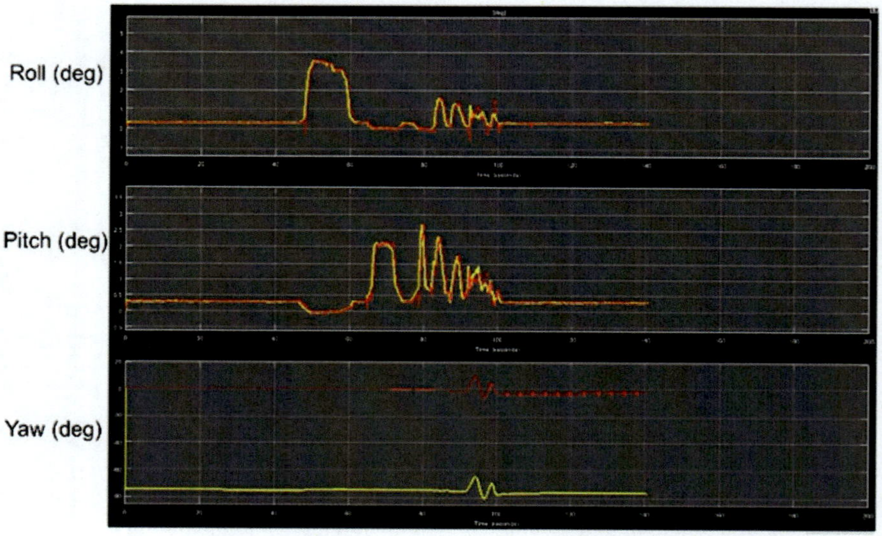

Fig. 7 Estimated angles of the test setup

Attitude estimation algorithm using the system model and sensor measurement models is designed. The algorithm uses the inputs of:

- Angular velocity estimated values in the previous step
- Attitude angle estimated values in the previous step
- Angular velocity measurements
- Angular velocity sensor error covariance value
- The offset value between the accelerometer position and the rotation center
- Accelerometer measurements
- Accelerometer error covariance value
- Moment of inertia matrix
- The offset value between rotation and the mass centers
- The total mass of the system
- Kalman estimation filter initial covariance matrices (tuning values)

As a result of the examinations on the test assembly, the results of the estimation filter are instantly taken in Fig. 7. The estimation filter is a conventional Kalman filter (Lefferts et al., 1982; Markley, 2003). The panels in the figure are representing the angles of roll, pitch, and yaw, respectively. The constant bias on the yaw angle can be calibrated easily. The theoretical magnetic field model outputs and magnetometer measurements from the test setup are obtained. By using these two vector results in the filter, the bias on the magnetometer measurements can also be estimated. Magnetometers can be calibrated by the bias estimation in the augmented states (Soken & Sakai, 2020; Zhang et al., 2015).

The magnetometer data of the test setup and the model-based sun sensor vector results are then used in the nontraditional filter SVD/EKF (Hajiyev & Cilden, 2017).

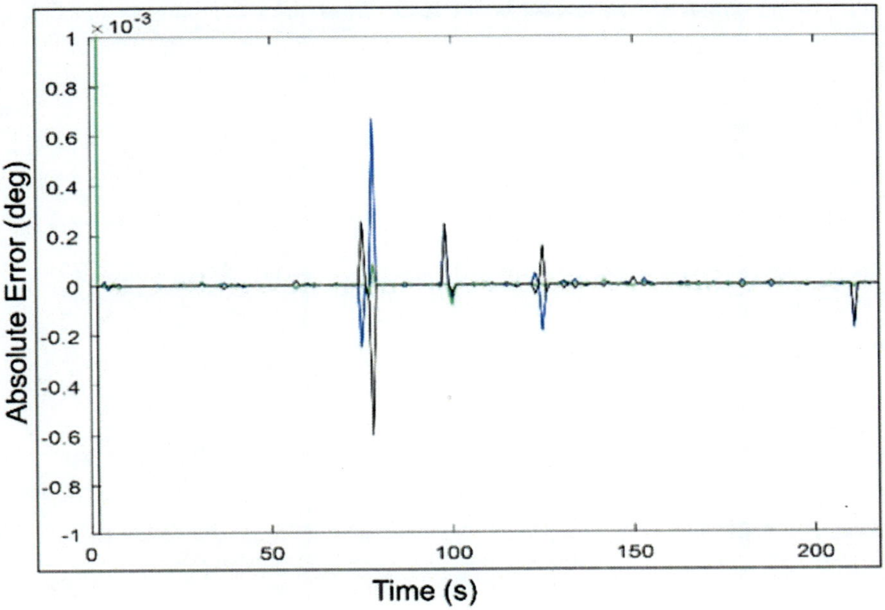

Fig. 8 SVD/EKF algorithm estimation

The estimation results are presented in Fig. 8 as roll, pitch, yaw angles along *x*, *y*, *z* axes, respectively. As seen in Fig. 9, the magnetometer bias is also estimated in a short period of time. The filter converges within 40 s.

Especially after 30 s, the errors that are approaching to zero can be seen in Table 2. The relative estimation errors in Table 2 are given as percentages.

5 Conclusion

Test table and the structural systems are designed. The sensors and actuators are selected after a trade-off study in designing the test setup. The physical properties of the platform are obtained (mass, moment of inertia) after the design. The algorithms designed for attitude estimation of the small satellites are verified using the test platform. For this purpose, traditional and nontraditional approaches of EKF are used.

Most of the small satellites in low orbit use magnetometers to determine their attitude angles. Whereas magnetometers have lightweight, reliable, and low power consumption, they are very convenient to be used as an attitude sensor in small satellites. Magnetometer bias caused errors must be overcome in order to determine the attitude correctly. Magnetometer measurements in the initial stages of satellite tasks are indispensable for attitude determination and control systems. In this study, Kalman filter-based algorithm is used to determine the magnetometer bias.

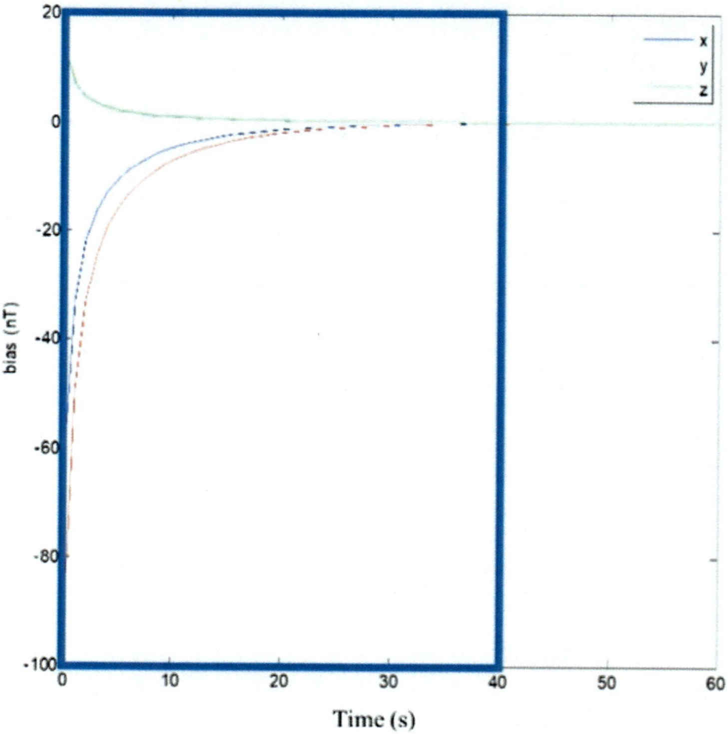

Fig. 9 The absolute error values of the bias estimation

Table 2 The relative estimation errors of the magnetometer bias

Time (s)	Relative error Rb_x (%)	Rb_y (%)	Rb_z (%)
10	−0.0027	−0.0027	0.0011
20	−0.7476e-3	−0.7476e-3	0.2990e-3
30	0.2226e-3	0.2226e-3	0.0890e-3
40	0.6674e-4	0.6674e-4	0.2669e-4
50	0.2002e-4	0.2002e-4	0.0801e-4
60	0.6004e-5	0.6004e-5	0.2402e-5

References

Al-Majed, M. I., & Alsuwaidan, B. N. (2009). A new testing platform for attitude determination and control subsystems: Design and applications. In *IEEE/ASME international conference on advanced intelligent mechatronics* (pp. 1318–1323). AIM.

Cardoso da Silva, R., Alves Rodrigues, U., Alves Borges, R., Sampaio, M., Beghelli, P., Gomes Paes da Costa, S., et al. (2016). A test-bed for attitude determination and control of spacecrafts. In *II Latin American CubeSat workshop* [Internet], Florianopolis, Brazil.

Hajiyev, C., & Cilden Guler, D. (2017). Review on gyroless attitude determination methods for small satellites. *Progress in Aerospace Sciences, 90*, 54–66.

Lefferts, E. J., Markley, F. L., & Shuster, M. D. (1982). Kalman filtering for spacecraft attitude estimation. *Journal of Guidance, Control, and Dynamics, 5*, 417–429.

Markley, F. L. (2003). Attitude error representations for Kalman filtering. *Journal of Guidance, Control, and Dynamics, 26*(2), 311–317. Available from: http://arc.aiaa.org/doi/10.2514/2.5048

Modenini, D., Bahu, A., Curzi, G., & Togni, A. (2020). A dynamic testbed for nanosatellites attitude verification. *Aerospace, 7*(3), 31. Multidisciplinary Digital Publishing Institute. Available from: https://www.mdpi.com/2226-4310/7/3/31/htm

Schwartz, J. L., Peck, M. A., & Hall, C. D. (2012). Historical review of air-bearing spacecraft simulators. *Journal of Guidance, Control, and Dynamics, 26*(4), 513–522. American Inst. Aeronautics and Astronautics Inc. Available from: https://arc.aiaa.org/doi/abs/10.2514/2.5085

Soken, H. E., & Sakai, S. (2020). Attitude estimation and magnetometer calibration using reconfigurable TRIAD+filtering approach. *Aerospace Science and Technology, 99*, 105754. Elsevier Masson SAS.

Tavakoli, A., Faghihinia, A., & Kalhor, A. (2017). An innovative test bed for verification of attitude control system. *IEEE Aerospace and Electronic Systems Magazine, 32*(6), 16–22. Institute of Electrical and Electronics Engineers Inc.

Zhang, Z., Xiong, J., & Jin, J. (2015). On-orbit real-time magnetometer bias determination for micro-satellites without attitude information. *Chinese Journal of Aeronautics, 28*(5), 1503–1509. Elsevier. Available from: https://www.sciencedirect.com/science/article/pii/S1000936115001430

Satellite Formation Flight via Thrusters and Proportional-Integral-Derivative Control Approaches

Tuncay Yunus Erkec and Chingiz Hajiyev

Nomenclature

KF	Kalman Filter
PID	Proportional-Integral-Derivative
ECI	Earth-centered Inertial
GPS	Global Positioning System
GNSS	Global Navigation Satellite System

1 Introduction

For adequate and appropriate design of satellite formation flight, user requirements must be carefully well understood and determined. User requirements can be sustained with the help of three basic elements.

- Design: Satellite relative navigation sensors and actuators
- Application: Number of stations
- Operation: Internal autonomous system level

In addition, for satellite mission subsystems and satellites in satellite formation architecture, precise and constant distances between satellites are needed to maintain

T. Y. Erkec (✉)
Turkish National Defence University, Hezarfen Aeronautics and Space Technologies Institute, Istanbul, Turkey
e-mail: terkec@msu.edu.tr

C. Hajiyev
Istanbul Technical University, Istanbul, Turkey
e-mail: cingiz@itu.edu.tr

T. H. Karakoc et al. (eds.), *New Achievements in Unmanned Systems*, Sustainable Aviation, https://doi.org/10.1007/978-3-031-29933-9_5

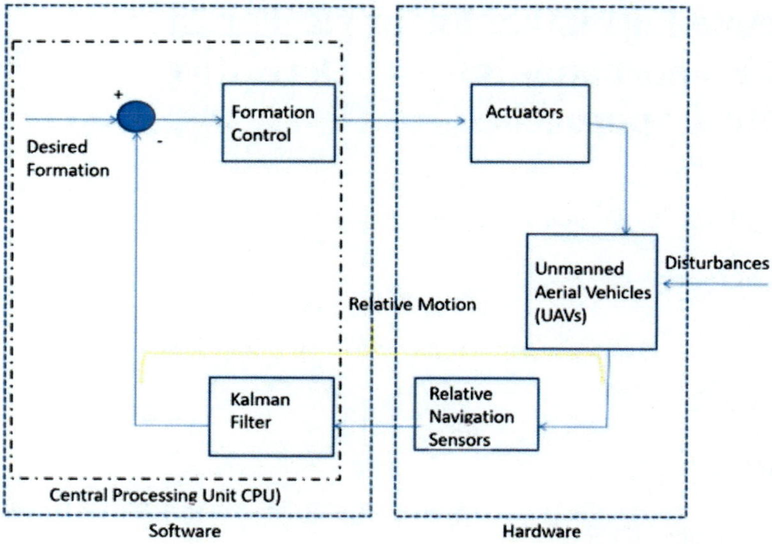

Fig. 1 Architecture of cluster satellite control. (Schilling, 2019; Erkec & Hajiyev, 2020)

formation geometry. The satellite and orbital information obtained later will not be valid and sufficient. Instead, continuous and simultaneous information acquired by sensors is required to establish the basis of 'real-time control'. In particular, for the positions of GNSS-based satellites within the formation, it may be necessary to establish an inter-satellite link for sensor and actuator control signals. As a result, there is no single general and valid approach required for satellite formation flight. Satellite formation flight control is a form of control in which performance is affected by the combination of information and control, availability and delay time, internal autonomous and ground station, precision and durability, and the preference ratio. The platform–sensor–control relationship in the satellite formation architecture is as shown in Fig. 1.

In this study, the relative vectors in the cluster satellite architecture are taken as an input to the control part of the estimations with different Kalman filters. According to the satellite thrusters used in the literature, the most suitable method was determined for the formation geometry preservation of the two satellites, which were compared with PID-based control algorithms.

2 Methodology

The target and follower satellite orbital positions determined by the Newton–Raphson model and EKF-based algorithm architecture stand out as the state estimation. The architecture created in this context is in two stages.

First stage:

Using the 'actual distance measurement model (Pseudo-ranging model)' approach with AKF/EKF based on GPS measurements, the target, and tracker satellite localization state parameters were roughly estimated.

Second stage:

Using the results of the first stage, the state vectors of the satellite relative motion model are estimated. Thus, the state vector estimates of the target and tracker satellites based on the AKF/EKF were used as the measurement values for the second stages AKF/EKF estimates of the position and velocity vectors of relative motion between satellites (Erkec & Hajiyev, 2021).

3 The Control Concept of Cluster Satellites Architecture

In the formation of satellite formation architecture, two phases are required, typically establishing and maintaining the formation.

The phase of providing the formation; within the operational concept, it is linked to how many launchers put satellites into orbit. Once the formation is formed, many different forces and accelerations slowly but continuously change the geometry and initial conditions. In accordance with the user's special requirements and needs, active control may be needed in the formation of the relative formation geometry according to the satellite relative navigation method to be selected.

In satellite manoeuvre planning, F is the thrust force, fuel flow m, specific thrust coefficient I_{sp} and fuel combustion injection rate and v_e are key terms associated with the custom-designed actuators. These terms are related to each other as in Eq. (1).

$$F = |\dot{m}|v_e = |\dot{m}|I_{sp}9.81\text{m/s}^2 \tag{1}$$

The actuators apply a certain acceleration a to the spacecraft, which increases the velocity Δv in the time interval Δt, which is the manoeuvre time, the application continues over time, the speed increase causes a position change Δr. The velocity change calculation is shown in Eq. (2).

$$\int_0^{t_0+\Delta t} a(t)dt = \int_0^{t_0+\Delta t} \frac{F(t)}{m(t)} = -\frac{F}{|\dot{m}|}\ln\left(1-\frac{|\dot{m}|\Delta t}{m_0}\right) \tag{2}$$

m_0 is the satellite mass at the start of the manoeuvre. In the equation on the right side of Eq. (2), constant fuel flow is assumed. Since the fuel consumed in many types of manoeuvres is much smaller than the satellite mass, the velocity change calculation is shown in Eq. (3).

$$\Delta v \approx -\frac{F\Delta t}{m_0} \tag{3}$$

$F\Delta t$ is the thrust transfer by the actuator.

3.1 PID Control Algorithm Designed in Satellite Formation Architecture

Each of the PID controller component parts is governed by a coefficient. These coefficients (K_p, K_d, K_i) take separate values for each system. The block diagram of the internal structure of the PID controller is given in Fig. 2. As can be seen from this block diagram, the structure of the PID controller consists of a combination of proportional gain, integrator, and derivative circuits (Coskun & Terzioğlu, 2007).

The output of the PID controller and the control law are expressed in Eq. (4) (Coskun & Terzioğlu, 2007). In the designed team satellite relative navigation architecture, the gains are made according to the manual tuning method. The control gain coefficients, K_p, K_i, K_d required for the close formation geometry preservation in each axis within the targeted 30-day period in the cluster satellite architecture, were determined by manual tuning method. While determining the gain coefficients, propellant properties, satellite properties, desired sensitivity, and response times were taken into consideration.

$$u_{(t)} = K_p e_{(t)} + K_i \int_0^t e_{(t)} dt + K_d \frac{d}{dt} e_{(t)} \tag{4}$$

It is expressed as in this equation, $e\_((t))$ specifies the error value. The error value is shown in Eq. (5).

$$e_{(t)} = R_t - b_t \tag{5}$$

In the constellation, relative navigation architecture designed in this study, the actual velocity value obtained from the relative satellite dynamic model is

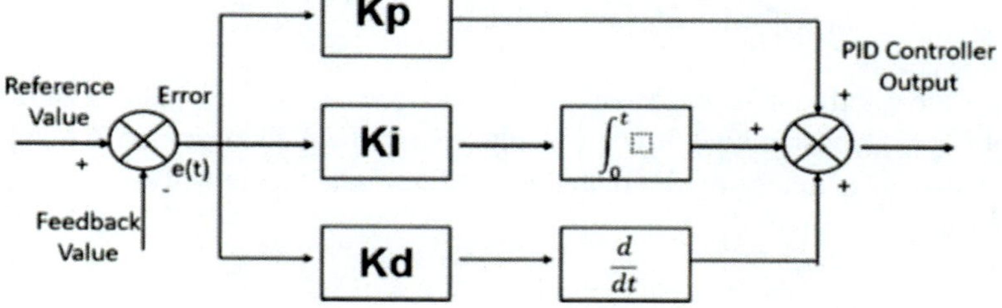

Fig. 2 Block diagram of the PID controller. (Coskun & Terzioğlu, 2007)

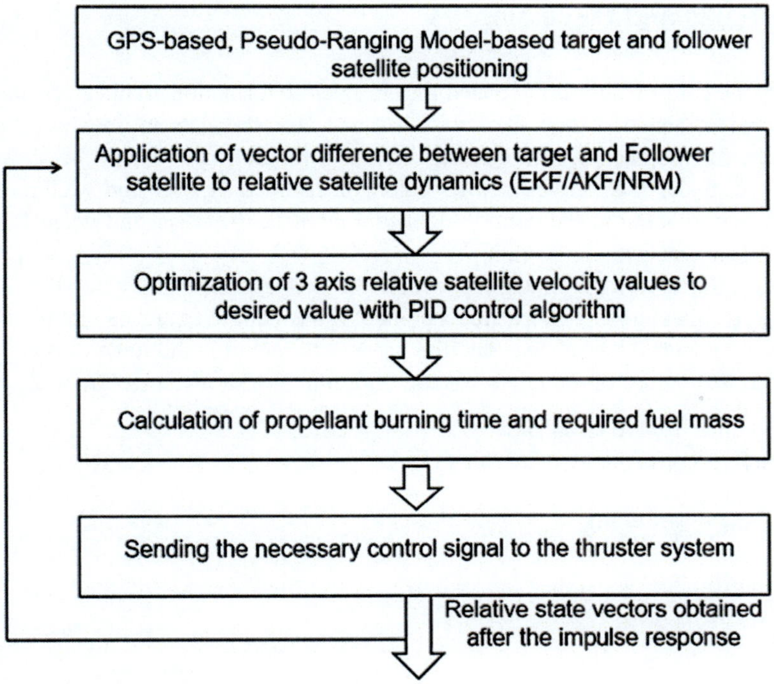

Fig. 3 The designed cluster satellite control algorithm diagram

considered as the reference value R_t. If the feedback value is b_t, the velocity outputs of the relative Kalman filter part included in the measurements are considered.

With the help of the control part outputs of the team satellite relative navigation architecture and Eqs. (4) and (5), the fuel mass and combustion time required for the targeted 30-day, close formation geometry preservation were calculated. The burning time supports the system frequency of 1 Hz and its repulsive properties.

The block diagram of the 1 Hz proportional-integral-derivative (PID) control algorithm for the propulsion system designed for the Follower Satellite (50 kg) is shown in Fig. 3. When the PRISMA mission implemented in the literature was examined, 1N thrusters were used for formation satellite containment. The control algorithms were evaluated for both 1N and 10N thrusters. When thrusters are evaluated, it is considered that 1N thrusters are more suitable for the designed satellite relative navigation mission.

After checking the Kalman filter outputs used for the relative satellite vector estimations at each iteration step, they are included in the AKF input again. After 100 s, PID control starts in the step where Kalman filter oscillation is balanced. The thrust control signal is sent to the reverse axis thruster for errors in satellite relative navigation state estimations.

4 Results and Discussion

Within the control algorithm created for the satellite formation architecture, the fuel mass and combustion time spent with 1N and 10N thrusters in obtaining the Δv required for maintaining the desired satellite formation parameters are shown in Figs. 4, 5, 6 and 7. Relative satellite vectors determined by EKF and AKF methods were taken as input to the control algorithm after the 100th second when Kalman filter values reached equilibrium.

When the algorithms designed within the scope of the study for the cluster satellite architecture enclosure are examined, it has come to the fore that the use of 1N thruster with AKF relative satellite vector estimation is the most cost-effective method. The values of the relative states algorithm are shown as follows:

- Rise time: 1.94163e-04 sn
- Block response Time: 0. 10805 sn
- Settling time: 0. 09735 sn
- Accuracy: 0.005 m/s

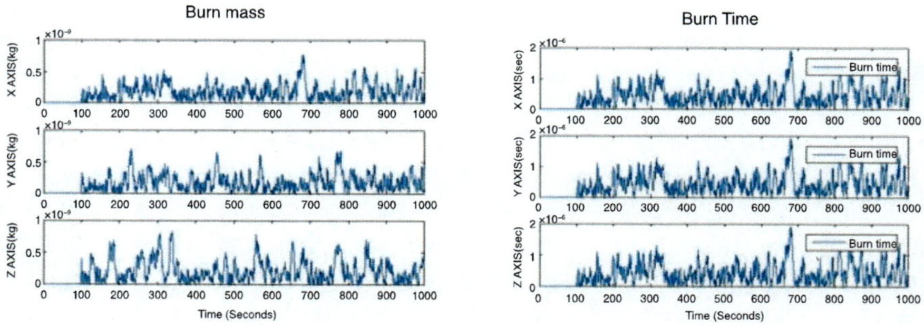

Fig. 4 1N thruster values when using AKF estimation algorithm

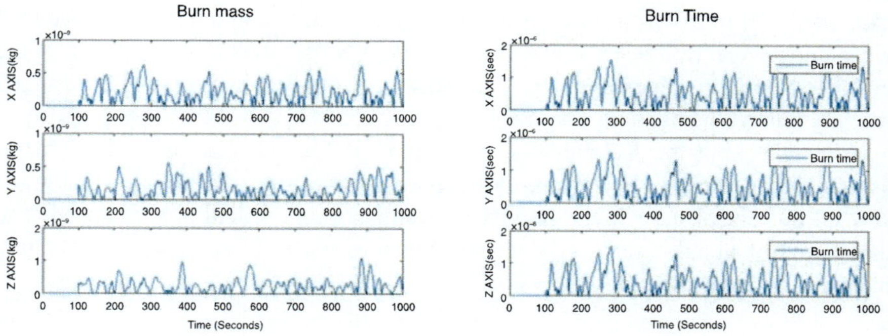

Fig. 5 1N thruster values when using EKF estimation algorithm

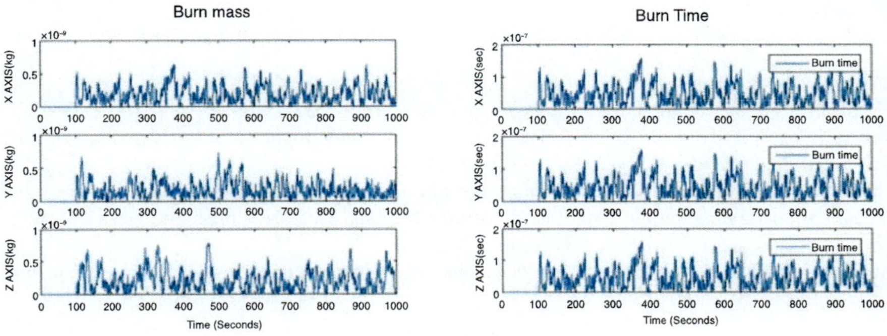

Fig. 6 10N thruster values when using AKF estimation algorithm

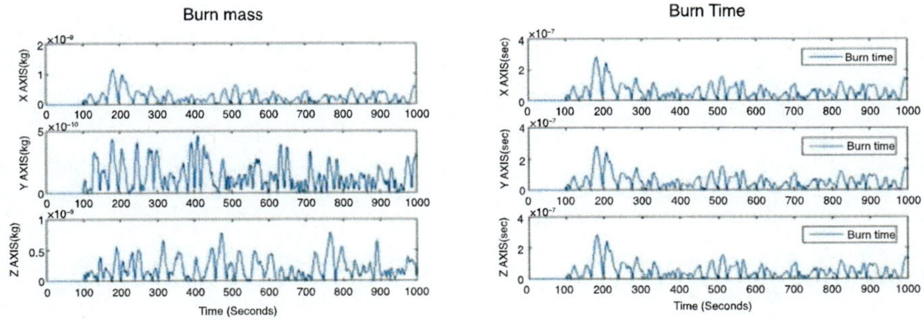

Fig. 7 10N thruster values when using EKF estimation algorithm

Within the scope of 1-month satellite formation architecture, the total Δv required for three-axis control is 75.278 m/s. However, approximately 6.5 kg of fuel is consumed in a 1-month formation satellite mission.

5 Conclusion

In the study, the control of formation satellite geometry vector values obtained by AKF and EKF relative vector estimation methods was made with PID approach. The designed control algorithm has been analysed within the scope of 1N and 10N thrusters.

The burn mass and burn time have been calculated for the 1-month mission period, and 3-axis precision satellite formation flight requirements have been analysed. In the relative satellite geometry enclosure, the AKF method has been estimated and the architecture using a 1N thruster emerged as the most appropriate approach.

Within the scope of the follower satellite propulsion system, ION thrusters were also used for experimental purposes. Despite its high Isp value and its low thrust value, studies are increasing in both theoretical and real missions for its use in the space environment, especially in formation and relative satellite flights.

References

Coskun, İ., & Terzioğlu, H. (2007). Hız performans eğrisi kullanılarak kazanç (PID) parametrelerinin belirlenmesi. *Selcuk-Technic, 6*(3), 180–205.

Erkec, T. Y., & Hajiyev, C. (2020). Relative navigation in UAV applications. *International Journal of Aviation Science and Technology, 1*(2), 52–65.

Erkec, T. Y., & Hajiyev, C. (2021). *Relative navigation of cluster satellites using kalman filters and global positioning system*. PhD thesis, Turkish National Defense University, Hezarfen Aeronautics and Space Tech. Institute, Istanbul, Turkey.

Schilling, K. (2019). *Mission analyses for low-earth-observation missions with spacecraft formation* (pp. 1–20). NATO.

Adaptive Kalman Filter-Based Sensor Fault Detection, Isolation, and Accommodation for B-747 Aircraft

Akan Guven and Chingiz Hajiyev

Nomenclature

OLKF	Optimal linear Kalman filter
AKF	Adaptive Kalman filter
AFEKF	Adaptive fading extended Kalman filter
MMNSF	Multiple measurement-noise scale factor

1 Introduction

Based on Kalman filtering technique, fault accommodation is variously analyzed in the literature. Sensor fault detection and isolation based on Kalman filtering technique (Gheorghe et al., 2013) is referenced in various studies. Kalman filter innovation sequence mean is an approach to detect and isolate aircraft sensor and control surface faults.

A faster converging Kalman filter for sensor fault detection and isolation is brought (Fang et al., 2018). The faults are detected and isolated in a shorter time with initialization of the covariance matrix and using an EKF compared to classical Kalman filter method. A modified AFEKF is suggested for satellite fault detection and isolation process which includes three stages and it differs fault detection and isolation, also a type of fault recognition.

Bias and noise increment fault cases are examined with the aim of detection and isolation of sensor faults for wind turbine systems (Zhang et al., 2018).

A. Guven (✉) · C. Hajiyev
Faculty of Aeronautics and Astronautics, Istanbul Technical University, Istanbul, Turkey
e-mail: akan.guven@itu.edu.tr; cingiz@itu.edu.tr

T. H. Karakoc et al. (eds.), *New Achievements in Unmanned Systems*, Sustainable Aviation, https://doi.org/10.1007/978-3-031-29933-9_6

Sensor fault detection and isolation is well performed in (Hajiyev & Caliskan, 2005) based on the statistical function of Kalman filter innovation sequence.

Single faulty sensor (sideslip angle, roll rate) and also in double fault sensor conditions are examined. An OKF is used to detect and isolate the sensor faults, and AKF is used for fault accommodation. The estimation results are investigated through the root mean square error values and AKF algorithm has better estimation results compared to the conventional OKF method as granted by having less error magnitude.

2 Method

The European languages are members of the same family. Their separate existence is a myth. For science, music, sport, etc., Europe uses the same vocabulary.

According to Newton' s second law of aircraft EoM:

Outer forces are thrust, gravity, aerodynamic forces and engine moments, aerodynamic moments are the external moments. Equations of external forces are shown in Eqs. (1), (2) and (3), and the equations of outer moments are shown in Eqs. (4), (5) and (6) with respect to (1) based on aerodynamic coefficients.

Equation of motion (EOM) is derived from the following assumptions:

- The aircraft is rigid title, author details, a short summary, keywords.
- The earth is flat and non-rotating.
- Constant aircraft mass during flight.
- Symmetry in the X_bZ_b plane which means the moment of inertia I_{xy} and I_{yz} are equal to zero. Symmetry in the X_bZ_b plane, which means the moment of inertia I_{xy} and I_{yz} are equal to zero.

$$F_x = \frac{1}{2}\rho V_t^2(-c_D\cos\alpha + c_L\sin\alpha) + F_{T_x} - mg\sin\theta \tag{1}$$

$$F_y = \frac{1}{2}\rho V_t^2 S c_Y + F_{T_y} - mg\cos\theta\sin\varphi \tag{2}$$

$$F_z = \frac{1}{2}V_t^2 S(-c_D\sin\alpha - c_L\cos\alpha) + F_{T_x} - mg\cos\theta\cos\varphi \tag{3}$$

$$M_x = \frac{1}{2}\rho V_t^2 Sbc_{lb} + M_{\text{engx}} \tag{4}$$

$$M_y = \frac{1}{2}\rho V_t^2 Sb\bar{c}c_{mb} + M_{\text{engy}} \tag{5}$$

$$M_z = \frac{1}{2}\rho V_t^2 Sb\bar{c}c_{nb} + M_{\text{engz}} \tag{6}$$

2.1 Kalman Filter for Estimation of Lateral States of the B-747 Aircraft

In the trimmed model of LTI Boeing 747 in steady state flight, the lateral axis is simplified which is performed in the simulation that has four main states (x) that consist of sideslip angle (β), yaw rate (r), roll rate (p), and roll angle ($\varnothing$) as given Eq. (7) and two inputs (u) that include the rudder deflection (δr) and aileron deflection (δa) given in Eq. (8).

To run the simulation of the whole model, the system matrices below are accurately gained as below:

State vector:

$$x = [\beta, r, p, \varnothing]^T \tag{7}$$

Control input vector:

$$u = [\delta a, \delta r]^T \tag{8}$$

System matrix:

$$A = \begin{bmatrix} -0.08895 & 0.06282 & -0.9795 & 0.04362 \\ -2.419 & -0.6024 & 0.3438 & 0.01244 \\ 1.491 & -0.2827 & -1.19 & -0.2719 \\ 0 & 1 & 0.05824 & 0 \end{bmatrix} \tag{9}$$

Control distribution matrix:

$$B = \begin{bmatrix} 0 & 0.01024 \\ -0.1967 & 0.09127 \\ -0.0138 & -0.5503 \\ 0 & 0 \end{bmatrix} \tag{10}$$

Measurement matrix of the system:

$$C = \begin{bmatrix} 1 & 0 & 0 & 0 \\ 0 & 1 & 0 & 0 \\ 0 & 0 & 1 & 0 \\ 0 & 0 & 0 & 1 \end{bmatrix} \tag{11}$$

The performance of the estimation or estimation error is dedicated by the covariance matrix. K is the value of the Kalman filter gain matrix also P is considered as estimation error covariance. The equations are established for prediction stage as in Eqs. (12), (13), and (14).

$$\widehat{x}_k = A\tilde{x}_{k-1} + Bu_{k-1} \tag{12}$$

$$\widehat{P}_k = A\tilde{P}_{k-1}A^T + Q \tag{13}$$

$$K_k = \widehat{P}_k C^T \left(C\widehat{P}_k C^T + R \right)^{-1} \tag{14}$$

Here, T and ^ symbols imply transpose and estimate value. $\widehat{x}_k$ state estimation refreshes its value as time spend and it is based on sentient. In correction stage, correction function presents the value to enhance the regained state estimation that comes from measurement. The noise covariance is directly proportional to the measurement noise, and Kalman gain matrix K will diminish the moment reliability is not steady on the measured output y during the new iteration to compute the next state $\widehat{x}_k$.

The residual is described in Eq. (15).

$$\Delta_k = y_k - C\widehat{x}_k \tag{15}$$

Inside the correction section, updated state estimate of the observer is given in Eq. (16)

$$\widehat{X}_{k+1} = \widehat{x}_k + K_k(y_k - C\widehat{x}_k) \tag{16}$$

And updated error covariance matrix is given in Eq. (16).

$$\widehat{P}_{k+1} = \widehat{P}_k - \left(K_k C\widehat{P}_k \right) \tag{17}$$

To acquire a great success from the Kalman filter, the values that are given, dynamic model and probabilistic data values, must be right. Therefore, a convenient model must match probabilistic model to reveal alterations in dynamics of the aircraft and environment circumstances. Kalman filter desires to have civil aircraft's initial state value and initial covariance matrix, and this property of Kalman filter differs from other types of filters.

2.2 *Statistical Test for Fault Detection*

The states of the Boeing 747 model are estimated through a linear optimal KF.

The following two hypotheses may be introduced (Hajiyev & Caliskan, 2005):

$$H_0 = \text{No sensor fault}$$

$$H_1 = \text{Sensor fault occorance on the system}$$

Innovation approach is utilized on the detection of sensor fault conditions. Establishing detection of sensor failure by altered mean of the innovation sequence of Kalman filter can be gathered as the statistical function below:

$$\beta_k = \sum_{j=k-M+1}^{k} \tilde{\Delta}_j^T \tilde{\Delta}_j \tag{18}$$

where $\tilde{\Delta}_j$ is the normalized innovation sequence of the OLKF and M is the width of the sliding window. The Kalman filter normalized innovation sequence is:

$$\tilde{\Delta}_k = \left[CP_{k-1}C^T + R_k\right]^{-1/2}\Delta_k \tag{19}$$

P_{k-1} is the covariance matrix of the extrapolation error and R_k is the covariance matrix of measurement error, Δ_k is OLKF innovation sequence.

This statistical function has χ^2 distribution with s degree of freedom where s is the dimension of the measurement vector. The level of significance α is selected as,

$$P\left\{\chi^2 > \chi^2_{\alpha,Ms}\right\} = \alpha; 0 < \alpha < 1 \tag{20}$$

The threshold value $\chi^2_{\alpha,Ms}$ can be determined from quantiles table of the χ^2 distribution. So, when the hypothesis H_1 is correct, the statistical value of β_k will have higher value compared to the threshold value $\chi^2_{\alpha,Ms}$, i.e.,

$$H_0 : \beta_k < \chi^2_{\alpha,Ms}, \forall k \tag{21}$$

$$H_1 : \beta_k > \chi^2_{\alpha,Ms}, \forall k \tag{22}$$

In Fig. 1, it can be implied that the statistical value β_k exceeds the threshold value during time $t = 400$ and $t = 600$, so the fault is detected.

2.3 *Sensor Fault Isolation Algorithm*

Fault isolation process is responsible to bring out which sensor is faulty. The algorithm is defined by the Eqs. (24) and (25), which are implemented for fault isolation. $S_{\tilde{\Delta}_k}$ is the sample covariance matrix and S_k^* is sample covariance matrix multiplied by all the elements $M - 1$. The dimensions of S_k are adjusted as sxs.

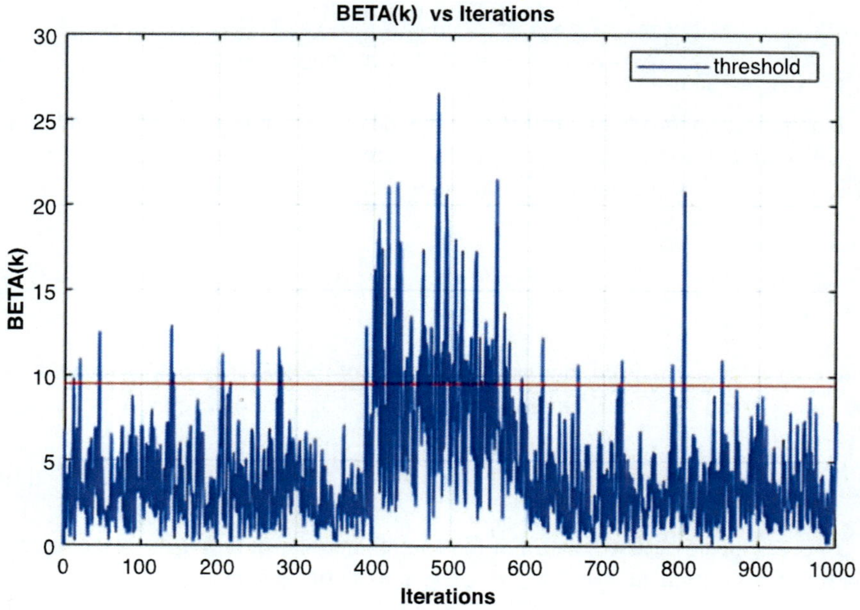

Fig. 1 β_k evolution for the case of continuous bias fault on sideslip angle sensor

Based on the statistics of the faulty sensor, it is assumed to be more affected compared to remaining sensors. Meanwhile, the fault has an influence on the variance of the innovation sequence

$$\tilde{\Delta}_k = \left(\frac{1}{M} \sum_{j=k-M+1}^{k} \tilde{\Delta}_j \right) \tag{23}$$

$$S_{\tilde{\Delta}_k} = \frac{1}{M-1} \sum_{j=k-M+1}^{k} \left[\tilde{\Delta}_j - \overline{\tilde{\Delta}}_k \right] \left[\tilde{\Delta}_j - \overline{\tilde{\Delta}}_k \right]^T \tag{24}$$

$$S_k^* = \sum_{j=k-M+1}^{k} \left[\tilde{\Delta}_j - \overline{\tilde{\Delta}}_k \right] \left[\tilde{\Delta}_j - \overline{\tilde{\Delta}}_k \right]^T \tag{25}$$

We denote the ith diagonal element of the statistic (25) by $S_k(i)$. It can be implied from the calculations that, max $\{S_k(i) \mid i = 1, 2, \ldots, s\} = S_k(m)$ for $i \neq j$, and $S_k(i) \neq S_k(j)$, therefore the mth sensor set is diagnosed as faulted. The process of decision making is run to isolate the faulty sensor.

From Fig. 2, the first row demonstrates that the sideslip angle sensor's KF normalized innovation function value has reached three sigma level between the time interval of $t = 400$ till $t = 600$ s, which means fault occurred and the remaining sensor KF innovation function values seem normal. Therefore, the sideslip angle sensor fault is isolated successfully.

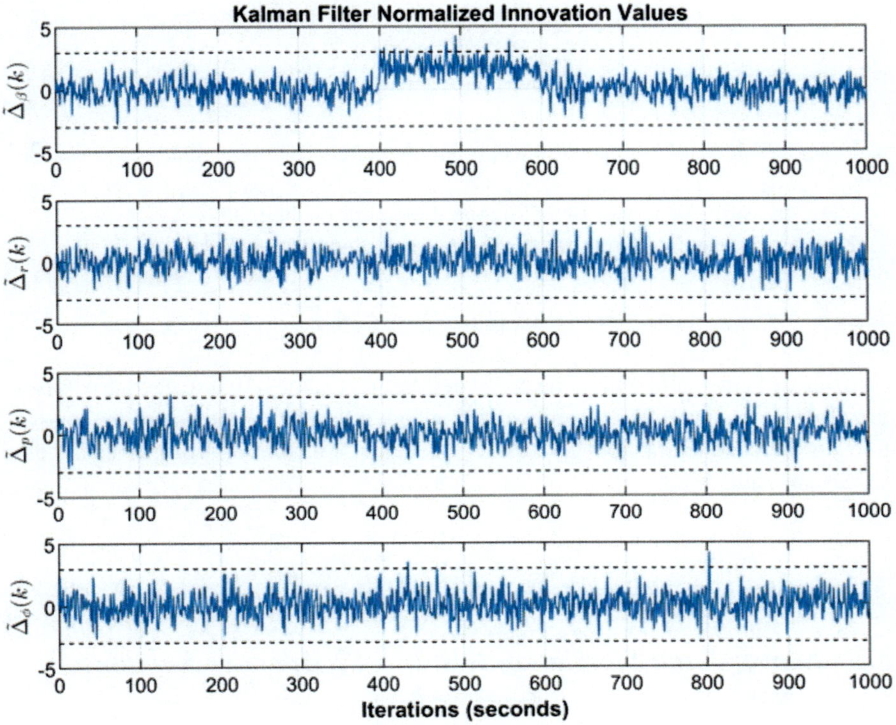

Fig. 2 Normalized innovations of OLKF for β, r, p, $\varnothing$ measurement channels in case of sideslip angle sensor fault

2.4 Adaptive Kalman Filter

Sensor faults can be detected and isolated in adequate method: yet, the estimation of the filter is not satisfying due to the faults. To overcome this problem, fault accommodation approach is compatible for enhancement of the estimation values.

A method called Adaptive Kalman filter is utilized to have better results in estimation of the states that are closer to the actual values even if a sensor fault is observed on the system.

The AKF is embedded with multiple MMNSF method. The target is to correct solely the correlated parts of the measurement noise-covariance, and eventually the Kalman gain. The scale factor Λ_k can be designated using the formula:

$$\Lambda_k = \left(\frac{1}{M} \sum_{j=k-\mu+1}^{k} \Delta_j \Delta_j^{\ T} - C_k P_{k-1} C_k^T \right) R_k^{-1} \tag{26}$$

In this case, rather than Λ_k a matrix $\Lambda_k^{\ *}$ is utilized. To establish the scale matrix $\Lambda_k^{\ *}$, the following calculation is performed (Hajiyev, 2016):

$$\Lambda_k^* = \mathrm{diag}\left(\lambda_1^*, \lambda_2^*, \ldots, \lambda_n^*\right) \tag{27}$$

In which

$$\lambda_i^* = \max\{1, \Lambda_{ii}\}\ i = 1, \ldots, n \tag{28}$$

Here, the sign Λ_{ii} serves as the diagonal element of the matrix Λ_k. Kalman filter gain should be run and re-evaluated by the equation:

$$K = P_k C_k^T \left[C_k P_k C_k^T + \Lambda_k^* R_k\right]^{-1} \tag{29}$$

In case of any malfunction, the correlated element value of the scale matrix raises; therefore, a smaller Kalman gain is found which reduces the impulse of the innovation on the state estimation process. Thus, more satisfying and proper estimation values are gathered.

3 Results and Discussion

3.1 Sensor Fault Simulation Results with Adaptive Kalman Filter

The technique has been executed in one state parameter faulty (sideslip angle, roll rate) and also in double state parameter faulty (roll rate, roll angle). It is proposed that there is a continuous bias fault and measurement noise increment fault between the given time period.

The fault is implied at time interval $t = 400$ to $t = 600$ s on lateral sensors which are sideslip angle, roll rate, and double sensor faults. The first fault is continuous bias fault on sideslip angle for 7.5 degrees and the remaining faults are both measurement noise increment faults of approximately 10 degrees per second and 20 degrees per second on roll rate and roll angle sensors, respectively.

An Optimal Linear KF is used to detect and isolate the sensor faults and Adaptive KF is utilized for fault accommodation.

The first part of the figures show the Kalman filter state estimation values and real state values comparatively. Second part illustrates the estimation error related to actual values of the Boeing-747 aircraft. The OLKF is not accurate after the fault took place at time $t = 400$ s till $t = 600$ s where the fault is exerted on sideslip angle sensor.

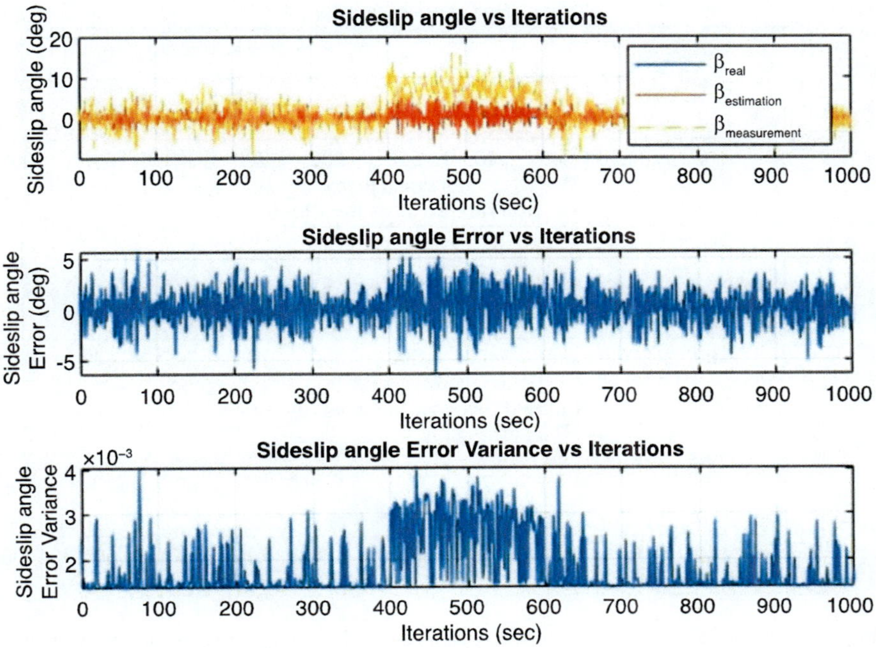

Fig. 3 Sideslip angle sensor fault evolution and AKF estimation value with continuous bias fault

3.2 *Adaptive KF for Sideslip Angle Sensor Fault*

From the figure, it can be implied that, when just one parameter disabled, the adaptive Kalman filter has good estimation values of sideslip angle. Essential parameters to assess estimation performance are velocity of the error convergence, the evolution of the error absolute value and the error variance along with the iterations, and the absolute value of the stationary error. By performing the AKF, the estimations are getting closer to the real values.

Figure 3 shows that the faulty measurement values (yellow dashed line) are enhanced by the algorithm and estimation values (red line) got closer to real value (blue line). Therefore, the fault is accommodated well.

3.3 *Adaptive KF for Double Sensor Fault*

The measurement noise-increment fault took place at time $t = 400$ s till $t = 600$ s where the fault is exerted on both roll rate and roll angle sensors. The adaptive Kalman filter estimates the right values even if the double measurement channels are faulty. The broken sensor values are corrected by the algorithm. As illustrated in Figs. 4 and 5, the estimations are convenient as compared to the real value. The algorithm estimates well and both two faults are accommodated firmly.

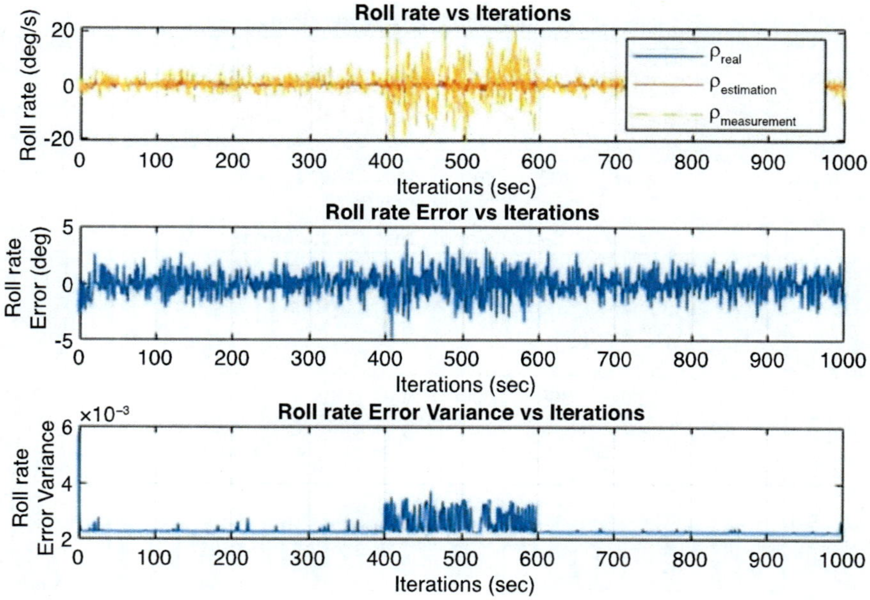

Fig. 4 Roll rate sensor evolution and AKF estimation value with measurement noise increment fault on roll rate and roll angle sensor fault

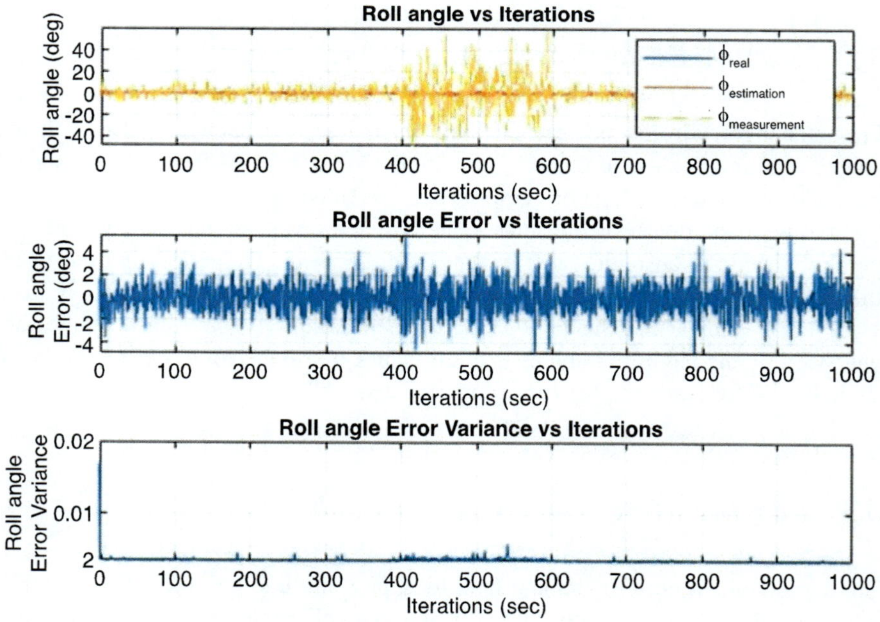

Fig. 5 Roll angle sensor evolution and AKF estimation value with measurement noise increment fault on roll rate and roll angle sensor fault

Both Figs. 4 and 5 demonstrate that simultaneous faults are accommodated well and reached good results.

3.4 *Root Mean Square Error Evaluation*

Root mean square error method is utilized to approve the results and that proves the estimations of AKF is better and RMSE values are seen in Tables 1 and 2.

4 Conclusion

In this study, Optimal Linear Kalman Filter (OLKF), Fault Detection, Fault Isolation, Adaptive Kalman Filter algorithms are embedded on the lateral dynamics of the Boeing-747 aircraft. In nominal cases, the OLKF gives fine estimation values. Anyways, if a malfunction on the measurement channels occurred, the accuracy of the filter estimations are unreliable and not firm.

Two scenarios of faulty case are implemented. First case includes a single sensor fault in sideslip angle sensor and the second case includes simultaneously double sensor fault on roll rate and roll angle. First, the fault detection process takes place and detects whether a fault occurred. Following, the fault isolation algorithm performs the statistics of rate of sample and theoretical variances to distinguish the faulty sensor. Lastly, after the fault isolation process, this study presents Adaptive Kalman filter algorithm to gather and enhance the estimation values for fault accommodation purpose.

Table 1 Root mean square error comparison for the sideslip angle sensor faulty case

Lateral states	OLKF RMSE value	AKF RMSE value
Sideslip angle	2.1039	2.0102
Yaw rate	3.2824	3.0142
Roll rate	2.1926	1.9773
Roll angle	2.7450	2.5547

Table 2 Root mean square error comparison for the roll rate and roll angle double sensor fault case

Lateral states	OLKF RMSE value	AKF RMSE value
Sideslip angle	1.1169	0.9447
Yaw rate	1.9365	1.6918
Roll rate	1.7616	1.1289
Roll angle	1.8693	1.4689

References

Fang, H., Tian, N., Wang, Y., Zhou, M., & Haile, M. A. (2018). Nonlinear Bayesian estimation: From Kalman filtering to a broader horizon. *IEEE/CAA Journal of Automatica Sinica, 5*(2), 401–417.

Gheorghe, A., Zolghadri, A., Cieslak, J., Goupil, P., Dayre, R., & Le Berre, H. (2013). Model-based approaches for fast and robust fault detection in an aircraft control surface servo loop: From theory to flight tests [applications of control]. *IEEE Control Systems Magazine, 33*(3), 20–84.

Hajiyev, C. (2016). An innovation approach based sensor fault detection and isolation. *IFAC-PapersOnLine, 49*(17), 420–425.

Hajiyev, C., & Caliskan, F. (2005). Sensor and control surface/actuator failure detection and isolation applied to F-16 flight dynamic. *Aircraft Engineering and Aerospace Technology, 77*(2), 152–160.

Zhang, J., You, K., & Xie, L. (2018). Bayesian filtering with unknown sensor measurement losses. *IEEE Transactions on Control of Network Systems, 6*(1), 163–175.

STEM Opportunities in Flight Testing Sunlight Reflector Ultralights

Narayanan Komerath, Ravi Deepak, and Adarsh Deepak

1 Introduction

Countering global warming is an urgent priority. It is now recognized (USNAS, 2021; Koch et al., 2021) that merely reducing anthropogenic emissions of infrared-absorbing Greenhouse Gases (GHG) is not quick enough to prevent tip-over into a regime of nonlinear interactions: reducing heat input to the atmosphere is essential. Reflecting sunlight into space is obvious but controversial (USNAS, 2021; Banerjee et al., 2021). In prior work (Komerath & Deepak, 2021), we have shown how our Glitter Belt architecture (Komerath et al., 2017; Komerath, 2021) of high-altitude ultralight reflector UAVs overcomes all cited objections. Implementing the architecture starts with scale model flight tests and swarms that provide unique measurement capabilities. In turn, this increases simulation and prediction confidence to safely reverse global warming. In (Komerath et al., 2021a), we discussed Eastern Hemisphere plans with meteorology and data transfer over the Indian Ocean region. The chapter looks at the Pacific Ocean and vehicle design/testing aspects.

Figure 1 (Komerath & Deepak, 2021) shows the Glitter Belt concept: swarms of solar-powered ultralight reflector sheets moving between daytime cruise at 30.48 km (100,000 feet) and night-time glide staying above 18.3 km (edge of controlled airspace at 60,000 feet). The swarms are generally in "glittering belts" roughly under the zone of peak summer sunshine, with measurement excursions to the Arctic and Antarctic Circles. Each Flying Leaf vehicle in a swarm is made up of 11 sheets, each 64 m × 64 m, assembled by high-altitude rendezvous from 11 Flying Leaflet

N. Komerath (✉) · R. Deepak
Taksha Institute, Hampton, VA, USA
e-mail: nkomerath@taksha.org; ravi@taksha.org

A. Deepak
Taksha Institute, Silicon Valley, CA, USA
e-mail: ravi@taksha.org

T. H. Karakoc et al. (eds.), *New Achievements in Unmanned Systems*, Sustainable Aviation, https://doi.org/10.1007/978-3-031-29933-9_7

Fig. 1 Flying leaf made of 11 sheets and 3 leaflet wings, at 30 km altitude. One trails a tethered measurement payload

carrier vehicles. Only 3 of the 11 carriers are retained, the rest returning to base within the same day to pick up new sheets to assemble another leaf. Each leaflet can carry 50 kg payload, but total payload on a flying leaf is limited to 300 kg. This accommodates meteorological instruments that are now carried on space satellites, as well as radio sondes and drop sondes carried on airplanes and balloons. Eastern Hemisphere routes span the Indian Ocean, generally staying within data range (500 km) of ground facilities. Initial Pacific Ocean meteorological flight testing (Fig. 2) touches Arctic Ocean and Antarctic coastline. It uses North and South American coastlines, Hawaiian Islands, Tahiti-Tonga, New Zealand coast, and Fijian Islands for ground station links, while staying clear of populated areas or large land masses for now.

2 Nature Credits, Public Participation

In (Komerath et al., 2021b), we argue for nature credits made available at a micro level, to recognize worldwide participation in activities that drive towards the UN's Sustainable Development Goals, broadened from carbon credits.

Credit for solar reflection was estimated to be between 0.068 and 0.356 Carbon Emission Units (tons of CO_2 equivalent in IR absorption) per square meter of full strength or AM0 sunlight, fully reflected into space, with a nominal 10-year lifetime expected. Aluminized Mylar sheets reflect over 98% of the solar IR spectrum (McNulty, 2016), even at a shallow 60° angle of incidence (30° azimuth). In contrast, reflectors at ground level can aspire to a maximum of 50% of this credit,

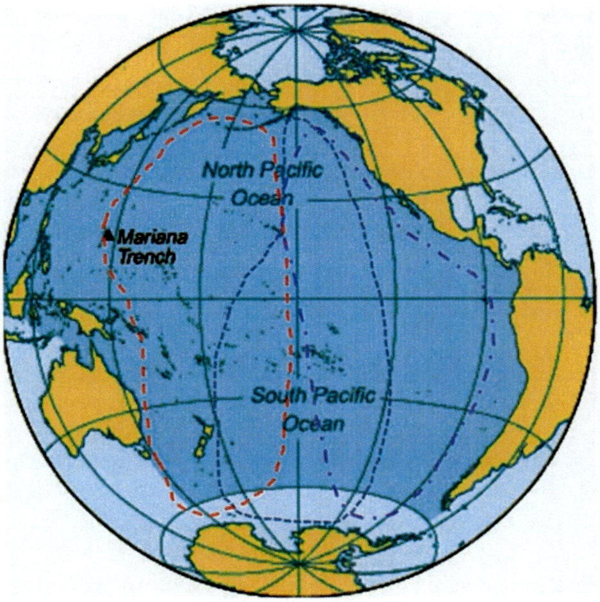

Fig. 2 Pacific Ocean routes for flying leaf summer follower swarms

even at midday on a clear summer day. Clouds and shorter daylight duration due to obstructions and seasonal variation as well as dust, dew and snow, degrade this much more. We further argued for micro credit computations, exploiting the prevalence of cashless electronic micropayment systems that use cellular or mobile telephones worldwide. Such microcredits enable mass participation in restoring nature and reversing global warming, far better than projects that are debated for decades to reduce "Carbon."

In this chapter, we show how sunlight reflection directly provides opportunities for students to participate, particularly STEM (Science, Technology, Engineering and Mathematics) students and teachers. This is evident from the technical development ladder. From an aeromechanics perspective, these vehicles are low-speed, laminar-flow ultralights with thin flexible sheets providing most of the lift. Because of the large sheet area, the wing loading is extremely low. This permits flight as slow as 9.5 m/s for the full-scale leaf at 30.48 km altitude and a similarly low rate of descent in night-time glide. Flight and structural parameters must be held within tight limits to avoid sheet instability. Aeroelastic stability of long flexible wings and joined leaf structure must be understood and controlled. Distributed payloads, propulsion and support wings all introduce unique aspects. Autonomous operation in swarms is no doubt an interesting problem.

Our approach to these problems is incremental, building scale models increasing span by factors of 2 (Table 1). We have come through 0.5 m, 1 m, 1.5 m, and now 2 m unpowered models. A 1.2 m powered model is in testing, towards building the

Table 1 Technical milestones relevant to STEM participation

Wing Span & (scale)	Objectives	Technical challenges
1 m (1/64)	Static stability & glide; sheet tension	Conceptual design scaling; gliding descent, dynamic response
2 m (1/32)	Wing loading; static stability	Construction methods. Control surfaces. Launch and landing
Powered 2 m (1/32)	Steady flight; drag budget. Radio control	Motors, controls, R/C communications, launch and recovery
Powered 4 m, 8 m (1/16; 1/8)	Takeoff, climb, autonomous flight; rendezvous	Flight control system; sensor feedback and rendezvous latching, unlatching, return to landing
Powered 16 m (1/4)	Autonomous, meteorology payloads, long distance night survival	Launch & recovery, Aeroelasticity; remote monitoring; airspace & ATC; autonomy; wind patterns.
32 m (1/2)	Sunlight reflection	Aeroelasticity; laminar flow control. Precision for astronomy; produce, deploy & monitor

first 2 m powered model, followed by 4, 8, 16, 32 before the full 64 m leaflet vehicle. At each size range, what can be learned must be exhausted before investing heavily in the next. Up to 16 m span, focus will be on meteorology and solving problems. The 32 m and 64 m will emphasize sunlight reflection and serve as night-time distributed telescope array elements..

3 Results and Discussion

3.1 STEM Participation

We conceptualize a Glitter Belt Consortium, based on Komerath et al. (2007) which is already well underway through the Taksha Institute. The structure is shown in Fig. 3. We are already working with a few universities in the United States and India and have received technical support and advice from government and industry experts. We invite worldwide participation. Participation opportunities extend into many fields of human interest and endeavor. The public policy debate on climate change is entangled in many issues. Everyone has an opinion which must be respectfully and objectively considered. The on-going debate about personal precautions and vaccination during a pandemic illustrates the range of imagination, communications, and emotions that come into play. Educational institutions should take the lead in openness to ideas as well as in developing and exploring useful solutions.

Conceptual design follows the initial top-down steps outlined in a textbook aimed at first-semester students (Komerath, 2018). This verified that the fully assembled Flying Leaf would stay above 20 km in night glide at speed for minimum drag, and

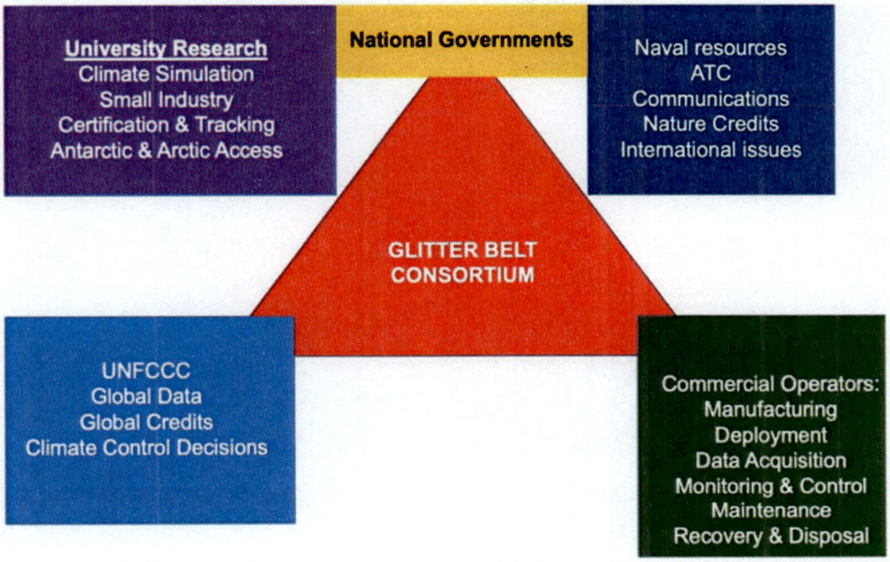

Fig. 3 Consortium structure used to develop the glitter belt

higher if lift coefficient were raised to 1.2, which is as high as we dare venture with flexible sheets. In our process, we iterate on a starting guess of gross weight and come up with a structural weight that we think we can meet. Closing the design consists of showing what is necessary to meet that structure weight, while having enough strength to survive the rough-and-tumble of flight testing. As our knowledge improves, the database of material properties is getting more accurate and detailed, while essential structural design innovation needs are encountered and overcome. In time, this will move into dynamic simulation using both commercial off-the-shelf (X-Plane11) software and our own software development. The electronic controls knowledge is being integrated both from fixed-wing hobby aircraft and the more advanced toy drone industry which performs precise position hold in winds under inertial and GPS control.

4 Flight Test Model Development

The low speed ultralight UAV domain is benign and offers low-threshold (baby) steps, while being rich in problems suitable to be broken down into STEM team problem-solving. We hope to depart from the usual competition model where N-1 teams must go away disappointed to one of a collaborative race to beat the looming threat. Every minute of reflecting sunlight at any level is creditable to delay global warming; every day gained in deploying a given area of high-altitude reflector is a significant step in slowing down sea-level rise, droughts, and other disasters. We seek to energize STEM and other youth in this global team endeavor. Microcredits

Fig. 4 1-m span model in stable glide

for reflecting sunlight should be pursued to get public participation. For instance, the 2 m × 1 m model shown here, even left outside for 6 h on a sunny day, reflects 12 KiloWatt Hours of sunlight at the AM1 nominal value assumed for solar panel design. Of this only about 8.4 kWh will reach space, to be credited. If 10,000 such models were used worldwide by students, that is, 8.4 MWh per day. That is a head start on 5 Flying Leaflets. The knowledge gained is a priceless bonus.

The following figures show how very inexpensive materials and amateur construction (constrained by COVID lockdowns denying access to good tools, facilities, and skilled help) combined with some innovation have been used to advance along our roadmap.

Figure 4 shows a 1-m span reflective Mylar sheet supported on a frame, with a styrofoam wing supporting it, which exhibits steady smooth flight. The model was built and flown by undergraduates after extensive wind-tunnel tests also done by undergraduates to determine how to hold thin sheets in tension at angle of attack (Smith-Pierce et al., 2018). These sheets, intended for party decorations, are reflective on both sides unlike the desired lower surface black sheets.

Figure 5 shows our first 2 m model. The support wing is made of balsa spars, drinking-straw leading edges and control surface axis tubes reinforced with bamboo party skewers, election-advertisement cards from the 2020 US Elections, and shrink-wrap plastic, during the COVID lockdown of 2020. A wooden longitudinal beam supports the landing gear (made of wire and bottle-caps) and a low-aspect ratio canard which was the original 0.5 m wing model (not shown in Fig. 5).

The 2 m model weighed more than 1.2 kgf (Figs. 6 and 7). With experience of building and testing it, carbon fiber (CF) based structure is used for subsequent models. Pultruded carbon fiber (CF) tubes are being used for main spars and a CF vertical truss for longitudinal support of the sheet frame. A 1 m-span canard has been

Fig. 5 2-m propulsive wing with landing gear and part of the sheet support structure. A roll of 25-micron aluminized Mylar sheet is shown

Fig. 6 2 m × 1 m model in stable glide

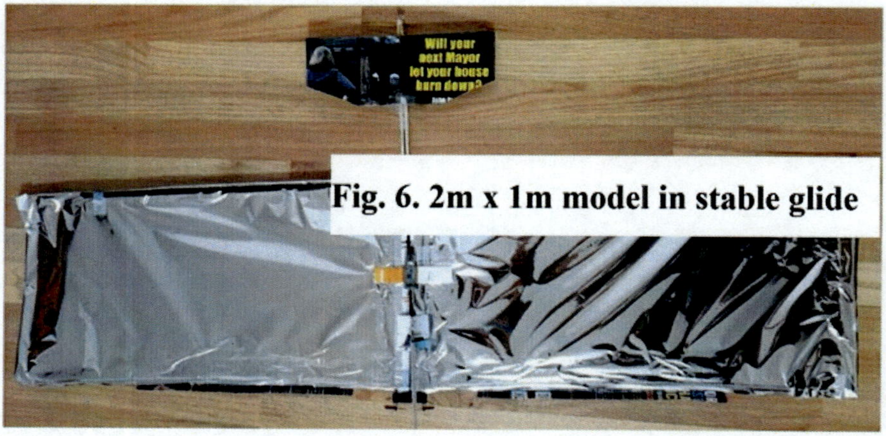

Fig. 7 2 m × 1 m model in stable glide

Fig. 8 Models propelled by PowerUp 4.0 R/C cellphone-BlueTooth controlled kit. Left: paper airplane. Right: 1.2 m × 0.3 m flying wing/ canard

flight-tested using a quadrotor drone. A rugged aluminum-spring landing gear has been built for glide tests to be augmented later with skids for grass landing.

Figure 8 shows the first powered model, modified from a paper airplane kit (PowerUp 4.0) operated by a phone-based app. This model was expanded to 1.2 m × 0.3 m at the present writing to fly stably at 4.73 m/s. Open-source flight control systems available on the internet and in publications enable student teams to build ever more sophisticated control systems. We propose to set up the knowledge retention system to permit continuous improvement as student cohorts cycle through curricula.

5 Conclusions

- A western hemisphere initial flight test and meteorological plan is described for high-altitude solar aerodynamic reflectors.
- Unpowered models with reflective sheets at 1 m and 2 m scale are shown to fly stably at low altitude, even with Aspect Ratio as low as 2.
- Powered model testing is feasible at 0.3 m scale.
- Autonomous flight systems from aeromodelling community already provide strong capabilities.
- Carbon fiber truss construction makes it feasible to meet conceptual design targets.
- Scale model and technology development invites participation with strong STEM opportunities.
- Launch speeds are low enough for playgrounds.
- Flying leaf swarms with autonomous control may enable high-altitude astronomy with large distributed array telescopes.

References

Banerjee, A., Butler, A. H., Polvani, L. M., Robock, A., Simpson, I. R., & Sun, L. (2021). Robust winter warming over Eurasia under stratospheric sulfate geoengineering–the role of stratospheric dynamics. *Atmospheric Chemistry and Physics, 21*(9), 6985–6997.

Koch, A., Brierley, C., & Lewis, S. L. (2021). Effects of Earth system feedbacks on the potential mitigation of large-scale tropical forest restoration. *Biogeosciences, 18*(8), 2627–2647.

Komerath, N. (2018). *Design-centered introduction to aerospace engineering* (1st ed.). SCV Inc. SBN-13: 978-1949335002.

Komerath, N. (2021, September 15). *Glitter Belt: Atmospheric reflectors to reduce solar irradiance*. US Patent 10,775586.

Komerath, N., & Deepak, A. (2021). Payload design for global continuous atmospheric mapping. In *Virtual global monitoring annual conference (eGMAC)*, May 24–28. https://gml.noaa.gov/annualconference/abs.php?refnum=34-210424-C. Viewed 7 Oct 2021.

Komerath, N., Nally, J., & Tang, E. Z. (2007). Policy model for space economy infrastructure. *Acta Astronautica, 61*(11–12), 1066–1075.

Komerath, N., Hariharan, S., Shukla, D., Patel, S., Rajendran, V., & Hale, E. (2017). *The flying carpet: Aerodynamic high-altitude solar reflector design study (No. 2017-01-2026)* (SAE Technical Paper).

Komerath, N., Sharma, A., & Deepak, R. (2021a). A. Glitter belt global measurement system: Indian Ocean component preparation. In *Proceedings of IEEE ICECCME*, Mauritius, October 7–8, 2021.

Komerath, N., Deepak, R., & Deepak, A. (2021b). Credit for reflecting sunlight. In *Proceedings of the SOLAR2021*, American Solar Engineering Society, Boulder, CO, August 3–6.

McNulty, D. (2016). *Reflectivity measurements*. Master's thesis, Idaho State University. https://www2.cose.isu.edu/mcnudust/publication/presentations/

Smith-Pierce, M., et al. (2018, July). *High altitude aerodynamic reflectors to counter climate change*. AIAA Paper, Applied Aerodynamics Conference, Atlanta, Georgia.

USNAS. (2021). *Reflecting sunlight: Recommendations for solar geoengineering research and research governance*. The National Academies Press. https://doi.org/10.17226/25762

Modelling and Simulation of Vertical Landing Dynamics of an Aircraft Based on a Model System

Selim Sivrioglu

1 Introduction

A short take-off and vertical landing (STOVL) aircraft has great advantages, especially for navy aircraft carriers. The STOVL fighter gives a significant edge to a military due to its capability to operate from small surfaces. A state-of-the-art vertical lift and landing technology for fighter jets is currently available in F-35B aircraft (Boyce & Burt, 2009).

Studies concerning STOVL aircraft attract much attention due to possessing some critical technology of such systems. Most of the published material concerning with STOVL aircraft technology written by engineers of F-35 program give overview of F-35 aircraft rather than an academic study. An earlier work written by NASA specialists, a nonlinear aircraft model based on wind tunnel data is studied in terms of powered and unpowered aerodynamics for a STOVL system without any given detail (Mihaloew & Drummond, 1989). Planar vertical take-off aircraft dynamics using AV-8B Harrier aircraft data is studied for a robust gain scheduled control by (Wu et al., 2008). A recent study (Wiegand et al., 2018) presented some technical details of the technology used in these aircrafts.

This chapter focused on the landing dynamics of a STOVL aircraft in modelling perspective based on a proposed model system. Landing simulations are performed using model dynamic equations and assuming the lift forces as linear and nonlinear function of time.

S. Sivrioglu (✉)
Antalya Bilim University, Antalya, Turkey
e-mail: selim.sivrioglu@antaya.edu.tr

T. H. Karakoc et al. (eds.), *New Achievements in Unmanned Systems*, Sustainable Aviation, https://doi.org/10.1007/978-3-031-29933-9_8

2 Aircraft Landing Structure

In this study, a model system is proposed based on the landing structure of F-35B aircraft. As shown in Fig. 1, the aircraft has four jet streams for vertical landing (Wiegand et al., 2018). These are the forward air jet, rear jet stream, right wing jets, and left wings jets. While the rear and wing jets are directly provided by the main engine of the aircraft, the front air jet is generated with a separate lift fan located behind the cockpit. Note that power of the lift fan is also provided by the main engine. The rear jet stream has an adjustable angle with a swivel module when the vertical landing operation occurs.

The basic vertical landing principle of the aircraft is to control the lift forces created by the jet streams to meet the weight of the aircraft. In a vertical landing, external wind loads may destabilize the aircraft. For a safe landing operation, the external effects should be considered in modelling and analysis. Therefore, the position of the aircraft should always be horizontal by controlling the air jet in a vertical landing height.

3 Vertical Landing Model System

The STOVL technology developed for F-35B aircraft is a very advanced technology and very few company has such capability. For research purposes, it is possible to create a model system to analyse the vertical landing dynamics in an academic perspective. For this aim, the model system shown schematically in Fig. 2 is proposed. It is assumed that the thrust in the model system is maintained by the axial lift fans. As seen in the model system, there are two main axial fans to generate lift forces F_{front} and F_{rear}, respectively. The wing fans provide stabilizing forces F_{wing1} and F_{wing2} for rolling motion of the model system. We assume that the model

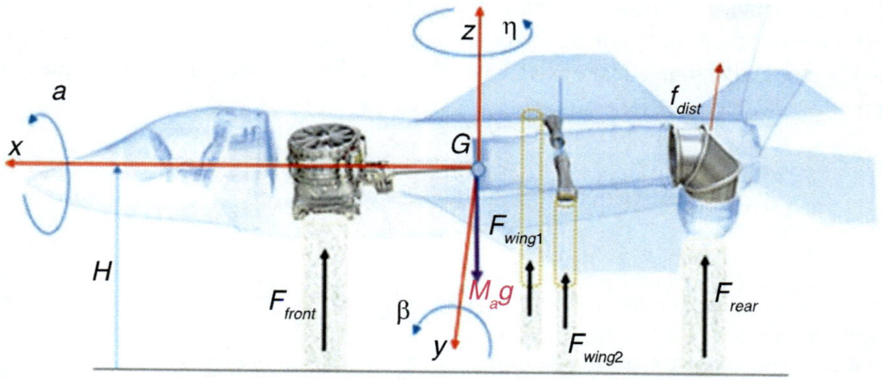

Fig. 1 The STOVL aircraft lift forces

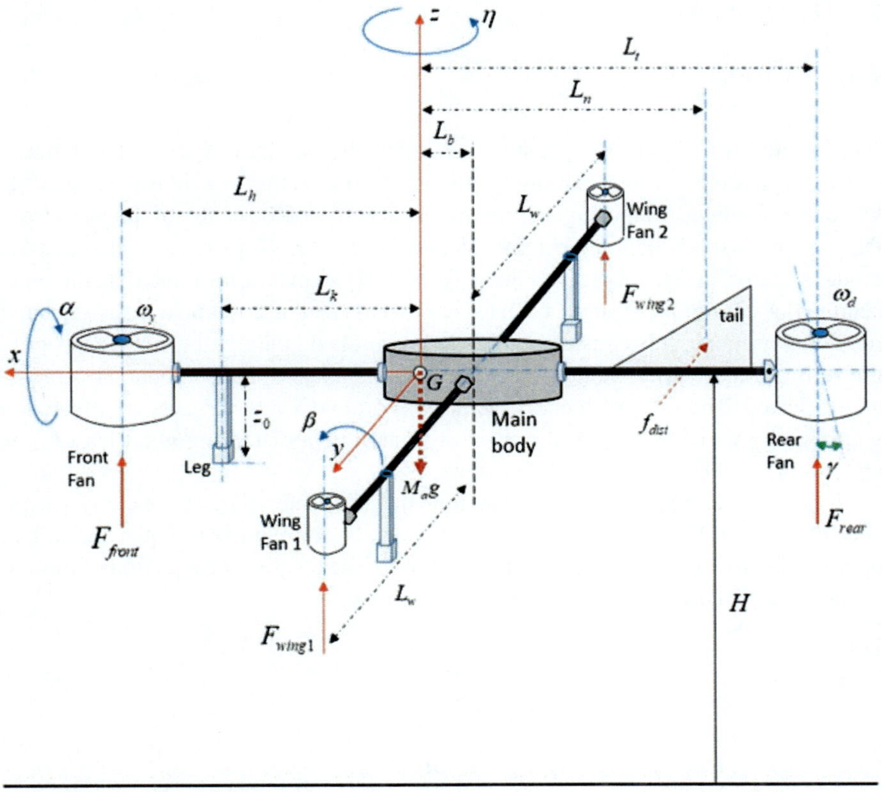

Fig. 2 Vertical landing model system

system has a short take-off and vertical landing capability provided by the fan lift forces. The force due to axial thrust of a fan is theoretically defined as

$$\sum F = \frac{1}{g_c}\frac{d(m\mathrm{V})}{dt} = \frac{1}{g_c}\frac{d\mathrm{M}}{dt} \tag{1}$$

where *dm*/*dt* is the mass flow rate of the air, *V* is the air speed and *M* is the momentum. Also, $1/g_c$ is the fan-related constant. For an experimental application, a controllable thrust force is needed for landing or take-off. Therefore, it is important to measure the mass flow rate and speed of the air for a fan to use in the model system.

4 Dynamic Modelling

4.1 In Air

The structure of the model system is schematically shown in Fig. 2. The landing process is modelled from a landing height of h with affecting weight force and air jet lift forces of the fans. The height H represents distance of the centre of the mass from the ground. Also, the landing height is decided by $h = H - z_0$. It is assumed that the model system is a rigid object and the mass of the system is concentrated on the mass centre G. A coordinate system is located at the mass centre to define translational and rotational motions. It is considered that during vertical landing, the aircraft has only the vertical translational motion. The other front and lateral translational movements are neglected. The roll, pitch and yaw motions around x, y and z axes are illustrated with angular variables α, β and η. It is assumed that a disturbing wind load f_{dist} is affecting the model system at the tail side.

In the ideal landing condition, we assume that the centre of mass is exactly known and lift forces affecting on the model system body and wings are equal. Also, any external disturbance force is not considered in the ideal case. The equation of motion of the aircraft model in the air ($h > 0$) is obtained as

$$\begin{aligned} M_a\ddot{z} &= -M_a g + \sum F_z \\ I_{p\beta}\ddot{\beta} &= \sum M_\beta \\ I_{p\alpha}\ddot{\alpha} &= \sum M_\alpha \\ I_{p\eta}\ddot{\eta} &= \sum M_\eta \end{aligned} \tag{2}$$

where the lift forces and moments are given as

$$\begin{aligned} \sum F_z &= F_{\text{front}} + F_{\text{rear}} + F_{\text{wing1}} + F_{\text{wing2}} \\ \sum M_\beta &= F_{\text{rear}} L_t - F_{\text{front}} L_h + \left(F_{\text{wing1}} + F_{\text{wing2}}\right) L_b \\ \sum M_\alpha &= F_{\text{wing1}} L_w - F_{\text{wing2}} L_w \\ \sum M_\eta &= 0 \end{aligned} \tag{3}$$

The ideal case requires a moment balance realization both in roll and pitch motion. In the moment equation $\sum M_\beta$, the moments $F_{\text{rear}} L_t$ and $F_{\text{front}} L_h$ should be equal values in pitch dynamics. Also, the moment term $(F_{\text{wing1}} + F_{\text{wing2}})L_b$ generated by the wing air jet forces assumed to be negligible since the air jets in the wings are active only when necessary for stabilizing rolling motion. Therefore, we may neglect this moment effect in the ideal case.

4.2 Lift Force Model

In the modelling, it is assumed that the landing of the aircraft model begins with an initial lift force that suspends the aircraft in the air. It is considered that the aircraft model approaches to the ground with the gradual decrease of the total lift force. The lift forces are modelled as linear and nonlinear function of time. The linear force is proposed as follows:

$$f_L = f_{\text{ini}} - \delta f_l \cdot t \tag{4}$$

Also, the nonlinear lift force is defined as

$$f_N = f_{\text{ini}} - \delta f_n \cdot t^2 \tag{5}$$

where f_{ini}, δf_l and δf_n show the initial lift force magnitude and reduction amount of the force per second and per second square, respectively.

4.3 After Landing

In the real case, when the aircraft touchdowns the ground, the landing gears support the aircraft. We assume that the model system touches the ground on simple legs and also think that these legs behave like springs with damping, ignoring their masses. The derived model switches the lift jet forces to spring and damping forces when the height $h = 0$. The dynamics after landing can be derived using the model given in Fig. 3. The equations of motion for the model body coordinate is obtained as

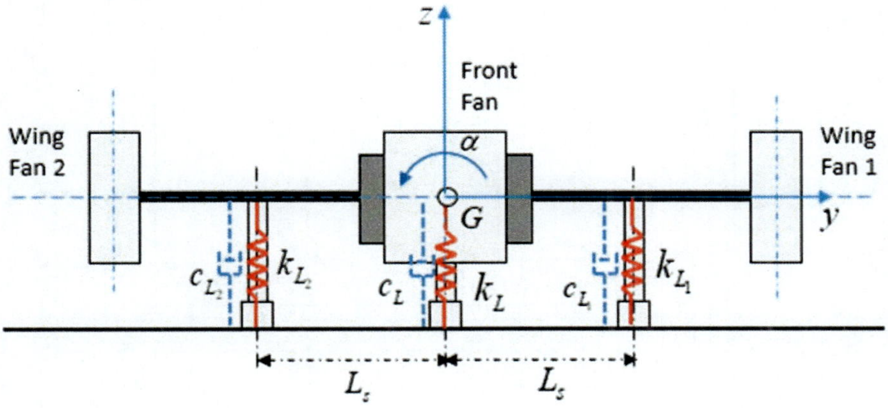

Fig. 3 Front view of the model system

$$
\begin{aligned}
M_a\ddot{z} &= -(k_L + k_{L_1} + k_{L_2})z - (c_L + c_{L_1} + c_{L_2})\dot{z} \\
I_{p\beta}\ddot{\beta} &= k_L(z - L_k\beta)L_k - (k_{L_1} + k_{L_2})(z + L_b\beta)L_b \\
&\quad + c_L(\dot{z} - L_k\dot{\beta})L_k - (c_{L_1} + c_{L_2})(\dot{z} + L_b\dot{\beta})L_b \\
I_{p\alpha}\ddot{\alpha} &= -k_{L_1}(z + L_s\alpha)L_s + k_{L_2}(z - L_s\alpha)L_s \\
&\quad - c_{L_1}(\dot{z} + L_s\dot{\alpha})L_s + c_{L_2}(\dot{z} - L_s\dot{\alpha})L_s
\end{aligned}
\tag{6}
$$

Using above equations, the dynamics of the model system is analysed after touchdown.

5 Simulation

A simulation file of the model system is built using Matlab/Simulink as shown in Fig. 4. The model system consists of force switch subsystems between the air and on the ground conditions. Main subsystems show the aircraft model system equations. Two separate files are generated for linear and nonlinear reaction force cases. The following initial conditions are applied:

$$
\begin{aligned}
z_0 = \quad & h = 15\,\text{m}, \quad \beta_0 = 5^\circ, \quad \alpha_0 = 5^\circ \\
\delta f_n = \quad & \qquad\qquad \delta f_l = 50\,\text{N}
\end{aligned}
$$

Simulation results of the landing process in the case of the linear and nonlinear lift forces are compared in Fig. 5. The landing time in nonlinear case is shorter than linear case. The variations of the forces are also illustrated in Fig. 6a, b. Note that while the initial force values are same at the beginning, the final force values at the landing to ground are different. The initial lift force should satisfy the condition

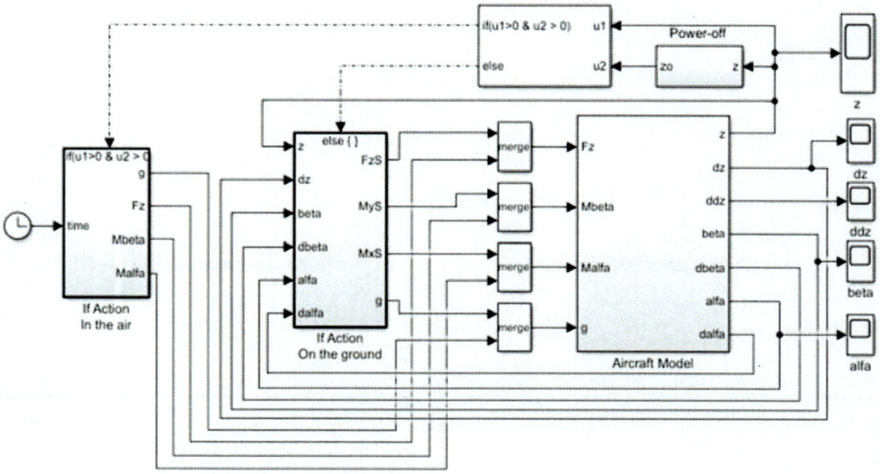

Fig. 4 Simulation model

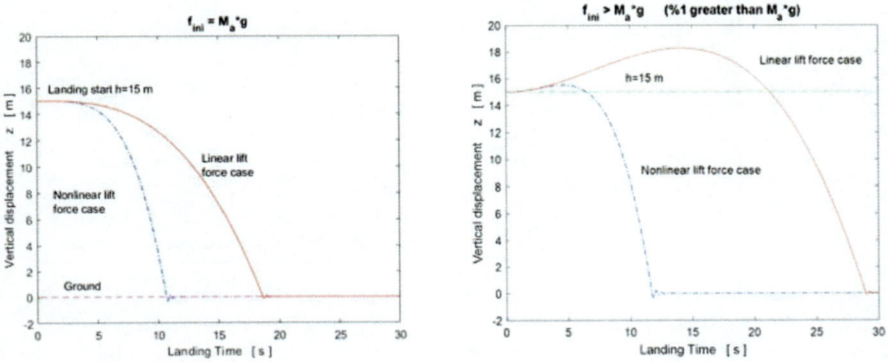

Fig. 5 Landing simulation results (**a**) $f_{\text{ini}} = M_a g$, (**b**) $f_{\text{ini}} \leq M_a g$

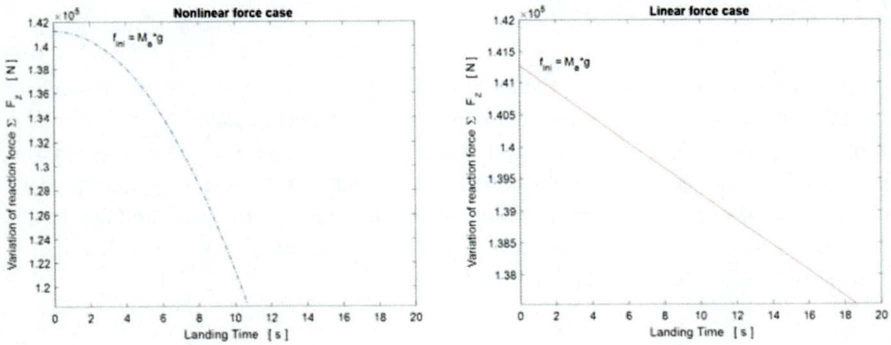

Fig. 6 Variation of lift forces (**a**) nonlinear force (**b**) linear force

$f_{\text{ini}} \leq M_a g$ to start descending without rising from the landing altitude h of the aircraft model. The correction of the model is tested by two different initial force values such as $f_{\text{ini}} = M_a g$ (Fig. 5a) and $f_{\text{ini}} > M_a g$ (Fig. 5b). As seen in Fig. 5b, when the initial force is greater than the weight force in 1% $f_{\text{ini}} > M_a g$, the aircraft model starts to rise in the air. The descending of the aircraft begins when the total lift force approaches the weight force due to the reduction of the force. The dependency of the model to initial lift force is much higher for linear case.

Variation of acceleration for both cases is given in Fig. 7a. Although nonlinear lift force provides a shorter landing time, the acceleration of the model at the touchdown is greater than linear case. Finally, variations of pitch angle are shown in Fig. 7b. similar results are obtained for roll motion. It is observed that pitch and roll angles approaches zero from an initial landing angle.

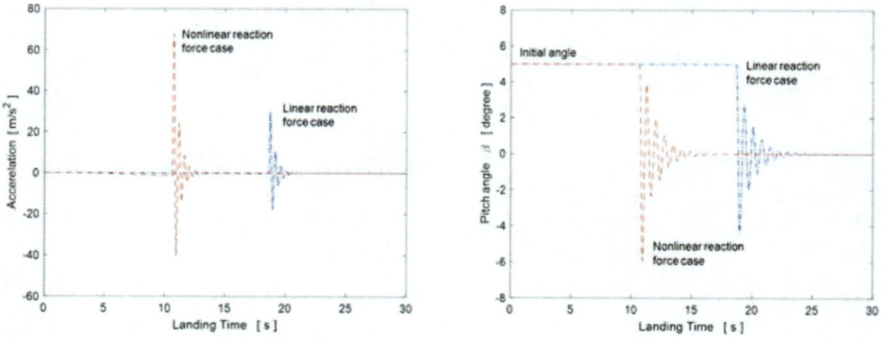

Fig. 7 Variation of (**a**) acceleration (**b**) pitch angle

5.1 Results and Discussion

In the proposed model system, the axial fans play a central role. The success of the model system in the landing process depends on the performance of the fans. The simulations are performed using reaction force models. In practice, the biggest issue is how to generate and control these forces using fans. If we test a fan and table its mass flow and output speed at different speeds, we can use this look-up table in the control algorithm. If the fan speed is adjusted by the motor voltage of the fan, it makes more sense to control it by voltage input.

6 Conclusion

In this study, a model test system is introduced for analysing the STOVL aircraft dynamics. Some of the simulation results were presented based on the derived model. The proposed lift force model is an open loop approach and a feedback control of these forces will also be studied. The obtained simulation results are promising for a real test system application.

References

Boyce, D. A., & Burt, R. J. (2009). Aircraft level dynamic model validation for the STOVL F-35 lightning II. Aircraft Structural Integrity Program (ASIP) conference, December 1–3, Jacksonville, FL, USA.

Mihaloew, J. R., & Drummond, C. K. (1989). STOVL aircraft simulation for integrated flight and propulsion controls research. 10th applied dynamics international users society annual conference, Colorado.

Wiegand, C., et al. (2018). F-35 air vehicle technology overview. 2018, AIAA aviation forum, aviation technology, integration, and operations conference, Atlanta, Georgia, USA.

Wu, S.-L., Chen, P.-C., Chang, K.-Y., & Huang, C.-C. (2008). Robust gain-scheduled control for vertical/short take-off and landing aircraft in hovering with time-varying mass and moment of inertia. In *Proceedings of IMechE Vol. 222. Part G: Journal Aerospace Engineering*.

Examination of Supercapacitors in Terms of Sustainability in Aviation

S. Cansu Gorgulu, Isil Yazar, and T. Hikmet Karakoc

1 Introduction

The development in the aviation sector progresses in proportion to the development in the propulsion systems used in aircraft. The performance and efficiency increase of propulsion systems is of great importance for aircraft. While speed was the basic parameter in the development of aviation in the 1950s, capacity became important in the 1960s. Efficiency and affordability came to the fore between 1970 and 1980, and cost, capacity and noise in the 1990s. Since the 2000s, the environment has become an important issue besides income and cost (Karakoc et al., 2017).

In the historical process of aviation, gas turbine engines have been used most commonly in the traditional sense. Along with technological developments, traditional propulsion systems have started to be replaced by alternatives over time. As an alternative propulsion system, electric propulsion systems have come to the fore after the 2000s (Yiğit, 2018).

The first electric vehicle model was developed in the Netherlands in 1835 (Kerem, 2014). Operating an aircraft with electricity was first proposed in 1916 (Atkinson et al., 2012; Karakoç & Yıldız, 2017), but due to technological

S. C. Gorgulu (✉)
Graduate School of Natural and Applied Science, Eskisehir Osmangazi University, Eskisehir, Turkey

I. Yazar
Faculty of Engineering and Architecture Aeronautical Engineering, Eskisehir Osmangazi University, Eskisehir, Turkey
e-mail: iyazar@ogu.edu.tr

T. H. Karakoc
Eskisehir Technical University, Faculty of Aeronautics and Astronautics, Eskisehir, Turkiye

Information Technology Research and Application Center, Istanbul Ticaret University, Istanbul, Turkiye

T. H. Karakoc et al. (eds.), *New Achievements in Unmanned Systems*, Sustainable Aviation, https://doi.org/10.1007/978-3-031-29933-9_9

developments, electric aircraft came to use after the 2000s. Examples of the first aircraft to have a hybrid electric propulsion system are the Alatus, produced in 2010; the EcoEagle, produced in 2011; and the DA36 E-Star2, produced in 2013 with the cooperation of Diamond Aircraft Company Airbus and Siemens. E-Star2 has been shown to have a significant reduction in fuel consumption compared to the non-hybrid DA36 (Meyer & Wall, 2017). Many airline companies continue to work on aircraft projects with fully electric and hybrid electric propulsion systems.

According to some researchers, battery technology commonly used in current electric aircraft needs to be improved by 10 to 20 times in order for electric aircraft to compete with aircraft with conventional propulsion systems (Moore & Fredericks, 2014). Hybrid electric propulsion systems are seen as a good alternative on the road to electric propulsion due to limitations in battery technology. The definition of hybrid electric propulsion system is a propulsion system that occurs by using two or more power sources, one of which is an electric motor, together or independently (Yiğit et al., 2018). The purpose of hybrid systems is to combine the advantages of different systems in a single system (Hiserote & Harmon, 2010). The total CO_2 emissions of an electric vehicle are halved compared to a comparable internal combustion engine (Bakker, 2010). On the other hand, a battery-powered vehicle has a range problem compared to an aircraft powered by an internal combustion engine (Friedrich & Robertson, 2014a, b). Due to such differences, hybrid propulsion systems that combine the advantages of both systems are preferred. Hybrid electric drive: It provides advantages such as fuel savings, lower pollution and noise, and lower emissions (Friedrich & Robertson, 2014a, b).

2 The Concept of Sustainability

Population growth and technological developments show that worldwide energy consumption will increase exponentially every year. An increase of 2.5% was recorded in the consumption of world primary energy resources in 2012 compared to 2011. Many different sources can be used to meet the energy needs. Fossil fuels provided almost the entire supply of fuel needs in transportation sector (Ozdemir & Mutlubas, 2019). Today, the fossil fuels that meet most of the energy needs are coal, natural gas and oil (Erdinç et al., 2011). Conventional fuel use removes limitations, but increases global warming and reduces natural resources. Conventional fuel is limited in nature and harmful to the environment (Türk et al., 2018). Using these fuels leads to increase in gas emissions and air pollution.

In recent years, many studies have been conducted on the effects of air pollution on health. Exposure to particulate matter and pollutants in the air, mortality from respiratory diseases and increased hospitalizations have been associated with each other (Brunekreef & Holgate, 2002).

With the outbreak of the COVID-19 pandemic in 2020, there was a pause in many sectors, including the transportation sector, and mandatory quarantines disrupted the transportation sector. As a result, the use of oil worldwide has decreased drastically, and as a result, a significant reduction in air pollution was observed (Hosseini, 2020).

Considering the role of the transportation sector in air pollution, it can be seen that it is necessary to take environmentalist steps. Environmentalist steps are taken in the aviation sector as well. One of them is the orientation towards energy storage systems, which is an important step for reducing gas emissions and increasing the orientation towards renewable energy (Kusdogan, 2017). Due to the nature of electrical energy, it must be stored as well as produced in order to be used throughout the flight (Yıldız & Karakoç, 2017; Kocaman, 2013). Energy needs may vary at different times. Energy storage technologies are also very important in this respect (Hadjipaschalis et al., 2009).

In hybrid systems, the part that provides the electrical propulsion is usually a battery, but due to the limitations in the current battery technology, different searches are on-going. One energy storage technology that stands out due to its many advantages is supercapacitors.

3 Supercapacitor Technology

Capacitor technology is much older than batteries, existing since 1745. Supercapacitors, on the other hand, are a newer technology. In 1966, engineers at Standard Oil patented an unexpected supercapacitor technology while working on fuel cells (Schindall, 2007).

By definition, a supercapacitor is a device that stores electrical energy at the interface between an electrolytic solution and a solid electrode (Sharma & Bhatti, 2010). Supercapacitors basically consist of a liquid electrolyte, two opposite electrodes and a separator. Although the separating surface acts as a barrier between the electrodes, it allows ion passage (Çalıker & Özdemir, 2013). When voltage is applied to the supercapacitor, an electric field is created between the plates, and the capacitor's energy is stored in this electric field. The capacitance of a supercapacitor depends on the surface area of the electrodes, their distance from each other, and the dielectric constant of the separator. The electrode material of the supercapacitor is generally activated carbon with a large surface area. In addition, there are many studies on increasing the surface area of the electrodes. Figure 1 shows the construction of a conventional capacitor and Fig. 2 shows the construction of a supercapacitor.

Supercapacitors, such as batteries and fuel cells, consist of an electrolyte solution and two electrodes, but redox reactions occur inside batteries and fuel cells, while chemical reactions do not occur inside supercapacitors. Polarization occurs with the movement of ions at the electrode/electrolyte interface (Winter & Brodd, 2004). Ions can also be quickly released again. Since no chemical reaction takes place inside the supercapacitor, it can be charged-discharged very quickly. In addition, thanks to this feature, they also have an infinite loop life. Although fuel cells and batteries have been used in electric vehicles for a long time, they are insufficient in applications that

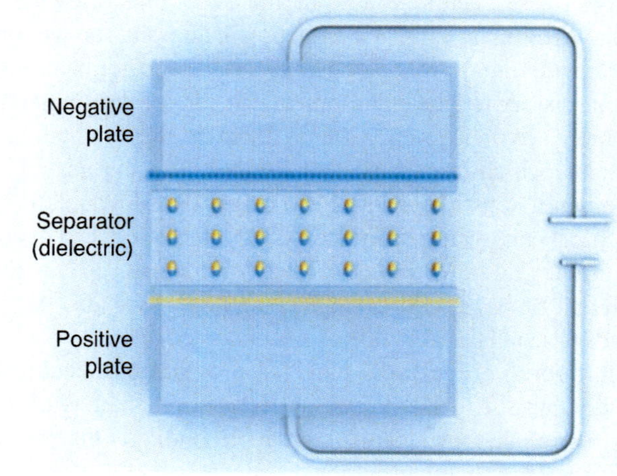

Fig. 1 The capacitor structure (Schindall, 2007)

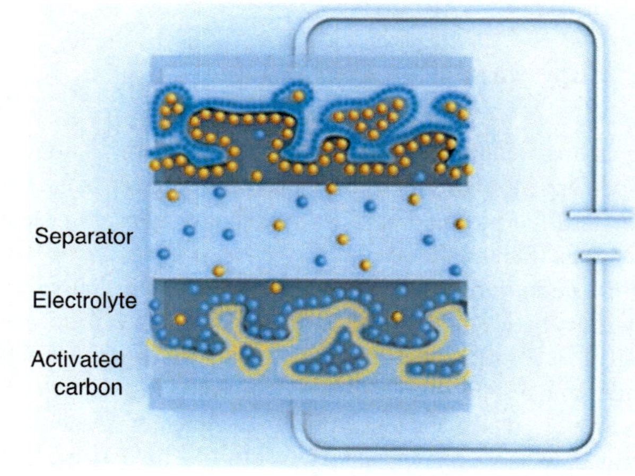

Fig. 2 The supercapacitor structure (Schindall, 2007)

require high specific power such as rapid acceleration or regenerative braking (Macro et al., 2005). While rechargeable batteries have low specific power and high specific energy value, the opposite is true for supercapacitors (Cericola et al., 2011). Figure 3 shows the specific energy vs specific power graph of different technologies.

Supercapacitors have higher specific power values than batteries (Bocklisch, 2016). The specific power value is related to capacity and indicates that supercapacitors can last longer than batteries. This is a very important advantage for the aviation industry. Another advantage of supercapacitors is that they can operate over a wide temperature range. While batteries can cause accidents such as burning, melting and explosion at high temperatures, such a problem does not occur with supercapacitors (Akgundogdu et al., 2017). Finally, the materials from which the supercapacitors are made are environmental-friendly, which is why electric propulsion is preferred.

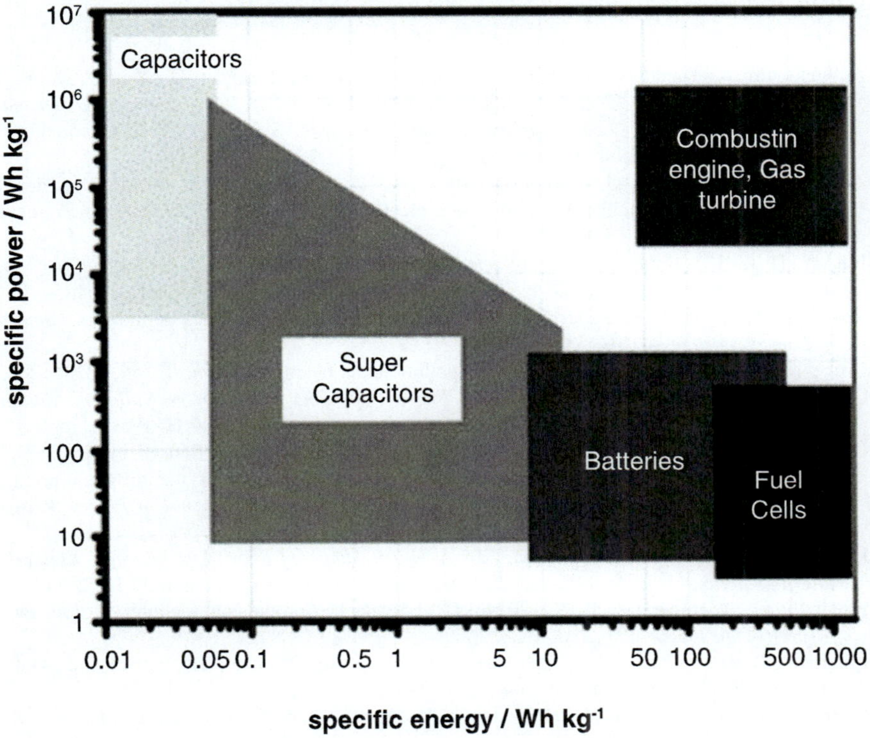

Fig. 3 Specific power – specific energy graph of different propulsion systems (Winter & Brodd, 2004)

4 Conclusion

The aviation industry also plays its part in the fight against the climate crisis and other environmental problems around the world. Environmentalist steps are taken in different areas throughout the sector. The use of electrical propulsion instead of traditional propulsion systems will reduce gas emissions and provide a cleaner atmosphere for living things. While the all-electric drive is a technology with limitations at the moment, hybrid systems are promising. In order for electrical energy to be used in the aviation industry, it must be stored. For this, batteries, which are electrochemical energy storage devices, can be used with fuel cells and supercapacitors. Each technology has different advantages and when used together, these advantages can be combined. Among them, the newest technology, supercapacitors, is a promising technology in many ways. Studies on supercapacitors continue.

References

Akgundogdu, A., Karadeniz, O., Şahin, U., İn, S., Tiryaki, H., Erdoğan, G., Yılmaz, M. Y., & Kocaarslan, İ. (2017). *Elektrikli Araçlar için Batarya Paketi ve Batarya Yönetim Sisteminin Gerçeklenmesi*. World electro mobility conference. Retrieved September 25, 2020, from https://www.emo.org.tr/ekler/ebfdbf02befde2a_ek.pdf

Atkinson, D. J., Atkinson, G. J., Bennett, J. W., Cao, W., & Mecrow, B. C. (2012). Overview of electric motor technologies used for more electric aircraft (MEA). *IEEE Transactions on Industrial Electronics, 59*(9).

Bakker, D. (2010). *Battery electric vehicles: Performance, CO2 emissions, lifecycle costs and advanced battery technology development*. Copernicus Institute University of Utrecht.

Bocklisch, T. (2016). Hybrid energy storage approach for renewable energy applications. *Journal of Energy Storage, 8*(2016), 311–319.

Brunekreef, B., & Holgate, S. T. (2002). Air pollution and health. *The Lancet, 360*, 1233–1242.

Çalıker, A., & Özdemir, E. (2013). *Modern Enerji Depolama Sistemleri ve Kullanım Alanları*. Retrieved October 17, 2020, from http://www.emo.org.tr/ekler/0a55200ff16175a_ek.pdf

Cericola, D., Novák, P., Wokaun, A., & Kötz, R. (2011). Hybridization of electrochemical capacitors and rechargeable batteries: An experimental analysis of the different possible approaches utilizing activated carbon, Li4Ti5O12 and LiMn2o4. *Journal of Power Sources, 196*(2011), 10305–10313.

Erdinç, O., Uzunoğlu, M., & Vural, M. (2011). *Hibrit Alternatif Enerji Sistemlerinde Kullanılan Enerji Depolama Üniteleri*. Electrical-electronics and computer symposium FEEB 2011.

Friedrich, C., & Robertson, P. A. (2014a). Hybrid-electric propulsion for aircraft. *Journal of Aircraft*. https://doi.org/10.2514/1.C032660

Friedrich, C., & Robertson, P. (2014b). Hybrid-electric propulsion for automotive and aviation applications. *CEAS Aeronautical Journal, 6*, 279–290.

Hadjipaschalis, I., Poullikkas, A., & Efthimiou, V. (2009). Overview of current and future energy storage technologies for electric power applications. *Renewable and Sustainable Energy Reviews, 13*(2009), 1513–1522.

Hiserote, R. M., & Harmon, F. G. (2010). *Analysis of hybrid-electric propulsion system designs for small unmanned aircraft systems*. 46th AIAA/ASME/SAE/ASEE joint propulsion conference & exhibit, 25–28 July 2010, Nashville, TN.

Hosseini, S. E. (2020). An Outlook on the Global Development of Renewable and Sustainable Energy at the time of COVID-19. *Energy Research Social Science, 68*(2020), 101633.

Karakoç, T. H., & Yıldız, M. (2017). Havacılıkta Kullanılan Bataryaların Tasarım Parametrelerine Göre Boyutlandırılması. *Journal of Sustainable Aviation Research, 2*(1).

Karakoc, T. H., Yazar, I., & Yigit, E. (2017). *A new trend for future aircraft propulsion: Electric propulsion*. International symposium on sustainable aviation, Kiev, Ukraine.

Kerem, A. (2014). Elektrikli Araç Teknolojisinin Gelişimi ve Gelecek Beklentileri. *Journal of Mehmet Akif Ersoy University Institute of Science and Technology, 5*(1), 1–13.

Kocaman, B. (2013). Akıllı Şebekeler ve Mikro Şebekelerde Enerji Depolama Teknolojileri. *BEÜ Fen Bilimleri Dergisi, 2*(1), 119–127.

Kusdogan, S. (2017). *Yenilenebilir Enerji Uygulamalarında Hibrit Enerji Depolama Teknolojileri ve Uygulamaları*. Retrieved February, 2021, from https://www.emo.org.tr/ekler/c66c76acabfcfa5_ek.pdf

Macro, S. W., Chan, K. T., Chau, C. C., & Chan. (2005). Effective charging method for ultracapacitors. *Journal of Asian Electric Vehicles, 3*(2), 771–775.

Meyer, R. T., & Wall, T. J. (2017). *A survey of hybrid electric propulsion for aircraft*. AIAA Propulsion and Energy Forum, 10–12 July 2017, Atlanta, GA.

Moore, M. D., & Fredericks, B. (2014). *Misconceptions of electric propulsion aircraft and their emergent aviation markets*. 52nd aerospace sciences meeting, National Harbor, Maryland.

Ozdemir, Z. O., & Mutlubas, H. (2019). Enerji Taşıyıcısı Olarak Hidrojen ve Hidrojen Üretim Yöntemleri. *Bartın University International Journal of Natural and Applied Sciences. JONAS, 2*(1), 16–34.

Schindall, J. (2007). The charge of the ultra-capacitors. *IEEE Spectrum, 44*(11), 42–46.

Sharma, P., & Bhatti, T. S. (2010). A review on electrochemical double-layer capacitors. *Energy Conversion and Management, 51*(2010), 2901–2912.

Türk, I., Yazar, I., Basaran, F. Ü., & Karakoc, T. H. (2018). *Simulation Based mathematical model of a solar powered DC motor for UAV applications*. Global conference on global warning.

Winter, M., & Brodd, R. J. (2004). What are the batteries, fuel cell and super-capacitors? *Chemical Reviews, 2004*(104), 4245–4269.

Yiğit, E. (2018). *Orta Sınıf Bir İnsansız Hava Aracının Elektrikli İtki Sisteminin Matematiksel Modellemesi ve Deneysel Doğrulanması*. M.Sc. thesis, Eskişehir Anadolu University Graduate School of Natural and Applied Sciences.

Yiğit, S., Yazar, I., & Karakoç, T. H. (2018). *Hibrit Elektrik İtki Sistemleri ve Sistem Tasarımında Komponent Seçiminin Önemi*. VII. National aeronautics and space conference, Ondokuz Mayıs University, Samsun.

Yıldız, M., & Karakoç, T. H. (2017). Havacılıkta Kullanılan Bataryaların Tasarım Parametrelerine Göre Boyutlandırılması. *Journal of Sustainable Aviation Research, 2*(1).

Improving the Risk Matrix

Sam Yoo, Dro Gregorian, Andrew Kopeikin, and Nancy Leveson

Nomenclature

FRWA	Future Rotary-Wing Aircraft
CMES	Combined Mitigation Effectiveness Score
CPMS	Combined Post-Mitigation Severity
MES	Mitigation Effectiveness Score
PMS	Pre-Mitigation Severity
PPMS	Post-Potential Mitigation Severity

1 Introduction

This chapter describes a potential alternative to the standard risk matrix that offers a more informative assessment of system risk. Risk is typically defined as a combination of probability/likelihood and severity/consequence of an event occurring. The risk matrix shown in Fig. 1 is a widely used analysis and reporting tool, often incorporated into risk management processes, to help decision makers address risk. It is a foundational tool of risk management in many industries, and program managers are often required to use it.

The strengths of the risk matrix that makes it so widespread are that (1) it is inherently simple to understand, (2) it promotes robust discussion, (3) it offers some consistency to prioritizing risks, and (4) it helps decision-makers to focus on the highest priority risks (Talbot, 2017).

S. Yoo (✉) · D. Gregorian · A. Kopeikin · N. Leveson
Massachusetts Institute of Technology, Cambridge, MA, USA
e-mail: sam.yoo@westpoint.edu; drogreg@mit.edu; kopeikin@mit.edu; leveson@mit.edu

T. H. Karakoc et al. (eds.), *New Achievements in Unmanned Systems*, Sustainable Aviation, https://doi.org/10.1007/978-3-031-29933-9_10

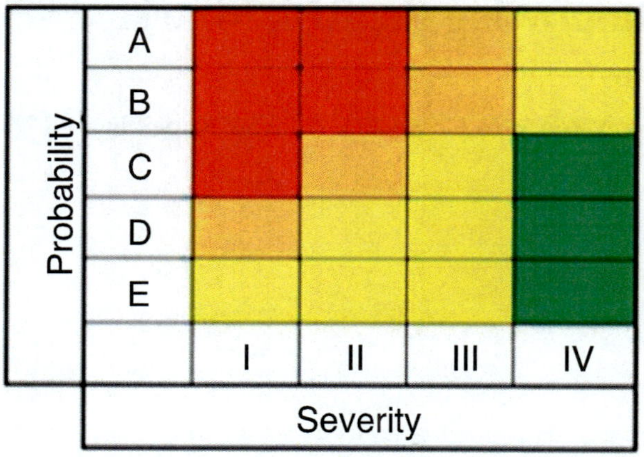

Fig. 1 MIL-STD-882 risk matrix (DoD, 2012)

However, there are serious problems with the use of risk matrices. The authors conducted 23 informal interviews with subject matter experts across different programs within the Department of Defense (DoD) about their views on risk management practices and how to improve them (Gregorian & Yoo, 2020). The interviews highlighted three significant shortfalls within the current DoD risk management process, which we suspect may apply to many, if not most, industries. First, there exists the organizational and cultural problem of treating risk management as a "check-the-box" effort rather than a critical component of program success. Second, there is a clear lack of a methodology in terms of identifying risk items. Third, there is a failure to prioritize risks properly due to a lack of experience, knowledge, or effective assessment methods.

There are additional fundamental shortfalls. Risk matrices usually focus on failures rather than hazards, which provides a reliability estimation rather than a comprehensive safety assessment. Another problem is related to how severity and likelihood are defined. In most uses, the ordinal scales for rating likelihood and consequence are qualitatively determined. This fails to consider more advanced methods to calculate likelihood, consequence, and the nuances between categories within the risk matrix (Leveson, 2019). Other weaknesses include: (1) prioritization of risk based on overall score instead of mitigation ability, (2) inaccurate quantitative analysis, (3) subjectivity to various psychological biases, (4) lack of skilled assessors (Leveson, 2019). Several different approaches have been proposed to improve on these weaknesses (Ni et al., 2010), but they still leave significant room for improvement in risk matrix implementation.

These problems drove the creation of a new risk matrix informed by the System Theoretic Process Analysis methodology (Leveson & Thomas, 2018), which represents risk as a form of hazard mitigation instead of arbitrary probabilities. The remainder of this chapter describes the methodology called the Scenario-Based Approach for the STPA-Informed Risk Matrix. Its feasibility is then demonstrated with an example applied to the risk assessment of a hypothetical Future Rotary-Wing Aircraft (FRWA) system design.

2 Methodology

The proposed risk assessment process is done in four steps. First, an STPA analysis is performed to derive causal scenarios that could lead to a loss unacceptable to the stakeholders. These scenarios systematically identify the hazards that need to be mitigated. Second, the effectiveness of potential mitigations to each of these risks are objectively evaluated and combined to formulate a *Combined Mitigation Effectiveness Score* (CMES). Third, the severity of the mitigated solution is assessed to form a *Combined Post-Mitigation Severity* (CPMS) score. Fourth and finally, the results are plotted onto a STPA-Informed Risk Matrix (SRM). Assuming all mitigations are applied, CMES represents the quality, rather than quantity, of mitigation effectiveness while CPMS represents the overall reduction in severity. The following subsections expand on these details.

2.1 System Theoretic Process Analysis (STPA)

STPA is a hazard analysis method that identifies potential causal factors in accidents, based on an extended accident causation model rooted in systems theory. STPA considers the widespread causes of mishaps (losses) today, including human operator error, organizational and managerial flaws, unsafe or inadequate software behavior, requirements and system design flaws, and component interactions (with failures included as a subset). Compared to traditional methods that focus on component failure and human error, STPA treats accidents as a control problem. This method scales with system complexity and allows for hazard analysis in the early stages of system development. Without addressing hazards that arise from interactions among system components, the accuracy of the overall risk assessment is significantly reduced, which can lead to major losses once the system is operational.

There are four steps in executing STPA, which are explained in the STPA Handbook (Leveson, 2011). The first step of STPA is to define the system boundary, unacceptable losses to the stakeholders, and hazards that could lead to these losses. The second step involves creating a control structure to provide a hierarchical, top-down view of the various controllers and controlled processes within the system boundary. The third step systematically identifies control actions within the control structure (see Fig. 3) that could lead to a hazard within a specific context. Finally, the last step is to identify loss scenarios, which are specific ways that unsafe control actions can occur.

2.2 Mitigation Effectiveness Assessment

Probability estimation for new systems is difficult to quantify objectively because it depends on historical data, which likely does not exist for new designs. It is especially difficult for designs involving new technology with complex interactions between operators, experimental components, software, and hardware.

An alternative is to use *mitigation effectiveness* as a proxy for likelihood. Mitigation effectiveness estimates how well a hazard can be controlled, rather than estimating the probability of its occurrence. The two are obviously related, but the former can be more readily estimated earlier in a system life cycle. Mitigation effectiveness is informed by the results of STPA, a structured and repeatable analysis method that can start at the conceptual development stage before a concrete design exists.

Mitigation effectiveness is derived from the *Safety Order of Precedence* (FAA, 2000) and an assessment called "strength of potential controls" proposed in (Leveson, 2019), which introduces the following ranking preference of mitigation strategy: (1) eliminate causal factors through design, (2) reduce or control causal factors through design, (3) detect causal factors through design, and (4) use training and procedures. The process in this chapter introduces a *Mitigation Effectiveness Score* (MES) that can be assessed for recommendations for each STPA causal scenario according to Table 1.

2.3 Mitigation Risk Severity Assessment

The Pre-Mitigation Severity (PMS) of each causal scenario, i.e., before any mitigation is applied, can be determined from the worst-case severity of the resultant loss the scenario traces to the STPA. Next, the *Post-Potential Mitigation Severity (PPMS)* evaluates the potential impact of each individual mitigation's change to severity. Once PPMS has been determined for each mitigation, the combined impact of all mitigations upon severity can be assessed in a *Combined Post-Mitigation Severity (CPMS)* using Eq. 1.

Table 1 Mitigation effectiveness scores (MES)

Mitigation level	Mitigation description	MES
Eliminated	Causal factor eliminated through design or combination of below mitigations	X
Reduced	Causal factor reduced or controlled through design	3
Detected	Causal factor detected and requires response to mitigate	2
Training & procedures	Causal factor mitigated through more training and procedures	1
None	No possible mitigation exists, or mitigation never applied	0

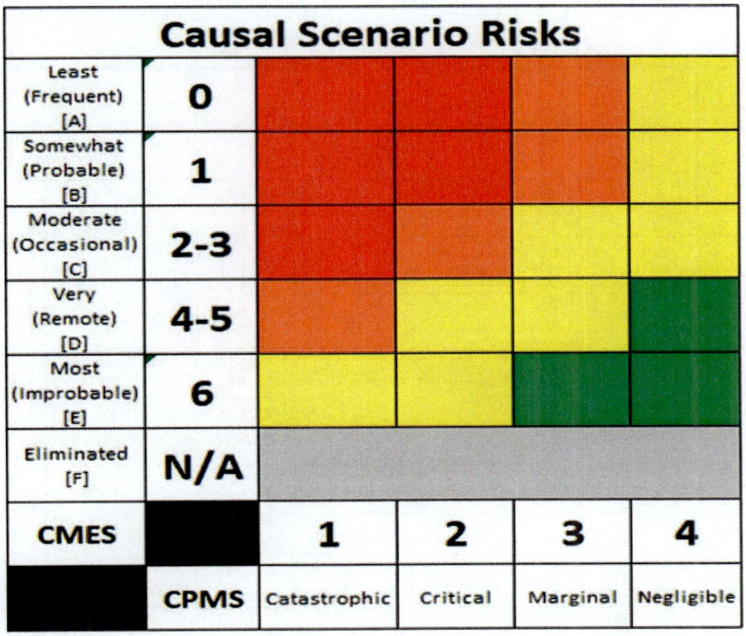

Fig. 2 STPA-informed risk matrix with two risks plotted from the results section

$$\text{CPMS} = \text{floor}\left(\frac{\sum_{1}^{N}\text{PPMS}}{N}\right) \tag{1}$$

2.4 Plotting the STPA-Informed Risk Matrix

The *STPA-Informed Risk Matrix (SRM)* is an improved version of the standard risk matrix, shown in Fig. 2. This matrix is designed with standard risk matrix formatting so that it can be seamlessly implemented in risk planning workflows. Causal scenarios from STPA are plotted on the matrix according to their mitigation effectiveness and resultant severity.

3 Results

This section describes a demonstration of the feasibility of the Scenario-Based Approach for the STPA-Informed Risk Matrix. The process was applied to the hypothetical conceptual design of a new Future Rotary-Wing Aircraft (FRWA)

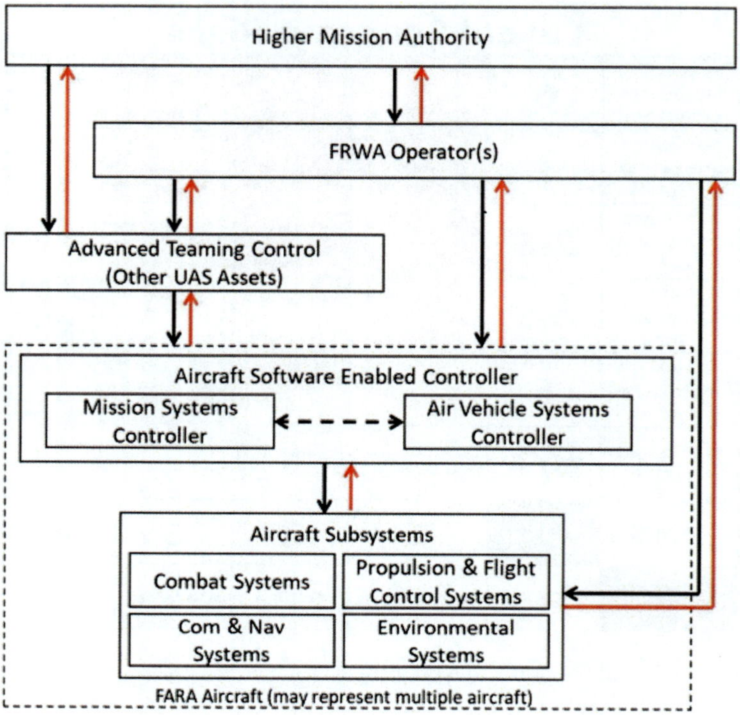

Fig. 3 FRWA high-level control structure

intended to conduct combat missions. The FRWA will be an optionally manned aircraft capable of operating fully or semi-autonomously. Without consideration of detailed design specifications, the controls that would enable the safe operation of the FRWA can be identified. The focus of this example is on system safety risk while excluding program risk factors including cost and schedule-related risks.

The first step of the process is to perform an STPA analysis of the system. For the sake of brevity in this chapter, the reader is referred to (Yoo et al., 2021) which describes this process in detail with the same FRWA example. An abstracted representation of the FRWA aircraft applicable for this analysis is presented in Fig. 3, and a more detailed view is available in the aforementioned paper. The following are example causal scenarios that are identified from the STPA analysis.

- Causal scenario – CS 2.0.1: Operator is incapacitated by enemy fire, injury, illness, and leans onto the controls accidentally activating them. As a result, the aircraft can become uncontrollable
- Causal scenario – CS 2.0.2: Operator mental model of aircraft flight control systems conflicts with reality and operator does not make the necessary manual inputs during a critical mode of flight. As a result, the aircraft becomes uncontrollable.

Table 2 Causal scenario risk calculations

RM	Recommended mitigation	MES	CMES	PPMS	CPMS
RM 01	Aircraft health systems monitor aircraft performance and alert operator to unsafe scenarios	2	3	4	3
RM 02	Emergency response system detects problems and alerts operator, awaits input, and auto engages if outside the time window allotted for operator action	2		3	
RM 03	Operator trained on proper preflight/flight/emergency procedures and techniques	1		2	

The risk assessors produce mitigations for each scenario and score the severity of outcome for each scenario. Next, they create recommended mitigations and score those according to the MES table. After applying each mitigation, a post-mitigation severity score is calculated. Lastly, CMES and CPMS are calculated and then each scenario is plotted on the SRM. The assessed CMES and CPMS are shown in Table 2.

With CMES and CPMS calculated, the scenarios can be plotted on the SRM, as shown in Fig. 2. These scenarios are considered medium risks per the SRM. This format affords risk planners a consolidated view of which scenarios to address based on resources available during the design process. It is important to note the assumption that all mitigations are applied simultaneously. Also, another iteration of risk assessment may need to be performed if the proposed mitigations cause additional hazards. This is to be expected. The goal is to iterate the risk assessment process until the system design is in a state acceptable for the design to move forward to the next review milestone.

4 Conclusions

Creating useful and accurate risk matrices is difficult. This chapter describes a new approach that uses results from STPA to provide a detailed assessment of risk within the context of unsafe causal scenarios. The concept of mitigation effectiveness provides an assessment based on the quality of the controls of those risks. Because STPA is an iterative process, the losses, hazards, and constraints may be refined as the designers mitigate risks. Eventually, all system designs need to move forward in development, at which point some types of design changes may no longer be feasible.

The scenario-based approach provides multiple strengths in its identification and mitigation of hazards. By redefining risk in terms of the causal scenarios generated by STPA, risk is captured far more completely than other forms of analysis. Additionally, other methodologies focus on component reliability, which is not the same as safety in accidents that do not result from component failures.

One potential weakness of this approach is that scenarios need to be generated. However, they will be generated later in the development process anyway if any hazard analysis is performed. In addition, there is research that shows STPA can be far less time-intensive than other methodologies (Yahia & Fawzy, 2013).

Another aspect of the scenario-based approach that serves as both a strength and weakness is how MES and CMES are calculated. A linear scale is used for mitigation effectiveness levels as opposed to other mathematical distributions. The primary reason for this choice was to simplify the understanding and application of the methodology for the project planner. While the use of other scales for calculating MES and CMES could have been applied (e.g., logarithmic, exponential, etc.), they were less intuitive.

STPA and the risk assessment approach presented can help provide important information to decision makers before detailed system designs are created. Additional approaches may also be possible and should be explored. Given the problems in producing accurate and useful risk matrices today, future improvements could greatly improve safety and project management across all industries.

Acknowledgments The authors would like to acknowledge the broader FRWA safety and security assessment team from MIT that participated in the STPA analysis.

References

Department of Defense. (2012). *System safety*. MIL-STD 882E.

Federal Aviation Administration. (2000). Chapter 3: Principles of system safety. In *System safety handbook*.

Gregorian, D., & Yoo, S. (2020). *Collection of risk management SME interviews*. MIT Research Notes.

Leveson, N. (2011). *Engineering a safer world: Systems thinking applied to safety*. The MIT Press.

Leveson, N. (2019). *Improving the standard risk matrix: Part 1*. [Online]. Available: www.sunnyday.mit.edu/Risk-Matrix.pdf

Leveson, N., & Thomas, J. (2018). *STPA handbook*. [Online]. http://psas.scripts.mit.edu/home/get_file.php?name=STPA_handbook.pdf

Ni, H., Chen, A., & Ning, C. (2010). Some extensions on risk matrix approach. *Safety Science, 48*(10), 1269–1278.

Talbot, J. (2017). *What's right with risk matrices?* [Online]. Available: www.juliantalbot.com/post/2018/07/31/whats-right-with-risk-matrices

Yahia, H., & Fawzy, E. (2013). *Range extender system for electric vehicles*. Valeo. [Online]. Available: http://psas.scripts.mit.edu/home/wp-content/uploads/2013/04/02_Yahia_STAMP-STPA-case-study-on-Electric-vehicle-range-extender.pdf

Yoo, S., Kopeikin, A., Gregorian, D., Munekata, A., Thomas, J., & Leveson, N. (2021). *System-theoretic requirements definition for human interactions on future rotary-wing aircraft*. International Symposium on Aviation Psychology.

The Artificial Immune System Paradigm for Generalized Unmanned Aerial System Monitoring and Control

Ryan McLaughlin and Mario Perhinschi

1 Introduction

It is of critical importance that an autonomous unmanned aerial vehicle (UAV) operates properly both under normal and abnormal conditions (AC) (FAA, 2018). The artificial immune system (AIS) paradigm has emerged as a promising tool for achieving such an objective (Dasgupta, 1998; Perhinschi & Moncayo, 2018). This chapter presents the envisioned general development process of an integrated and comprehensive methodology for monitoring and control of an autonomous UAV inspired by the beneficial properties of the biological immune system (BIS) (Coico & Sunshine, 2015). This includes outlining for each phase of the AC monitoring and control process of the biological sources of inspiration and the implementation methods with their potential benefits and drawbacks.

2 Method

As a monitoring and control system, the AIS would be tasked with addressing the operation of the UAV with respect to ensuring maximum safety and performance characteristics. The dynamic system itself, the UAV in this case, would be expected to maintain certain desirable or at least acceptable levels of performance in completing a mission under both normal and abnormal operational conditions. In general, it is expected that this monitoring and control system must be able to handle the presence of ACs, a process that would be broken down into the detection, identification, evaluation, and finally accommodation (DIEA) of the AC in question.

R. McLaughlin (✉) · M. Perhinschi
West Virginia University, Morgantown, WV, USA
e-mail: rgmclaughlin@mix.wvu.edu; Mario.Perhinschi@mail.wvu.edu

T. H. Karakoc et al. (eds.), *New Achievements in Unmanned Systems*, Sustainable Aviation, https://doi.org/10.1007/978-3-031-29933-9_11

Additionally, to ensure maximum effectiveness and efficiency, the monitoring and control system should consist of one single entity that is properly structured but simple, flexible, adaptive, and extendable. It should also be able to adapt to ACs that the dynamic system has not been exposed to before.

The AIS paradigm is an artificial intelligence technique that takes great inspiration from the BIS for each step of the DIEA process. Beginning with the detection step of the DIEA process, the AIS takes inspiration from the self/non-self discrimination that is performed by specialized immunity cells. When exposed to some antigen, the BIS can quickly determine that there is some invader into the body by comparing antigen-specific and own-specific chemical markers. In terms of the AIS, the "body" would be represented by nominal function of the dynamic system, in this case the UAV, and the antigens would be some AC that is affecting the system, which would be represented by some function of the system that would fall outside of nominal conditions. Through similarity to the chemical compounds that define both the antigen and the "body," a set of "features" in terms of UAV significant dynamic variables must be defined and used to discriminate between normal and abnormal operation.

The mechanisms by which the BIS is able to determine whether an antigen-fighting T-cell is capable of functioning as desired are through a two-step positive and negative selection process. Through the positive selection process, the T-cells that are capable of recognizing and binding to the antigen peptides are kept, while those who fail to recognize them are eliminated. In the following negative selection process, the T-cells that bind to the body peptides are removed because those T-cells could cause a detrimental autoimmune response, so only T-cells that do not bind to the self-peptide are kept. Both positive and negative selection-type processes have applications when building, structuring, and determining the functionality of the AIS.

The AIS itself will be structured by creating a body of data that represents the functioning of the dynamic system under all nominal conditions. To do this, one must define a feature set, that is, the list of all variables that will be critical to determining how the system is functioning. In the case of a UAV, these variables include, but are not limited to, aircraft dynamics, aircraft state, and control system outputs. These variables are then collected through either simulation, actual flights, or some combination of each. These raw data will be normalized and clustered to create a body of data that will represent the self. Additionally, data can also be collected under the effect of an AC to attempt to structure the non-self in a similar way to the self, which can be useful throughout the AC-DIEA process.

Positive/negative selection is key to how the AIS will determine whether to trigger a detection of an AC. By positive selection, an incoming data point during a flight will be compared to each "self" cluster as previously defined. If the data point is found to fit in no cluster, it is considered to not be a part of the self and therefore represents function under an AC. Conversely, if the non-self has also been structured, the incoming data point can be compared to each non-self cluster. Then, if the data point is found to fit inside one of these non-self clusters, the point will be negatively selected as being representative of function under an AC. The amount of

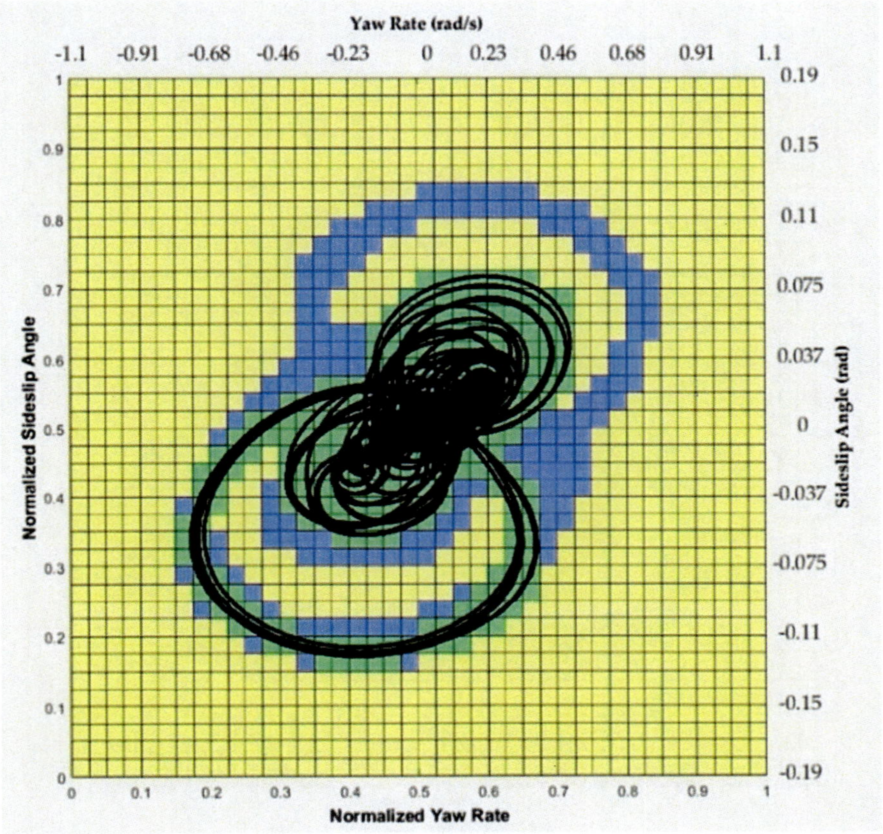

Fig. 1 Nominal test data

these data points that are abnormal are counted over a moving time window and, if the sum is higher than a particular threshold, an AC is declared. Under general operating conditions, the positive selection algorithm will be less computationally intensive, as the self cluster will be matched as opposed to using negative selection where all non-self clusters would need to be checked. Conversely, if some abnormal condition has occurred, the negative selection process will be able to react faster for the same reason as before. An example of this type of detection process can be seen in Figs. 1 and 2.

The next feature of the BIS that would be desirable for monitoring and control system is the ability to memorize previous specific antigen exposure. A small subset of T-cells is stored as memory T-cells after being used to fight an antigen. Then, if the BIS is exposed to the same antigen again, it quickly knows how to produce the T-cells to deal with the antigen. This ability is used in the AIS for the identification step of the DIEA process. In the case of the structured non-self, each of the non-self clusters can be associated with a label that represents the abnormal condition that was present when the data was collected. Then, if that abnormal condition is present

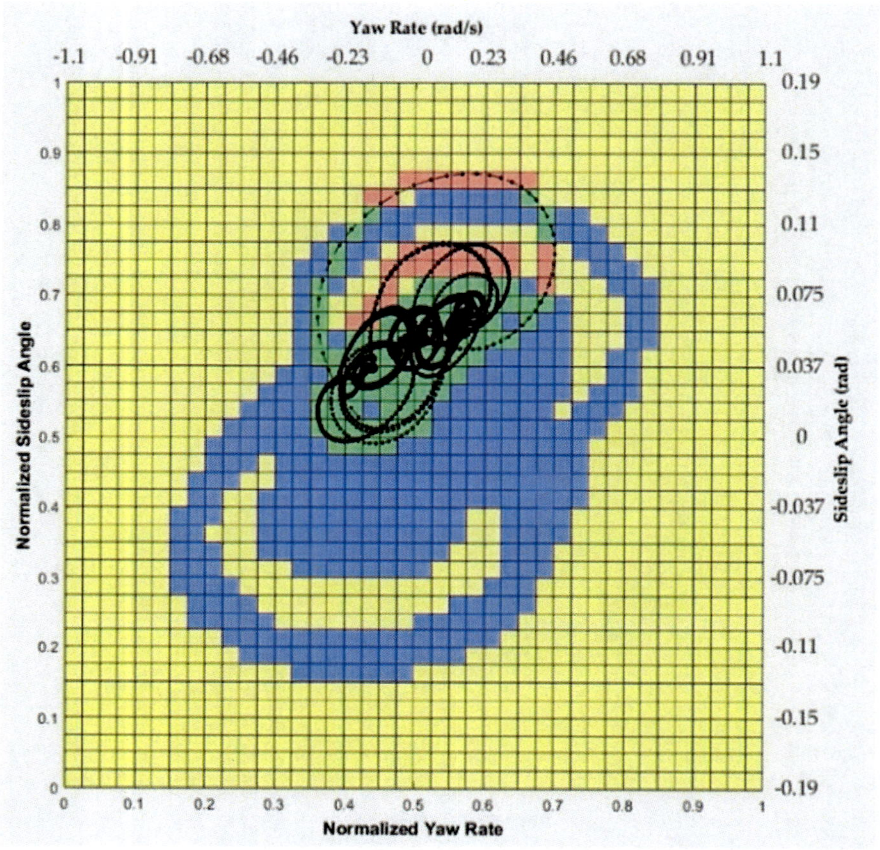

Fig. 2 Test data under the effect of an AC

again, the incoming data points will begin to fall inside that cluster. Finally, the label of the cluster serves as the AIS's way of remembering the AC that caused the dynamics of the system to change in that particular way. If no structuring of the non-self is available, another algorithm relating to the function of dendritic cells can be considered. This requires having patterns of dynamic behavior recorded for each AC and using a matching algorithm to compare how the current function of the aircraft compares to the recorded pattern.

Following this, the AIS's evaluation process takes inspiration from the ability of the BIS to both localize and evaluate the antigen invasion. In the BIS, this is handled by dendritic cells, which were also the inspiration for one of the above identification paradigms. When using the AIS, the evaluation step is divided into three main parts: qualitative evaluation, direct quantitative evaluation, and indirect quantitative evaluation. The qualitative evaluation step consists of determining the main type of the failure. For an actuator, this could include whether the actuator was locked, freely moving, or structurally damaged. This step of the evaluation process can be treated as an additional identifier and can be handled similarly to how the identification step itself is addressed. The direct quantitative evaluation step consists of determining the magnitude of the failure. When the non-self is structured, each non-self cluster could also have a severity associated with it. Approaches based on determining the distance of the abnormal points to the nearest self data cluster and dendritic cell-type algorithms have also shown success in direct evaluation. The indirect evaluation step will determine the effect that the directly evaluated AC will have on the flight envelope or equivalently the "self" of the AIS. The AC itself will limit attainable values for some features and based on the new limits of this variable, the amount of "self" clusters that the aircraft can now reach are reduced to only include those that have the involved variable within the new constrained range of values.

The accommodation step of the process can also be solved by taking inspiration from the BIS. The BIS is able to memorize, through exposure, the desired immune response to the presence of a particular antigen. For the AIS, this will translate to the ability of the system to, once it knows what fault is affecting it, deploy antigens in the form of control law changes to function appropriately despite the new limits on control authority. This step can be handled by using specialized artificial memory cells that store information on the aircraft's dynamic response when a certain command is issued. Then when a command is issued by the control laws, the AIS can search the memory cells and match the dynamics of the provided command to the dynamics of another state that can be reached with the limitations enforced by the AC.

Similar to the antigen handling by the BIS, the AIS should be capable of performing the entire AC-DIEA process in an integrated way with a single entity that is addressing each step of the process. The general BIS functionality is briefly summarized in Fig. 3. The general process of how the AIS handles AC-DIEA is summarized in Fig. 4.

One of the most important aspects to the proper function of the AIS is the process of feature selection. If a variable that is necessary to describe the flight envelope is missing, the DIEA process, particularly with respect to evaluation and accommodation, may be incomplete or unable to be performed. An example set of features that can be used for UAV AC detection and identification is: angle of attack, sideslip angle, velocity, acceleration on x, y, and z axes, roll and pitch attitude, roll, pitch, and yaw rate, estimated roll, pitch, and yaw rate, roll, pitch, and yaw acceleration, and estimation errors between measured and estimated roll, pitch, and yaw rates.

3 Results and Discussion

Using this set of features, an AIS was constructed that was able to detect and identify failures involving both sensors and actuators on board a UAV. When considering the detection of an AC, the three parameters of interest are the false alarm rate, detection rate, and time until a detection is triggered. The AIS is developed such that, for all

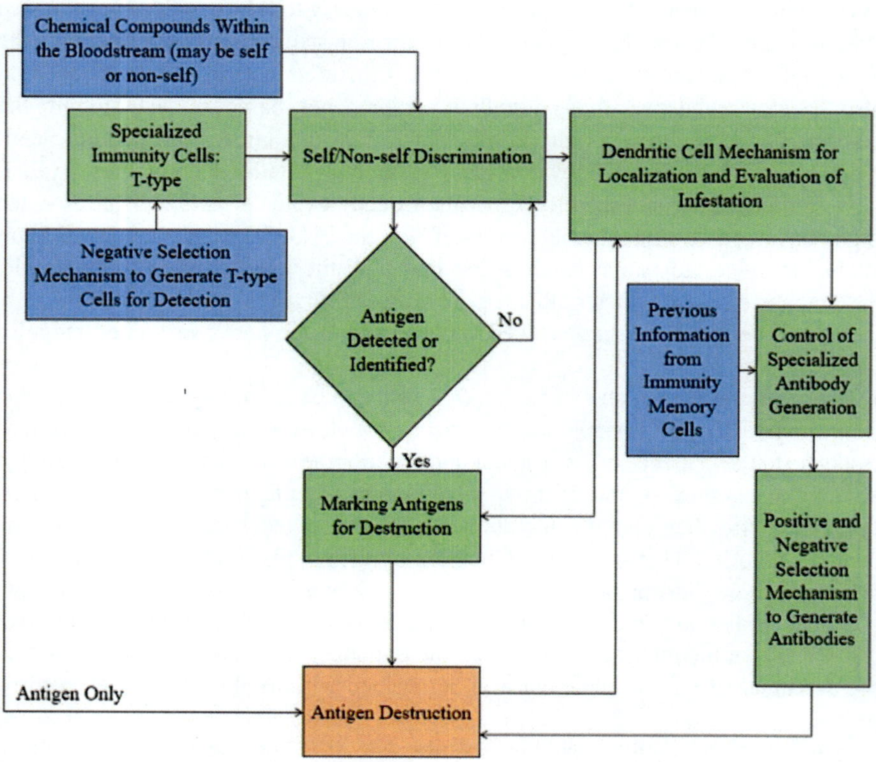

Fig. 3 General BIS functionality

ACs considered, was able to achieve 0 false alarms, that is, a detection was never triggered during nominal flight. For the other detection variables, we will consider in this example a single family of faults, which will be faults occurring on the left aileron. These faults will include both actuator locks and structural damage to the actuator. For this family of faults, an average detection rate of 87.20% was reached, with a detection time of 0.20 seconds for most cases. These results for six different flight trajectories with the left aileron locked at trim can be seen in Fig. 5.

To evaluate identification performance, the parameters of interest will be the rate of correct identifications, the identification rate, and the identification time. For the actuator faults considered, the AIS was able to produce the correct identification for each of the four trajectories considered. For specifically stabilator faults, the average identification rate over the four trajectories was 87.36%, with an average time to identification of 0.36 seconds. The identification rates for this case can be seen in Fig. 6.

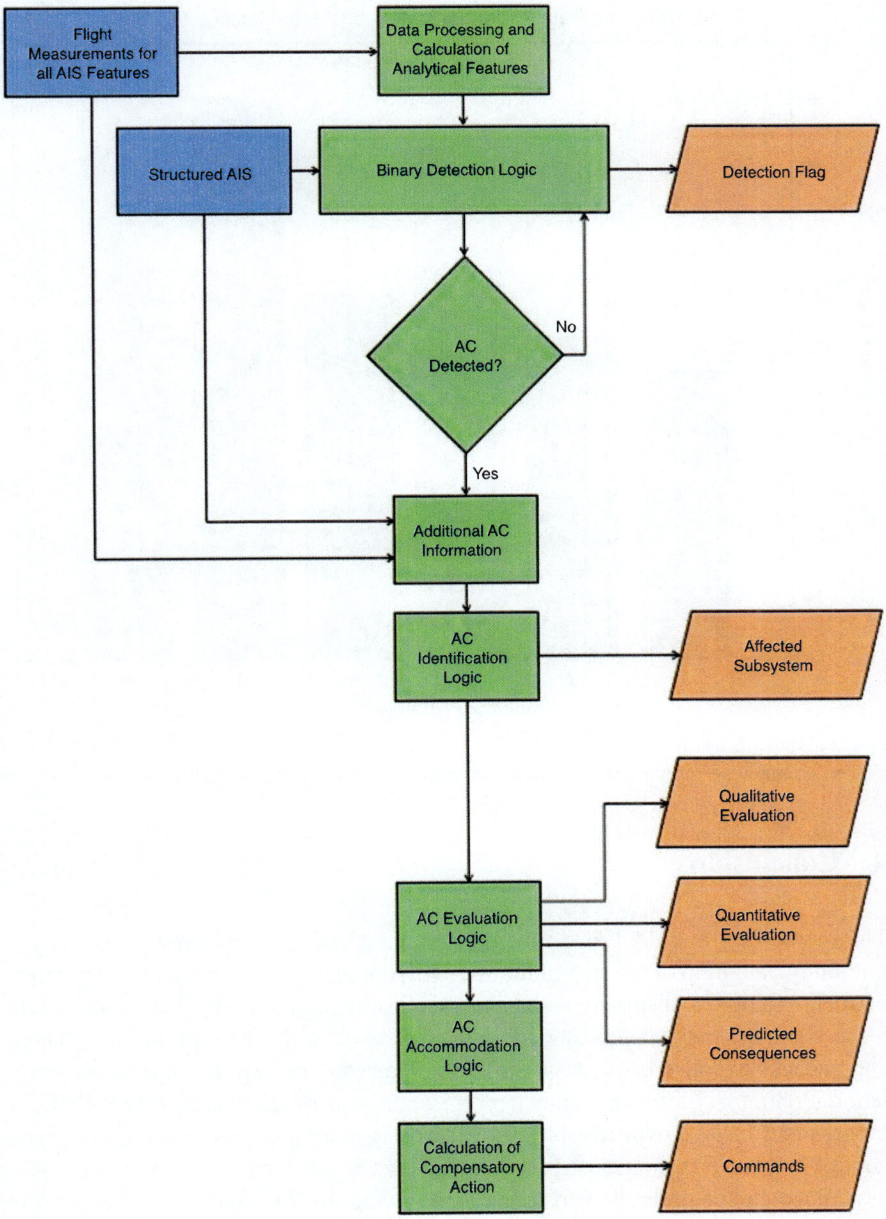

Fig. 4 General flowchart describing AC-DIEA process within the AIS paradigm

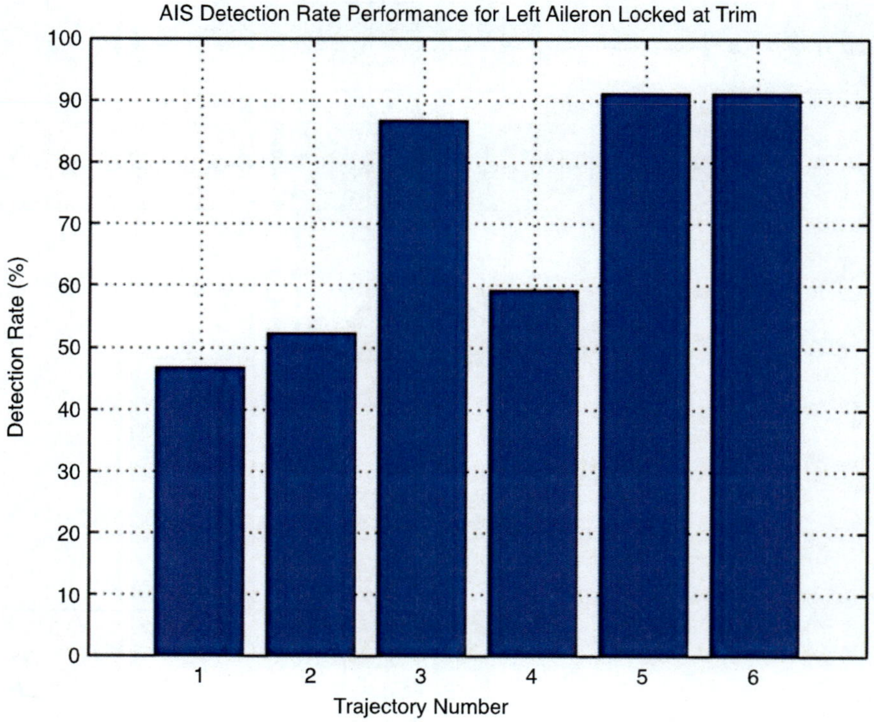

Fig. 5 Example performance of AIS detection scheme for left aileron locked at trim

4 Conclusion

The monitoring and control of an autonomous UAV is considered based on the immunity paradigm. Each step of the AC-DIEA process has been addressed in terms of their biological inspiration and the practicality of their implementation, with respect to the advantages and disadvantages of differing approaches. These approaches have also been successfully implemented and evaluated through simulation. Preliminary tests have shown promising results for each step of the AC-DIEA process and have shown capability for usage as a comprehensive monitoring and control system. The simulation results have demonstrated the ability of the proposed system to achieve desirable performance in solving the dynamic system monitoring and control problem. This research formulates the general framework for AC-DIEA using the immunity paradigm and creates the premises for developing a comprehensive and integrated solution for the monitoring and control of aerospace systems.

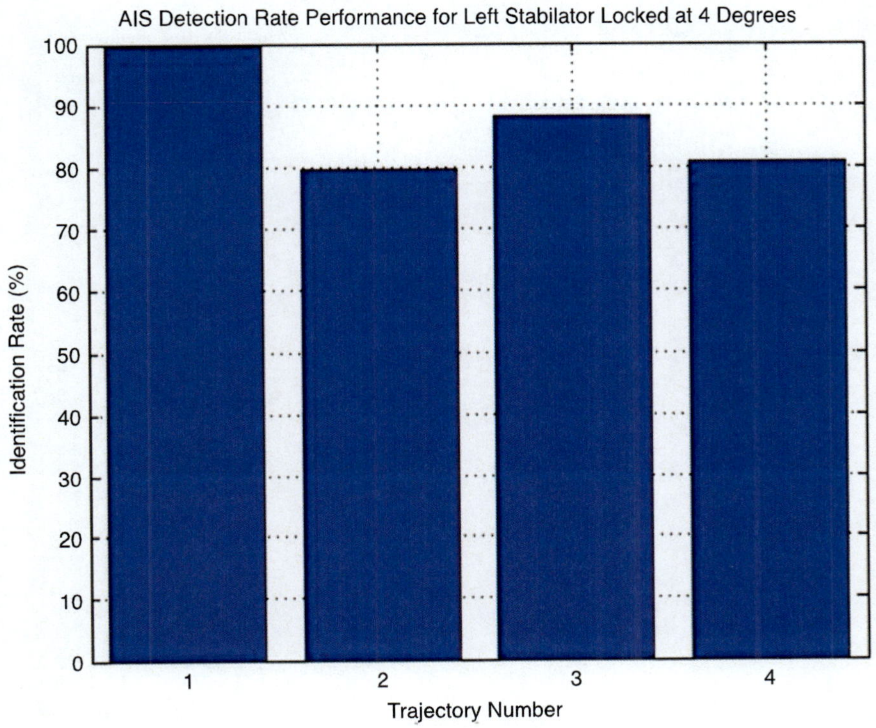

Fig. 6 Example performance of AIS detection scheme for left stabilator locked at 4 degrees

Acknowledgment Support for the first author was partially provided by the NASA West Virginia Space Grant Consortium. This research was made possible by the NASA West Virginia Space Grant Consortium, Grant # 80NSSC20M0055.

References

Coico, R., & Sunshine, G. (2015). *Immunology: A short course* (7th ed.). Retrieved August 21, 2021, from https://ebookcentral.proquest.com/lib/wvu/reader.action?docID=1936429&query

Dasgupta, D. (1998). *Artificial immune systems and their applications*. Springer.

Federal Aviation Administration. (2018). *FAA aerospace forecast, fiscal years 2019–2039*. Available at: https://www.faa.gov/data_research/aviation/aerospace_forecasts/

Perhinschi, M. G., & Moncayo, H. (2018). Artificial immune system for comprehensive and integrated aircraft abnormal conditions management. In J. Valasek (Ed.), *Advances in computational intelligence and autonomy for aerospace* (AIAA Progress in Aeronautics and Astronautics Series).

Nonlinear Six-Degree-of-Freedom Flight Modelling and Trimming of a Single-Propeller Airplane

Kasim Biber

Nomenclature

C	Mean aerodynamic chord (MAC)
C_L, C_D, C_m	Lift, drag, and pitching moment coefficients
C_{L1}, C_{D1}, C_{TX1}	Force coefficients for cruise flight
C_l, C_n, C_Y	Rolling and yawing moment and side force coefficients
C_X, C_Y, C_Z	Non-dimensional X, Y, Z body axis force coefficients
C_{Lq}, C_{Mq}	Lift and pitching moment coefficient due to pitch rate
C_{Yp}, C_{lp}, C_{np}	Side force and rolling and yawing moment coefficient due to roll rate
C_{Yr}, C_{lr}, C_{nr}	Side force and rolling and yawing moment coefficient due to yaw rate
$C_{L\delta e}$, $C_{M\delta e}$	Lift and pitching moment coef due to elevator [1/rad]
$C_{l\delta a}$, $C_{n\delta a}$, $C_{Y\delta a}$	Rolling and yawing moment and side force coef due to aileron [1/rad]
$C_{l\delta r}$, $C_{n\delta r}$, $C_{Y\delta r}$	Rolling and yawing moment and side force coef due to rudder [1/rad]
i_h	Horizontal incidence angle [deg]
I_{XY}, I_{XZ}, I_{ZX}	X-Y, X-Z, and Z-X body axis product of inertias [slug-ft^2]
m	Aircraft mass [slugs]
M	Mach number
p, q, r	Roll, pitch and yaw rates [rad/s]
S	Wing area [ft^2]
SHP	Engine shaft horse power [hp]
u, v, w	x, y and z velocity components [ft/s]
Vt	Flight velocity [ft/s]

K. Biber (✉)
Bartin University, Bartin, Turkey
e-mail: kbiber@bartin.edu.tr

T. H. Karakoc et al. (eds.), *New Achievements in Unmanned Systems*, Sustainable Aviation, https://doi.org/10.1007/978-3-031-29933-9_12

W	Aircraft weight [lb]
x, y, z	Coordinate axis
x_E, y_E, z_E	X, Y and Z earth axis positions

Greek Letters

α	Angle of attack
β	Sideslip angle
δ_a, δ_e, $\delta_{r:}$	Deflection angles for aileron, elevator and rudder
Θ, Ψ, Φ	Euler pitch, yaw and roll angle

1 Introduction

Modeling and simulation using computer tools have become an essential element to the mission success of air vehicles. The range of missions involves aircraft operations within the flight envelope, in which the aircraft has to follow a certain flight path. There has to be a reasonable confidence of flying such a flight path before the mission is undertaken.

Modelling the flight dynamics is very useful in planning the aircraft mission and in turn the aircraft design. The simulations based on the modelling provide aircraft response to commands and controls, allowing a careful evaluation of mission flight paths. The value and usefulness of the modelling is closely tied to the reliability and the type of modelling that represents the flight dynamics. There are analytical and experimental methods commonly used for such flight modelling. Analytical techniques are highly advanced with developments in computer modelling of fluid dynamics. Such that, flight dynamics modelling can be integrated with structural dynamics to investigate aeroservoelasticity and flight loads (Dussart et al., 2018; Nuguyen & Tuzcu, 2009) or propeller theory to account the effects of propeller wake on airframe aerodynamics (Kim et al., 2014). However, these methods have to be correlated to experiments before they can be used in confidence. A combination of analytical and experimental methods is commonly used today, as in (Kim et al., 2014; Chen et al., 2009) for flight dynamics modelling of unmanned air vehicles.

In this investigation, the methodology outlined in (Stevens & Lewis, 2003) was adopted for modelling the flight dynamics of an example aircraft. Because of its unique design features, wind-tunnel tests were conducted on a power-off and stick-fixed 1/12 scale model of the example airplane. The tests included the investigation of flap, elevator, rudder, and aileron deflection on longitudinal and lateral-directional aerodynamics. Using the wind tunnel data along with analytical estimates obtained from USAF DATCOM report (Hoek & Fink, 1960), the nonlinear differential equations for aircraft state vectors as presented explicitly in (Garza & Morelli,

2003) were solved numerically for six-degree-of-freedom flight. The solution was made by using MATLAB programming, including trim and numerical integration routines. Time history variations of all 12 state variables for trimmed flight are shown in plots. This type of work should be considered as a forward step towards the aircraft mission evaluation efforts in preliminary design.

2 Description of Airplane

The example aircraft, as described in (Biber, 2006, 2011, 2016) is a cost-effective and relatively short-range cargo with high-wing configuration. It is equipped with a twin-pack engine driven by a single-propeller through a unique combining gearbox. Figure 1 shows the airplane side view with its body axis. The diameter ratio of fuselage to propeller is 0.65, which is relatively high (Biber, 2011), as compared to other single propeller airplanes such as the trainer considered in (Milenkovic-Babic, 2018; (Stojakovic & Rasuo, 2016).

The airplane is in the commuter category with 19,000 lb of maximum take-off weight. Its twin-pack engine is rated at 2700 horsepower take-off setting, which is halved for one-engine inoperative condition. The propeller used is made of a composite with six blades and has a diameter of 12.9 ft. Its activity factor is 76 and design lift coefficient is 0.42. The propeller has 1200 RPM at take-off setting, and its blade aerodynamic data was available for performance calculations. The major airplane data for cruise flight is provided in Table 1 for flaps up case, as in (Biber, 2006).

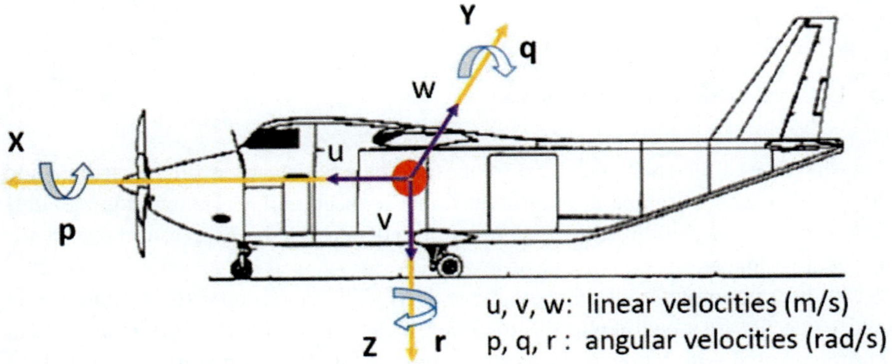

Fig. 1 Airplane side view showing linear and angular velocities on *X*, *Y*, *Z* body axis. Origin of axis is at the centre of gravity location

Table 1 The airplane data for cruise flight

Altitude, h (ft)	10,000
Mach number, M	0.25
True air speed, U_1 (ft/s)	280
c.g. location, fraction c	0.25
Angle of attack, α_1 (deg)	0
Weight, W (lbs)	19,000
I_{xx} (slug-ft^2)	50,000
I_{yy} (slug-ft^2)	45,000
I_{zz} (slug-ft^2)	90,000
I_{xz} (slug-ft^2)	0
C_{L1}	0.4
C_{D1}	0.0525
C_{TX1}	0.0525

Table 2 Static stability and control data

Wind tunnel data		Estimated data	
C_{L0}	0.303	$C_{L\dot{\alpha}}$	2.61
C_{D0}	0.0518	C_{llr}	9.665
C_{M0}	0	C_{Lu}	0.027
$C_{L\alpha}$	5.357	$C_{D\alpha}$	0.191
$C_{L\delta e}$	0.286	C_{Txu}	−0.158
$C_{M\alpha}$	−1.662	$C_{M\dot{\alpha}}$	−14.614
$C_{M\delta e}$	−1.833	C_{Mu}	0
$C_{Y\beta}$	−0.859	C_{Mq}	−54.126
$C_{Y\delta r}$	0.229	C_{Yp}	0
$C_{Y\delta a}$	0	C_{Yr}	0.937
$C_{n\beta}$	0.132	C_{np}	−0.038
$C_{n\delta r}$	0.132	C_{nr}	−0.214
$C_{n\delta a}$	−0.011	C_{lp}	−0.893
$C_{l\beta}$	−0.108	C_{lr}	0.127
$C_{l\delta r}$	−0.0229		
$C_{l\delta a}$	0.212		

3 Force and Moment Data

For the modelling analysis, the airplane coefficient data as available from wind tunnel tests was presented in two-dimensional look-up tables. The wind tunnel data was obtained in stability axis and transformed into the body axis for use in the modelling analysis.

A linear interpolation routine available in MATLAB was used to compute forces and moments for given angle of attack and side slip angle. The interpolation was complemented with a linear extrapolation in case the data temporarily exceeded the limits of look-up tables. Considering the effects of elevator, rudder and aileron as given in terms of derivatives in Table 2, total force and moment coefficients without the effects of flap, engine thrust and landing gear were calculated as follows:

Force coefficients	Moment coefficients
$C_L = C_L(\alpha) + \frac{dC_L}{d\delta_e}\delta_e$	$C_l = C_l(\alpha,\beta) + \frac{dC_l}{d\delta_a}\delta_a + \frac{dC_l}{d\delta_r}\delta_r$
$C_D = C_D(\alpha,\beta) + \frac{dC_D}{d\delta_e}\delta_e + \frac{dC_D}{d\delta_r}\delta_r + \frac{dC_D}{d\delta_a}\delta_a$	$C_m = C_m(\alpha) + \frac{dC_m}{d\delta_e}\delta_e$
$C_Y = C_Y(\alpha,\beta) + \frac{dC_Y}{d\delta_r}\delta_r + \frac{dC_Y}{d\delta_a}\delta_a$	$C_n = C_n(\alpha,\beta) + \frac{dC_n}{d\delta_r}\delta_r + \frac{dC_n}{d\delta_a}\delta_a$

The derivatives used for control surfaces in the above equations are given in Table 2. Additionally, damping derivatives were estimated for subsonic range using USAF DATCOM method provided in (Hoek & Fink, 1960). They are also presented in Table 2. Damping derivatives require the use of constants for the example airplane. The stability derivatives have per-degree units while the damping derivatives are in per radian per second.

Engine thrust force was assumed to act along the X-body axis through the centre of gravity. It was computed by linear interpolation of engine-shaft horse power database available, as a function of altitude and airspeed:

$$\text{SHP} = \text{SHP}(h, V_t)$$

The shaft horse power from two-engines was considered within a flight envelope defined by altitude and airspeed. The flight envelope has a maximum altitude of 25,000 ft. for engine powers available for cruise, climb and take-off settings. Cruise power setting at 10000 ft. altitude is considered for the present analysis.

4 Mathematical Model of Flight Dynamics

The methodology for developing the nonlinear aircraft equation of motion for six-degree-freedom flight is given in (Stevens & Lewis, 2003). The airplane is assumed rigid with constant mass density and symmetry about the X-Z plane in body axes. Forces and moments are acting on airplane from aerodynamics, propulsion and gravity. The nonlinear aircraft dynamics is modelled for both translational and rotational motion.

The equations given in literature are however transformed into a more suitable form for numerical integration with only one-time derivative on the left side of equation. These equations are given as follows for translational motion:

$$\dot{u} = rv - qw - g.\sin\theta + (\_qSC_X + T)/m$$
$$\dot{v} = pw - ru + g.\cos\theta.\sin\varphi + (\_qSC_Y)/m$$
$$\dot{w} = qu - pv + g.\cos\theta.\sin\varphi - (\_qSC_Z)/m$$

The translational equations are expressed in terms of V_T, α, β instead of u, v and w because V_T, α, β can be measured directly on real aircraft and have a more direct relationship to piloting and the aerodynamic forces and moments. The relationships between V_t, α, β and u, v, w are as follows:

$V_t = \sqrt{u^2 + v^2 + w^2}$,	$\alpha = \tan^{-1}\left(\frac{w}{u}\right)$,	$\beta = \sin^{-1}\left(\frac{v}{V_t}\right)$

and,

$u = V_t \cos\alpha \cos\beta$,	$v = V_t \sin\beta$,	$w = V_t \sin\alpha \cos\beta$

The equations for V_T, α, β can be differentiated with respect to time to give:

$V_t^{\bullet} = \frac{uu^{\bullet} + vv^{\bullet} + ww^{\bullet}}{V_t}$,	$\alpha^{\bullet} = \frac{uw^{\bullet} - wu^{\bullet}}{u^2 + w^2}$,	$\beta^{\bullet} = \frac{V_t v^{\bullet} - vV_t^{\bullet}}{V_t^2\sqrt{1-\left(\frac{v}{V_t}\right)^2}}$

For the rotational motion, equations can be arranged to give time derivatives of roll, pitch and yaw rates (p, q and r) as follows:

$$p^{\bullet} = \left(c_1 r + c_2 p + c_4 h_{eng}\right)q + \_qSb(c_3 C_l + c_4 C_n)$$
$$q^{\bullet} = \left(c_5 p - c_7 h_{eng}\right)r - c_6\left(p^2 - r^2\right) + \_qS\_cc_7 C_m$$
$$r^{\bullet} = \left(c_8 p - c_2 r + c_9 h_{eng}\right)q + \_qSb(c_4 C_l + c_9 C_n)$$

where the inertia terms are expressed as follows:

$c_1 = \frac{(I_Y - I_Z) - I_{XZ}^2}{I_X I_Z - I_{XZ}^2}$	$c_2 = \frac{(I_Y - I_Z + I_Z)I_{XZ}}{I_X I_Z - I_{XZ}^2}$	$c_3 = \frac{I_Z}{I_X I_Z - I_{XZ}^2}$
$c_4 = \frac{I_{XZ}}{I_X I_Z - I_{XZ}^2}$	$c_5 = \frac{I_Z - I_X}{I_Y}$	$c_6 = \frac{I_{XZ}}{I_Y}$
$c_7 = \frac{1}{I_Y}$	$c_8 = \frac{(I_X - I_Y)I_X - I_{XZ}^2}{I_X I_Z - I_{XZ}^2}$	$c_9 = \frac{I_X}{I_X I_Z - I_{XZ}^2}$

The nonlinear aircraft equations also include rotational kinematic and navigation equations. The rotational kinematic equations, which relate Euler angular rates to body-axis angular rates, are given in terms of time derivatives of pitch, roll and yaw (θ, Φ and ψ) angles as follows:

$$\varphi^{\bullet} = p + \tan\theta \; (q\sin\varphi + r\cos\varphi)$$
$$\theta^{\bullet} = q\cos\varphi - r\sin\varphi$$
$$\psi^{\bullet} = \frac{q\sin\varphi + r\cos\varphi}{\cos\theta}$$

The navigation equations relate aircraft translational velocity components in body axes to earth axis components, neglecting wind effects. These differential equations describe the time evolution of the position of the aircraft c.g. relative to earth axis and they are given in terms of *x*, *y* and *z* coordinates as follows:

$$\dot{x_E} = \begin{array}{l} u\cos\psi\cos\theta + v(\cos\psi\sin\theta\sin\varphi - \sin\psi\cos\varphi) \\ + w(\cos\psi\sin\theta\cos\varphi + \sin\psi\sin\varphi) \end{array}$$

$$\dot{y_E} = \begin{array}{l} u\sin\psi\cos\theta + v(\sin\psi\sin\theta\sin\varphi + \cos\psi\cos\varphi) \\ + w(\sin\psi\sin\theta\cos\varphi - \cos\psi\sin\varphi) \end{array}$$

$$\dot{z_E} = u\sin\theta - v\cos\theta\sin\varphi - w\cos\theta\cos\varphi$$

Assuming thrust acts along the *X* body axis, body axis accelerations a_x, a_y, and a_z are calculated from:

$a_X = \frac{qSC_X+T}{mg}$,	$a_Y = \frac{qSC_Y}{mg}$,	$a_Z = \frac{qSC_Z}{mg}$

Airplane state equations can be presented in the state vector form as follows:

$$\vec{x} = [V_t, \alpha, \beta, \varphi, \theta, \psi, p, q, r, x_E, y_E, z_E]^T$$

For the analysis, the airplane is considered at level flight and its control surfaces such as elevator, aileron, and rudder can be displayed with a control vector showing the angular movement of three control surfaces as follows:

$$\vec{u} = [\delta_e, \delta_a, \delta_r]^T$$

The set of first order ordinary differential equations expressed in time derivatives of 12 state variables represents the equations of motion for a rigid aircraft. Trim is required prior to the solution of such differential equations.

5 Steady State Trim

The nonlinear equations of motion can be linearized at an equilibrium flight condition known as trim, which can be used as an initial condition for flight simulation. At the trim, all translational and rotational accelerations are equal to zero. Time derivatives of state equations for *Vt*, α, β, *p*, *q* and *r*, augmented with a rate of climb and a coordinated turn constraint, can make up a total of eight trim equations (Stevens & Lewis, 2003). With eight trim equations to find the trim condition, any combination of eight state or control surface deflection variables if available can be free to vary so that the equations can be satisfied.

A convenient method of trim as described in (Stevens & Lewis, 2003) is made through a numerical algorithm in which a COST function is formed for the specific aircraft from the sum of squares of time derivatives of V_t, α, β, p, q and r mentioned above. Initially, the velocity and altitude is user defined, and all other state and control variables are set to zero. A function minimization algorithm called SIMPLEX can then be used to adjust the control variables and the appropriate state variables to minimize this scalar cost. With this algorithm, trimmed control variables for the airplane model are computed.

6 Time History of Aircraft Motion

Once the aircraft is trimmed and brought to the equilibrium, the first-order nonlinear differential equations are solved numerically. Though the solver "ode45" is available in MATLAB library for the numerical solution, the forth-order Runge–Kutta integration was implemented for clarity. The integration was made in 0.1 s time steps for 10 s total time as listed below:

```
% time history integration
dt = 0.1;
time = 0.0;
x = state;
u = control_trim;
for i = 1:100
   X(:,i) = x;
 t(i) = time;
[time,x] = aircraft_rkutta(dt,time,x,u,xcg);
   fprintf(' ');
end
```

Thus, the solution was obtained for time history of all 12 state variables. Table 3 shows initialization and trimmed data at 10 s for these state variables with the centre

Table 3 State variables for trimmed solution at 10 s

State variables, x	Initialization x_0	Trimmed solution
V_t, ft./s	269.345	362.8833
α, rad	0.0	−0.00442
β, rad	0.0	−0.32389
φ, rad	0.0	11.51163
θ, rad	0.0	−0.67673
ψ, rad	0.0	0.26547
p, rad/s	0.0	1.41009
q, rad/s	0.0	−0.46006
r, rad/s	0.0	−0.02197
x_E, m	0.0	2328.536
y_E, m	0.0	292.9915
z_E, m	10000.0	8439.974

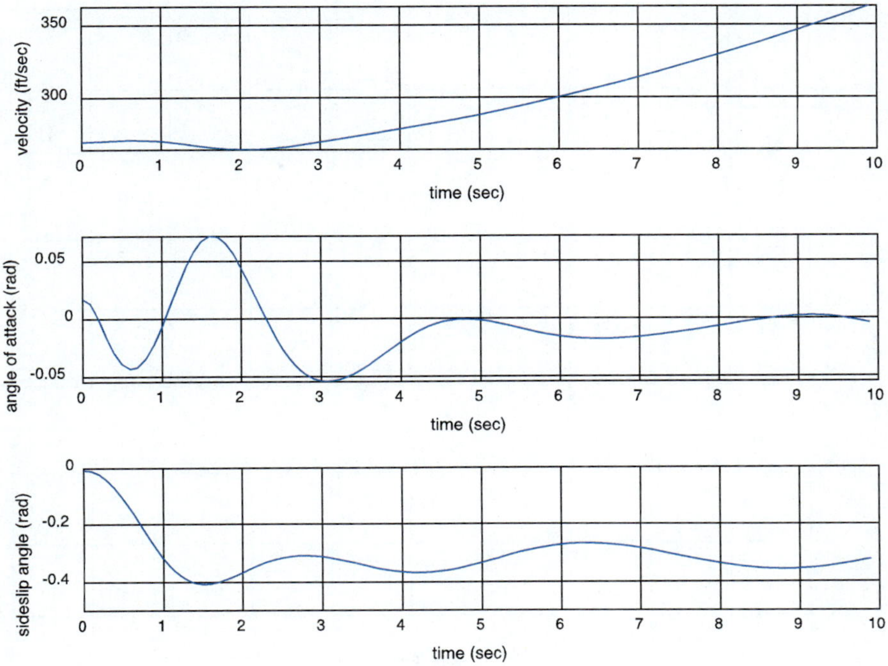

Fig. 2 Time history of velocity, angle of attack and side slip angle

of gravity at 0.25c location. The initialization was made for an airspeed, corresponding to Mach 0.25 at the cruise altitude of 10,000 ft.

Figures 2, 3, 4 and 5 show the instantaneous time history variation of aircraft state variables. Within 10 s of time span, state variables show deviations from their initial set values. Variations in velocity indicate acceleration in the cruise flight. Both angle of attack and side slip angle show an increase with time. Euler angle in roll has an increasing and pitch and yaw angles have decreasing attitudes. Navigation distances also show variation in time. North distance increases and altitude and east distance show decrease from their initial values.

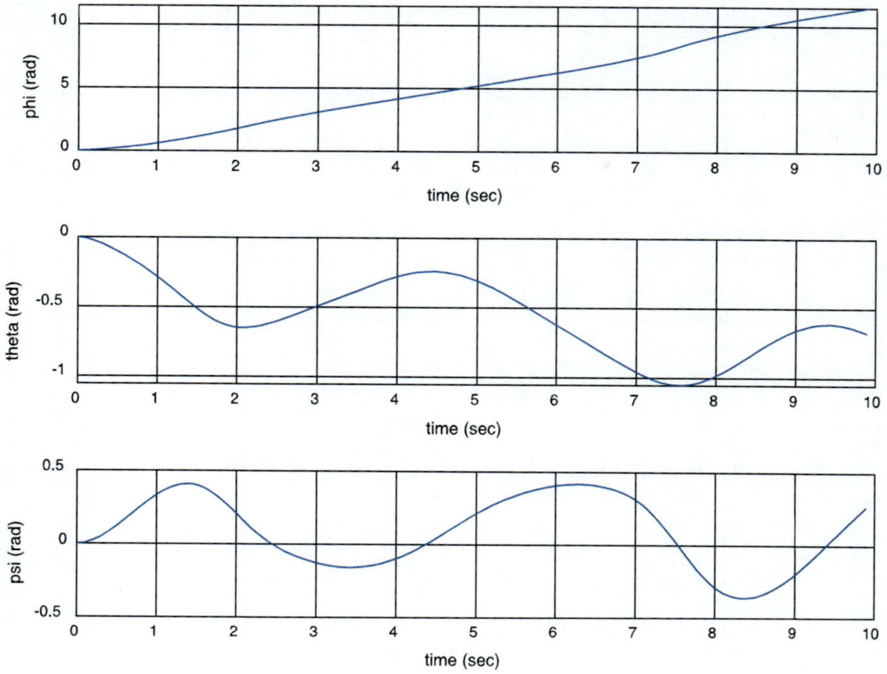

Fig. 3 Time history of roll, pitch and yaw angles

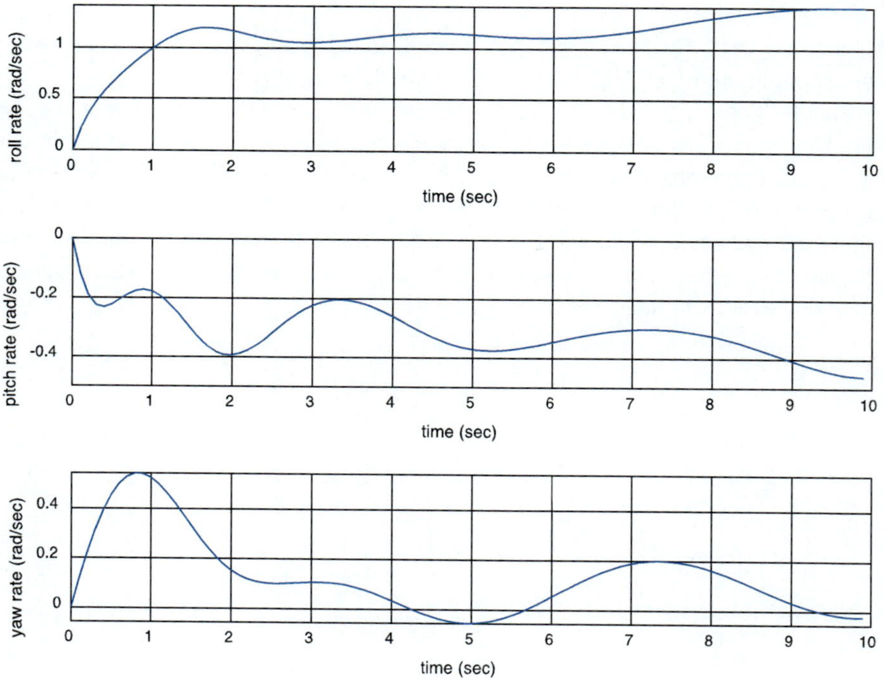

Fig. 4 Time history of roll, pitch and yaw rates

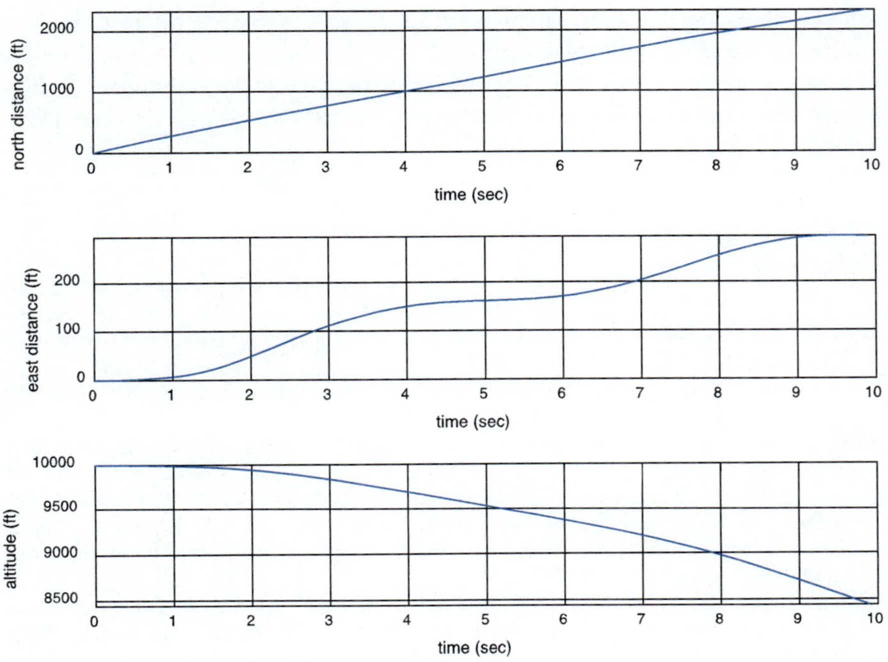

Fig. 5 Time history of north and east distances and altitude

7 Conclusion

The flight dynamics modelling is of importance because it represents the flight motion numerically for a given input, as close to the flight motion in the real world as the application requires. For this purpose, a mathematical model was developed by using MATLAB programming to solve the nonlinear aircraft equation of motion for six-degree-of-freedom flight. The input for the mathematical model includes force and moment coefficient data in look-up tables and engine power with varying altitude and airspeed. Linear interpolations are performed to evaluate the input data. The damping derivatives needed for the model can be obtained from empirical USAF DATCOM methods.

This mathematical model was adopted for evaluating the flight motion of an example single-propeller airplane equipped with a twin-pack engine. The aerodynamic input for the flight model was obtained from wind tunnel tests of power-off and stick-fixed 1/12 scale model of the airplane. The tests included the investigation of flap, elevator, rudder, and aileron deflection on longitudinal and lateral-directional aerodynamics. The propulsion input, on the other hand, was obtained from two engines used within a flight envelope defined by altitude and airspeed.

For the example airplane, the first-order differential equations of motion were solved numerically at trim condition for cruising velocity and altitude. The results of solution show time history variation of all 12 state variables for instantaneous airplane motion. The procedure described in this chapter should be useful for the mission evaluation of such a propeller airplane during its preliminary design. Improvement in modelling the flight dynamics is expected as the airplane design evolves.

References

Biber, K. (2006). Stability and control characteristics of a new FAR23 airplane. *Journal of Aircraft, 43*(5), 1361–1368.

Biber, K. (2011). Estimating propeller slipstream drag on airplane performance. *Journal of Aircraft, 48*(6), 2172–2174.

Biber, K (2016). Flight dynamics modeling of a propeller driven Cargo Airplane. In *AIAA-2016-0806, AIAA SciTech 2016*, San Diego, CA, USA.

Chen, X. Q., Ou, Q., Wong, D. R., Li, Y. J., Sinclair, M., & Marburg, A. (2009). Flight dynamics modelling and experimental validation for unmanned aerial vehicles. In *Mobile robots – State of the art in land, sea, air and collaborative missions* (Open access – Peer reviewed chapter 9). IntechOpen. https://doi.org/10.5772/6994

Dussart, G., Portapas, V., Pontillo, A., & Lone, M. (2018). Flight dynamic modelling and simulation of large flexible aircraft. In *Flight physics – Models, techniques and technologies* (Open access peer-reviewed chapter 3). https://doi.org/10.5772/intechopen.71050

Garza, F. R., & Morelli, E. A. (2003). *A collection of nonlinear aircraft simulations in MATLAB*. NASA/TM-2003-212145.

Hoek, D. E., & Fink, R. D. (1960). *USAF stability and control DATCOM* (Vol. 1–4). Global Engineering Documents.

Kim, C. J., Kim, S. H., Park, T. S., Park, S. H., & Lee, J. W. (2014). Flight dynamics analyses of a propeller-driven airplane (I): Aerodynamic and inertial Modeling of the propeller. *International Journal of Aeronautical and Space Sciences, 15*(1), 345–355. https://doi.org/10.5139/IJASS.2014.15.4.345

Milenkovic-Babic, M. (2018). Propeller thrust force contribution to airplane longitudinal stability. *Journal of Aircraft Engineering and Aerospace Technology, 90*(9), 1474–1478. https://doi.org/10.1108/AEAT-04-2017-0104

Nuguyen, N., & Tuzcu, I. (2009). Flight dynamics of flexible aircraft with aeroelastic and inertial force interactions. In *AIAA-2009-6045, AIAA atmospheric flight mechanics conference*, Chicago, IL.

Stevens, B. L., & Lewis, F. L. (2003). *Aircraft control and simulation* (2nd ed.). Wiley.

Stojakovic, P., & Rauso, B. (2016). Single propeller airplane minimal flight speed based upon the laminar Maneuver condition. *Journal of Aerospace Science and technology, 49*, 239–249. https://doi.org/10.1016/j.ast.2015.12.012

Transonic Airfoil Development for an Unmanned Air System

Kasim Biber

1 Introduction

An unmanned air system has been conceptualized to provide the context of a vehicle within which an energy-efficient airfoil design and optimization problem can be formed. The air vehicle concept is used to generate flight conditions and desired performance for an airfoil design process. There is no official requirement for this class of vehicle, but some notional vehicle performance objectives are formulated so that the new airfoil can be designed and optimized for its main wing.

The concept vehicle is assumed to operate for the majority of its flight time in an efficient cruising condition at transonic speed, and so one objective would be to maximize the endurance of the vehicle. Maximizing flight time should improve the quality of information collected and could also save mission costs. For this objective, the airfoil design parameter would be minimum drag coefficient at a given cruise lift coefficient or a range of cruise lift coefficients. This means that the new airfoil should be energy efficient.

The concept vehicle also has an objective to achieve a high transonic top speed. The ability to achieve a high dash speed would aid in vehicle responsiveness, for example, in relocation between areas of interest. A thinner airfoil may be desirable for maximizing top speed, but it could lead to increased structural weight and a heavier vehicle. For various reasons, but primarily for antenna integration, the aircraft may also require relatively thick wings. Therefore, a third objective is to design an airfoil that has a maximum thickness ratio of 16% chord.

Because of these factors, it is likely that basing the wing design on an existing airfoil will provide sub-optimal performance. Thus, it becomes necessary to develop a unique airfoil to use as the baseline for further development efforts. The

---

K. Biber (✉)
Bartin University, Bartin, Turkey
e-mail: kbiber@bartin.edu.tr

T. H. Karakoc et al. (eds.), *New Achievements in Unmanned Systems*, Sustainable Aviation, https://doi.org/10.1007/978-3-031-29933-9_13

specifications for the new 16% thick airfoil include a range of Reynolds numbers per foot from 1.7 million to 2.5 million and Mach numbers from 0.4 to 0.8.

The long-endurance flight capability requires the air vehicle to have maximum aerodynamic efficiency and minimum fuel consumption, as illustrated by their prominence in the Breguet endurance equations. For this reason, the objective has been to extend the laminar flow over the airfoil as much as possible, leading to a substantial decrease in drag and consequent reductions in fuel consumption and pollution. However, as evidenced from some other research, reported in (Biber & Tilmann, 2004; Cella et al., 2005; Drela, 1992), the laminar flow exhibits strong sensitivity to the leading edge sweep angle and to the environment conditions. Therefore, the current investigation focuses on two-dimensional transonic flow over a wing section having no sweep angle and taking full advantage of natural laminar flow extension.

With these given specifications, this chapter aims at presenting only the process of developing an aerodynamically optimal airfoil shape for application in an unmanned air system. For the same airfoil, the polar performance and its sensitivity to flow transition has also been reported in (Biber & White, 2019; Biber, 2019).

2 Computational Tools

The airfoil design included the use of a collection of FORTRAN computer programs called the MSES, version 3.12b and its optimizer called LINDOP, version 2.50. The MSES program consists of main and supporting programs, as described in its user's manual in (Drela, 1996; 2004). It is capable of design, analysis, and optimization of airfoils used for both high lift systems and transonic wings.

Boundary layer transition in an MSES solution, as in an XFOIL solution, is triggered by either a free transition where e^N criterion is met or a forced transition where a trip or the trailing edge is encountered as in (Drela & Youngren, 2001). The e^N method is only appropriate for predicting transition in situations where the growth of two-dimensional Tollmien–Schlichting waves via linear instability is the dominant transition-initiating mechanism. The e^N method is always active, and free transition can occur upstream of the trip.

The e^N method has the user-specified parameter "N_{cr}", which is the log of the amplification factor of the most-amplified frequency which triggers transition. A suitable value of this parameter depends on the ambient disturbance level in which the airfoil operates and mimics the effect of such disturbances on transition. For the present airfoil design and analysis, the standard N_{cr} value of 9 was used; however, the effect of changing the transition parameter to other values such as 4, 5, 6, and 12 was also investigated.

3 The Process of Airfoil Development

The notional air vehicle is envisioned to have a long endurance at medium altitudes, requiring its wing to be as aerodynamically efficient as possible. The primary requirements for the wing section considered in this study included 16% chord maximum thickness, over 50% chord laminar flow, low pitching moment, and operations at Mach numbers ranging from 0.4 to 0.8 and Reynolds numbers per foot ranging from 1.7 million to 2.5 million. Both Mach and Reynolds numbers are based on free-stream flow conditions. MSES flow parameters were defined for vortex + doublet far-field, isentropic except near shocks and free-transition for either specified angle of attack or Mach number. Reynolds number was specified for viscous analysis, but it was set to 0.0 for inviscid runs.

A compromise was made among all of the requirements to increase the operational Mach number while maintaining airfoil maximum thickness with a high L/D ratio and large drag bucket for given range of Reynolds number. Surface pressure distribution or surface geometry was changed in mixed or inverse design modules of the MSES software to meet the design objectives. The interactive and iterative work resulted in a preliminary 16% thick airfoil considered to be an initial or reference airfoil, as shown with a solid line in Fig. 1.

Once the initial or reference airfoil was determined, the optimization capability of LINDOP driver was implemented to obtain the new or final transonic airfoil with 16% chord maximum thickness ratio, as shown with a dotted line in Fig. 1. For the optimization process, global variables and their corresponding fixing constraints were set to 5 for sweeps in angle of attack, 15 for sweeps in Mach number, and 20 for making use of LINDOP optimizer. Incremental sweep values in angle of attack and Mach number were provided in a separate *spec.xxx* file. Geometry deformation modes were represented by some functions describing camber and upper and lower surface of airfoil. These functions have end points; 0 at the leading edge and 1.0 at the trailing edge, as indicated in the file *modes.xxx*. The particular geometric shapes are implemented in FUNCTION GFUN in the program package.

The reference airfoil geometry file, *blade.xxx* was used for all seven design points selected for climb, cruise, and maneuver flight conditions. MSET was run with the

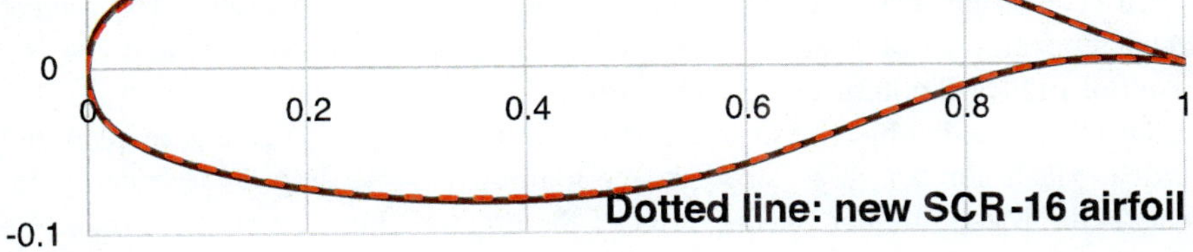

Fig. 1 Geometric comparison of 16% thick initial or reference airfoil with the optimized one

airfoil geometry file *blade.xxx* to create one *mdat.xxx*, and then the same file was copied for all seven design points. The *modes.xxx* file was also kept the same to ensure the same airfoil shape during optimization. The *mses.xxx* file had a different extension for each design point with conflicting requirements.

The optimization process was started by first converging *mses.xxx* files for all seven cases. MSES sets all the geometry mode amplitudes to zero during a calculation and calculates the sensitivities of various quantities such as CL, CD, etc. to the mode displacements. These sensitivities are written out to the unformatted file *sensx.xxx*, which is then read in by LINDOP optimizer. LINDOP reads all input files with some initialization and lists the available operating points and design parameters.

LINDOP in general is used to minimize an objective function $F(X_k)$ with respect to the parameters X_k as in (Drela, 1996, 1998). In this study, the objective function F was the drag coefficient C_D while the parameters X_k were the airfoil geometry deformation modes.

One optimization step consisted of the generation of design parameter changes via line minimization in LINDOP, followed by a nonlinear MSES solution calculation. This sub-cycle was executed toward the line minimum. The gradient vector and line-minimization direction vector were generated to find the optimum. The optimization process resulted in a new 16% thick transonic airfoil, designated as SCR-16 and shown in Fig. 1 in comparison with the reference airfoil.

The optimized airfoil clearly has better performance at transonic flow conditions. In order to investigate the cause of this improvement in design, the transition location of the optimized airfoil was compared with the reference one at $\alpha = -1$ deg, Re = 2.5 million, and $N_{cr} = 9$. It was shown that the optimization results in producing a relatively larger extent of laminar flow on the airfoil. The transition occurs at about 0.82c on upper surface and 0.60c on lower surface for the optimized airfoil. However, it gradually moves upstream as the Mach number nears its critical value for given flow conditions.

The critical Mach number is the free-stream Mach at which sonic flow is first achieved on the airfoil surface. It was determined for both reference and optimum airfoils by the method described in (Anderson, 2001). For the given airfoil, a minimum value of pressure coefficient was obtained for incompressible flow by running the MSES program. The pressure coefficient was corrected by using the Prandtl–Glauert rule and plotted against the free-stream Mach number. Another curve was obtained by the variation of critical pressure coefficient with Mach number. The intersection of these two curves represents the point corresponding to sonic flow at the minimum pressure location on the airfoil. The value of free stream Mach at this intersection is, by definition, the critical Mach number. The critical Mach number has a value of 0.710 for the reference airfoil, and it moves to 0.724 as a result of LINDOP optimization of the airfoil.

One of the main objectives of transonic airfoil design is to be able to increase the critical Mach number so as to obtain the highest possible drag divergence Mach. This is the Mach number for the onset of the dramatic increase in wave drag at a given angle of attack or lift coefficient, for a given maximum thickness ratio. With

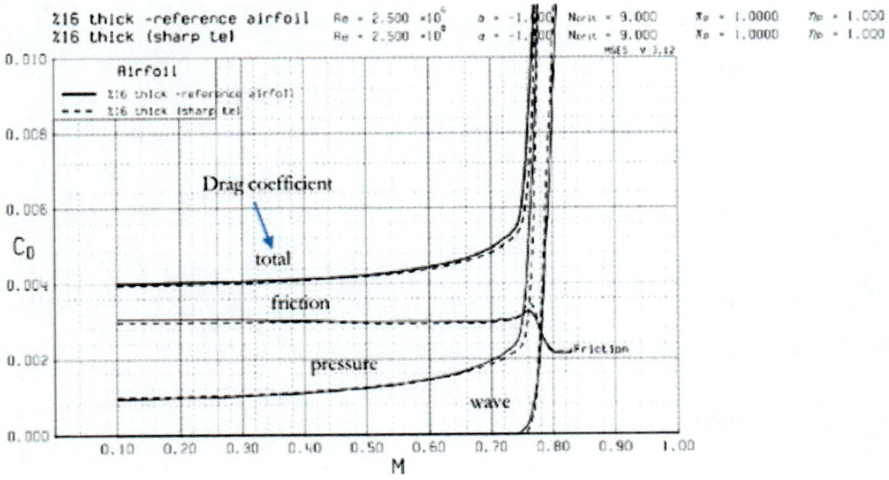

Fig. 2 Variation of drag coefficient with Mach number (drag divergence Mach) for the comparison of reference airfoil (solid line) with optimized one (dotted line)

this objective in mind, the MSES code was run with the Mach sweep option, and the airfoil drag was monitored.

Figure 2 shows a drag divergence plot comparing the optimized airfoil with the reference one at an angle of attack of – 1 deg and a Reynolds number of 2.5 million. The figure has a variation of total drag and its components such as friction, pressure, and wave drag coefficients with Mach number. With the airfoil optimization, there is clearly a reduction in drag coefficient, which is more significant for the pressure component during the Mach divergence. The start of wave drag is realized at Mach 0.74 for the reference airfoil, and it moves to Mach 0.75 with the optimized one. After this Mach, the drag coefficient starts diverging progressively from its profile value due to the increased compressibility effects. It is desirable to have the smallest possible initial rate of drag increase beyond the drag divergence Mach because the best cruise performance is obtained at a Mach number of 0.02–0.03 in excess of drag divergence Mach as in (Torenbeek, 1982).

Figure 3 shows polar performance data comparing the reference airfoil with the optimized one at Re = 2.5 million and $N_{cr} = 9$. The comparison was made at Mach 0.4. The airfoil initially has a drag bucket with a lift coefficient range of as much as 0.2. This bucket is shifted upward in the direction of higher lift coefficient with LINDOP optimization. The new optimized airfoil has an upper corner of drag bucket operating at a relatively higher lift coefficient. This shift of polar performance is also seen on the upper surface location of transition producing an increase in the extent of laminar flow.

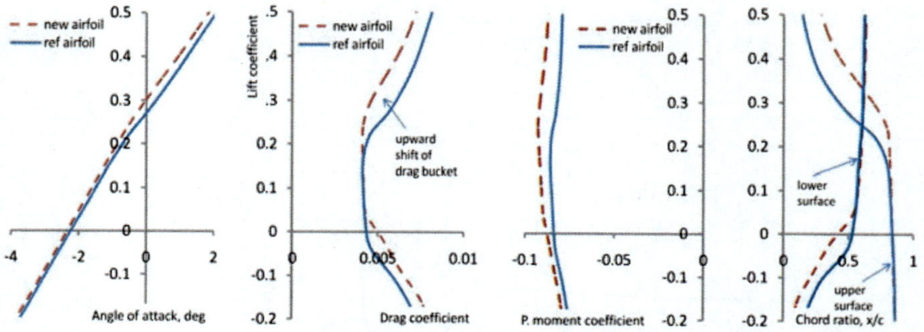

Fig. 3 Polar comparison of reference airfoil with the optimized one at Mach 0.4

4 Conclusion

The MSES/LINDOP computer program was used to develop a new 16% thick transonic airfoil. Airfoil specifications included a range of Reynolds numbers per foot from 1.7 million to 2.5 million and Mach numbers from 0.4 to 0.8, based on the climb, cruise, and maneuver flight conditions of an unmanned air system. The airfoil is to be used for a two-dimensional wing with no sweep angle. For the airfoil development, the flow transition was left free with $N_{cr} = 9$.

Shape optimization in geometry and inverse design modules of the MSES program was first used to design an initial (reference) 16% thick airfoil. The performance of the initial airfoil was then optimized for seven design points with conflicting requirements in Reynolds and Mach number by using LINDOP optimizer. For the free transition case at Reynolds number per foot of 2.5 million, the new optimized airfoil has the following design features:

- The flow transition occurs at approximately 80% chord upper surface and 60% chord lower surface of airfoil, taking full advantage of laminar flow as intended.
- The optimization results in an increase of critical Mach number. While the critical Mach is 0.710 for the reference airfoil, it moves to 0.725 for the optimized one.
- Drag divergence Mach plot shows a summation of friction, pressure, and wave drag components. For the plotted data, the start of wave drag occurs at Mach 0.74 for the reference airfoil and it moves to Mach 0.75 for the optimized one.
- The optimization also results in an upward shift of drag bucket in the direction of higher lift coefficient. The new optimized airfoil has an upper corner of drag bucket operating at a relatively higher lift coefficient.
- The endurance parameter, CL1.5/CD, shows an almost linear increase within the region of drag bucket, indicating the benefit of having a laminar flow airfoil.
- Increasing Mach number causes the drag bucket to move upward in the direction of both higher lift and higher drag coefficients. This shift is also seen on the upper surface location of transition producing an increase in the extent of laminar flow.

References

Anderson, J. D. (2001). *Fundamentals of aerodynamics* (3rd ed.). McGraw Hill Higher Education.

Biber, K. (2019). Transonic airfoil design and optimization for an unmanned air vehicle concept. In *AIAC-2019-061, 10th Ankara international aerospace conference*, Ankara, Turkey.

Biber, K., & Tilmann, C. (2004). Supercritical airfoil design for future high altitude and long endurance concepts. *Journal of Aircraft, 41*(1), 156–164.

Biber, K., & White, T. (2019). Transonic airfoil design and optimization for an unmanned air vehicle concept. *Journal of Mechanics Engineering and Automation, 9*. https://doi.org/10.17265/2159-5275

Cella, U., Quagliarella, D., & Donelli, R. (2005). Design and optimization of a transonic natural laminar flow airfoil. In AIDAA, *XVII Congresso Nazionale AIADAA*, Italy, VOLTERRA.2.

Drela, M. (1992). Transonic low-Reynolds number airfoils. *Journal of Aircraft, 29*(6), 1106–1113.

Drela, M. (1996). *A user's guide to LINDOP, version 2.50*. MIT Department of Aeronautics and Astronautics.

Drela, M. (1998). Pros and cons of airfoil optimization. In D. A. Caughey & M. M. Hafez (Eds.), *Chapter in Frontiers of computational fluid dynamics*. World Scientific. ISBN 981-02-3707-3.

Drela, M. (2004). *A user's guide to MSES, version 3.12b*. MIT Department of Aeronautics and Astronautics.

Drela, M., & Youngren, H. (2001). *A User's guide to XFOIL, version 6.99*. MIT Department of Aeronautics and Astronautics.

Torenbeek, E. (1982). *Synthesis of subsonic airplane design* (pp. 241–252). Delft University Press/Kluwer Academic.

Optimization of Energy Efficiency According to Freud's Disk Theory Depending on Propel Pitch Angles

Ukbe Ucar, Zehra Ural Bayrak, and Burak Tanyeri

Nomenclature

A	Pitch (degree)
α_{max}	Max pitch (degree) = 30
α_{min}	Min pitch (degree) = 18
v	Aircraft speed (m/s)
v_{max}	Max aircraft speed (m/s) = 20
v_{min}	Min aircraft speed (m/s) = 8
π	Pi number = 3.14
r	70% of the blade length(meter) = 0.089
σ	Stiffness ratio of blade element = 0.806
ro	Density of air(kg/m^3) = 0.909
ur	Resultant speed(rad/s) = 71.228
cl	Lift coefficient = 0.05
s	Total wing areas (m^2) = 0.006
thrust	Thrust (N)
us	The velocity of the air continuing to accelerate far enough from the propeller (m/s)
u	Speed of air in the plane of the propeller (m/s)
ee	Energy efficiency (%)

U. Ucar (✉) · B. Tanyeri
School of Aviation, Department of Aircraft Maintenance and Repair, Firat University, Elazig, Turkey
e-mail: uuucar@firat.edu.tr; btanyeri@firat.edu.tr

Z. U. Bayrak
School of Aviation, Department of Avionics, Firat University, Elazig, Turkey

T. H. Karakoc et al. (eds.), *New Achievements in Unmanned Systems*, Sustainable Aviation, https://doi.org/10.1007/978-3-031-29933-9_14

1 Introduction

Propeller systems move on a rotating shaft and are designed in line with aerodynamic and hydrodynamic conditions. These systems consist of blades of different lengths and numbers. The working principle of the propeller system is based on pushing and pulling power. Wind turbines, ships, airplanes, cars, and heating and cooling technologies are among the areas where propeller systems are used.

One of the vehicles in which propellers are used extensively is drones. Drones are constantly used in many areas such as agricultural applications, logistics, military operations, security, traffic, and sports. In addition to these, the number of civilian users is increasing day by day. It is of great importance that the drones, which are produced intensively, are designed in a reliable, efficient, and stable manner. One of the most important part of the drone, which consists of many components, is the propellers. In this study, ideal dimensions for propeller design for drones are determined by mathematical modelling, which is used for the first time in the literature, according to Freud's Disk Theory and for the purpose of energy efficiency, depending on the propeller pitch angles. In addition, the proposed method is analysed by comparing the simulation algorithm and the real propeller system. There are many studies on the subject in the literature, some of which are shown in Table 1.

In the next section, the solution methodologies used in solving the problem are explained. This section is followed by the result and discussion section. In the last part, general evaluations about the study are made.

Table 1 Literature survey

Author information	Considered system	Solution method
Gaggero et al. (2017)	Marine propellers	Multi-objective Numerical optimization
Gur and Rosen (2009)	Ultralight aircraft	Multidisciplinary Solution approach
Zhang et al. (2021)	Quadrotor fixed-wing hybrid UAV	Multidisciplinary Design optimization
Bayraktar and Güldaş (2020)	Quadrotors	Simulation approach
Lee et al. (2020)	Quadcopter	Genetic algorithm
Bacciaglia et al. (2020)	Ship boats	Particle swarm optimization
Magnussen et al. (2015)	Multicopter	Mathematical modelling
Kapsalis et al. (2021)	A tactical, fixed-wing UAV	CFD method
Mian et al. (2021)	UAV propeller	Space mapping surrogate modelling
Delbecq et al. (2020)	Electric multirotor drones	Generic and efficient sizing methodology
Gaggero et al. (2017)	Marine propellers	Multi-objective numerical optimization

2 Solution Method

In this study, energy efficiency is maximized using mathematical modelling method depending on the pitch angle and aircraft specifications. Matlab Simulink has been used to evaluate the effectiveness of the proposed method and to determine the relationships between energy efficiency and parameters affecting the energy efficiency. The details of the two proposed methods are given below.

2.1 *Optimization*

Optimization is a solution methodology based on mathematical modelling that will ensure that the resources of any structure or system (raw materials, machinery, time, budget, workforce, etc.) are optimally used for the relevant purposes under the constraints affecting the system. In this solution methodology, firstly the system or problem is modelled with mathematical and logical formulations. Then, ideal variable values are reached solving the developed mathematical model accordance with the relevant constraints and objectives. In general, the structure of a mathematical model is expressed in the following equations.

Objectives

$$\max/\min Z = f(x, y) \tag{1}$$

Constraints

$$g(x, y) = 0 : g(x, y) = a * x + b * y = 0 \tag{2}$$

$$h(x, y) \geq 0 : h(x, y) = c * x * y \geq 0 \tag{3}$$

$$x \in R^n \tag{4}$$

$$y \in \{0, 1, 2, \ldots., m\} \tag{5}$$

Based on this structure, a mathematical solution based on Non-Linear Programming has been developed for the solution of the problem and the information about the model is given below.

Objectives

$$\max = ee \tag{6}$$

Constraints

$$\pi * r * \sigma * ro * \left(ur^2\right) * cl * \cos(\alpha) = \text{thrust} \tag{7}$$

$$0.5 * ro * \left(us^2 - v^2\right) * s = \text{thrust} \tag{8}$$

$$\frac{(v + us)}{2} = u \tag{9}$$

$$ro * s * u * 0.5 * \left(us^2 - v^2\right) = ee \tag{10}$$

$$\alpha \geq \alpha_{\min} \tag{11}$$

$$\alpha \leq \alpha_{\max} \tag{12}$$

$$v \geq v_{\min} \tag{13}$$

$$v \leq v_{\max} \tag{14}$$

$$ee, \text{thrust}, \alpha, u, us, v \in R^n \tag{15}$$

In Eq. (6), the objective function is expressed in which the energy efficiency is maximized. In Eqs. (7) and (8), there are formulations for determining the thrust value. In Eq. (9), the 'u' value is also determined. Eq. (10) is the mathematical formula that determines the value of the 'ee' variable in the objective function. In Eqs. (11) and (12), the maximum and minimum value ranges that 'α' values can take are specified. In Eqs. (13) and (14), the lower and upper limits are expressed for the 'v' value. In Eq. (15), the type of decision variables and the values they can take are expressed.

2.2 Simulation

Simulation is a solution technology based on statistical methods that allows real systems to be transferred to the computer environment and made changes and experiments on them. Thanks to simulation, analyses can be made on systems that cannot be changed or that will create high costs if done. In addition, simulation methodology is used to determine how the systems will work during the design phase. In this study, simulation algorithm has been used to analyse the parameters affecting the propeller energy efficiency and to evaluate the effectiveness of the proposed mathematical model. The block diagram of the algorithm developed using Matlab Simulink is shown in Fig. 1.

In the next section, the two proposed methods will be compared and analysed on a real propeller system.

3 Results and Discussion

In the application study, alpha mini drone and 'AP-PR-008-21-7-H' type propeller have been taken into consideration and the proposed solution methodologies have been applied on these systems. Gams 22.5 program has been used for the optimization of the mathematical model, and the Matlab Simulink programme has been used for the simulation method. The variable and objective values obtained as a result of the analysis study are given in Table 2.

When the results in Table 2 are examined, it has been determined that an improvement of 1.64% for thrust and 11.84% for energy efficiency has been

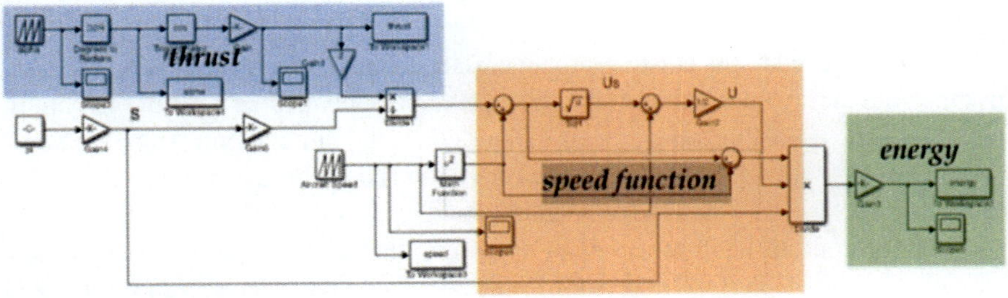

Fig. 1 Block diagram of the algorithm

Table 2 Experimental data

Variable name	Existing value	Optimal value	Simulation mean value
α	20.64	18	24
v	8	20	14
thrust	48.60	49.399	47.382
ee	3446.639	3854.836	3415.957

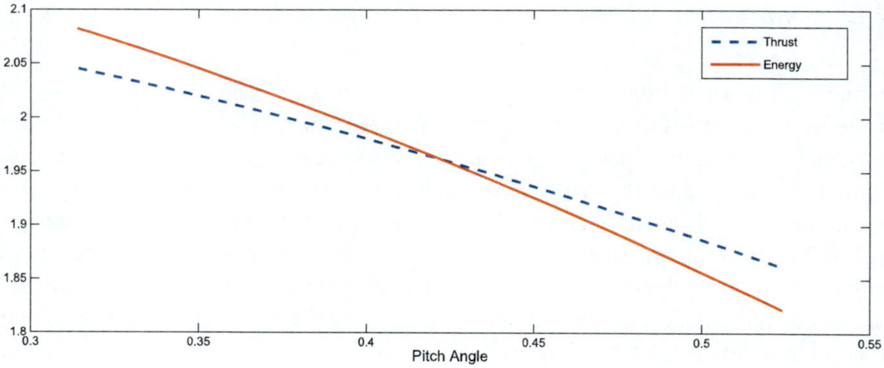

Fig. 2 Thrust and energy graph according to the pitch angle

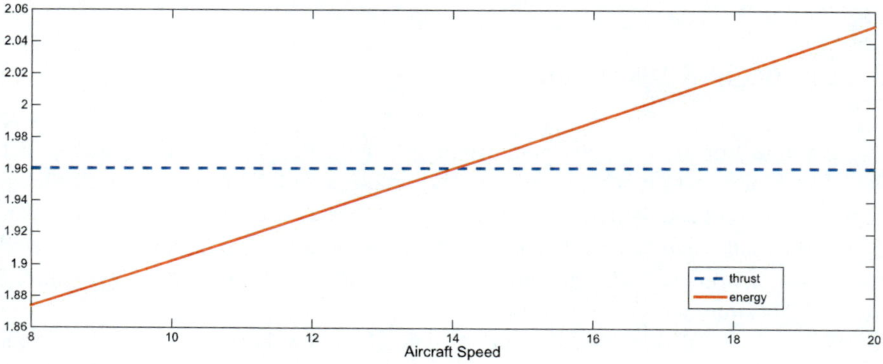

Fig. 3 Thrust and energy graph of according to the aircraft speed

achieved compared to the existing propeller data, thanks to the optimization method. In the simulation values column, the average values obtained as a result of running the algorithm 1000 times have been given. When the simulation results are compared with the optimal values, 4.08% lower values for the 'thrust' value and 11.39% lower for the 'energy' value have been obtained.

Matlab Simulink has been used to determine the relationship between 'energy efficiency', 'thrust', 'alpha' and 'speed', the specified values have been normalized and the results obtained are expressed in the graphs below.

Figure 2 shows the change in thrust and energy when the vehicle speed remains constant and the pitch angle is increased. According to this graph, it is determined that there are significant decreases in 'thrust' and 'energy' as the pitch angle increases.

In Fig. 3, the pitch angle has been kept constant and the changes on the 'thrust' and 'energy efficiency' have been observed by increasing the aircraft speed. In the graph, it has been determined that as the speed increased, the energy increased greatly, in addition, there has no significant change on the 'thrust'.

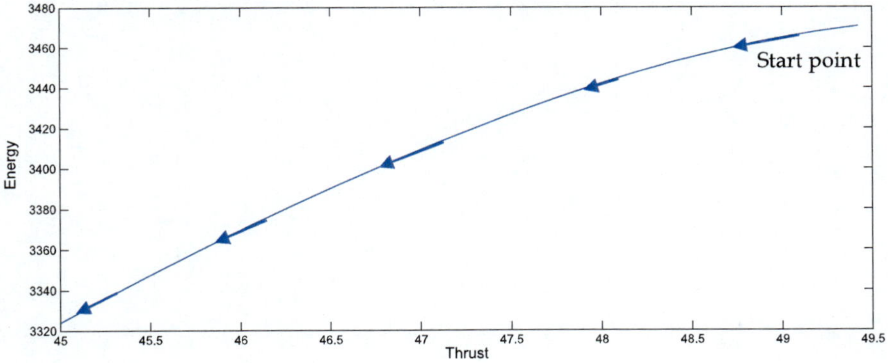

Fig. 4 Thrust energy graph as pitch angle increases as aircraft speed increases

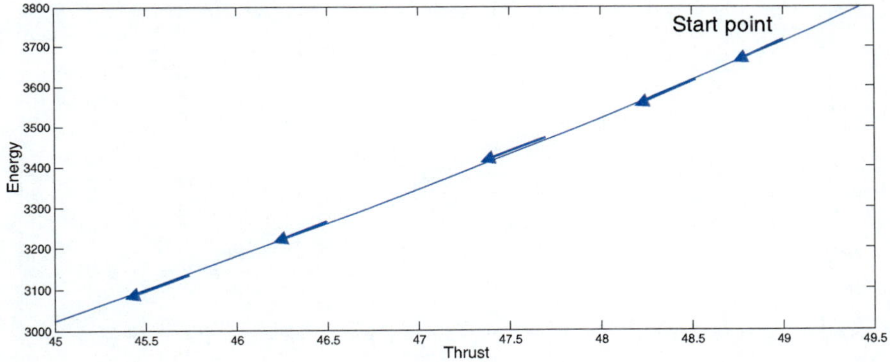

Fig. 5 Thrust energy graph as pitch angle increases as aircraft speed decreases

In Fig. 4, 'pitch angle' and 'aircraft speed' are increased simultaneously to understand the interaction in 'thrust' and 'energy'. The lower right corner shows the starting point of the graph. According to this graph, it has been observed that when the two input values increase, 'thrust' and 'energy' decrease logarithmically.

In Fig. 5, the change in 'thrust' and 'energy' is examined by increasing the 'alpha' value and decreasing the 'vehicle speed' value. The starting point of this graph is the bottom corner of the figure. According to the results in Fig. 5, it has been observed that the thrust and energy have decreased when the stated values have been changed as indicated.

In addition to this information, the results in Figs. 4 and 5 show that 'pitch angle' has a significant impact on system outputs. Moreover, it is understood from the results in Figs. 4 and 5 that the change in 'aircraft speed' has a lower effect on the system compared to the 'pitch angle'.

When Figs. 6 and 7 were examined, it was found that as the 'alpha' value decreased, 'thrust' and 'energy' increased significantly. In addition, it is understood from the results in Figs. 6 and 7 that an increase or decrease in aircraft speed causes a small change in system outputs.

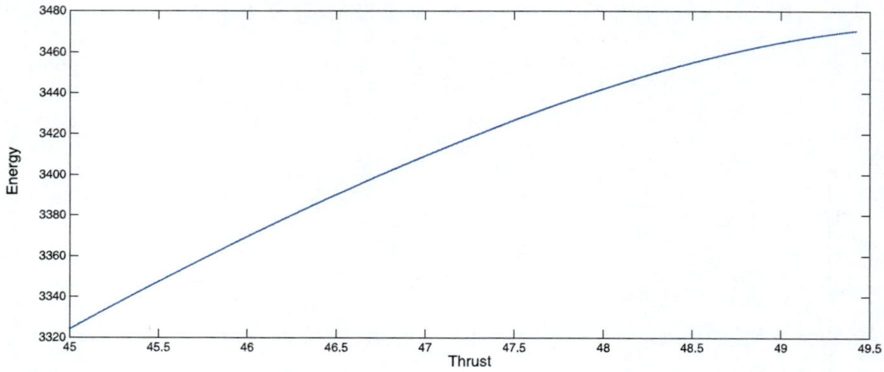

Fig. 6 Thrust energy graph as pitch angle decreases as aircraft speed decreases

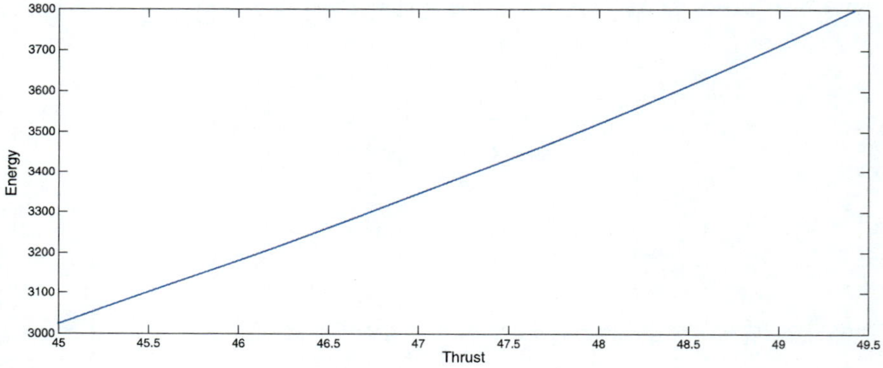

Fig. 7 Thrust energy graph as pitch angle decreases as aircraft speed increases

4 Conclusion

Propeller systems are mechanical technologies used in many vehicles and machines. Propeller technologies are used in many systems, especially wind turbines, aircraft, air conditioning systems, cooling systems, sea and land vehicles. Unmanned aerial vehicles (UAVs) are also one of the areas where these technologies are used, and propeller production is increasing exponentially with the developing technology. Effective and efficient design of propeller systems, which is one of the most important components of UAVs, is of great importance in many respects. The main importance of the propeller system is that the UAV stays stable in the air, moves in the desired direction and position, and performs landing and take-off operations reliably.

In this study, in line with the aim of maximizing energy efficiency, optimal propeller sizes for drones have been tried to be determined with a mathematical modelling approach based on non-linear programming in line with 'Freud's Disk

Theory'. There is no study in the literature in which this approach is used on propeller systems. The effectiveness of the proposed model has been tested by comparing it with the current propeller values and simulation method. As a result of the analysis study, it has been determined that the optimal values have been reached with the mathematical modelling approach, and 11.84% more energy gain was obtained compared to the current values. In addition, the interactions between the parameters affecting the design and the 'thrust' and 'energy' have been analysed with the simulation algorithm. The results of the analysis showed that the pitch angle has a great importance on the propeller design and as the pitch angle increases, the thrust and energy efficiency decrease exponentially. It has been understood from the analysis results that the aircraft speed has a lower importance on the propeller design compared to the pitch angle. In future studies, it is expected that more effective and efficient propeller systems will be designed by adding different parameters and constraints to the proposed method.

References

Bacciaglia, A., Ceruti, A., & Liverani, A. (2020). Controllable pitch propeller optimization through meta-heuristic algorithm. *Engineering with Computers, 37*, 2257–2271.

Bayraktar, Ö., & Güldaş, A. (2020). Optimization of Quadrotor's thrust and torque coefficients and simulation with Matlab/Simulink. *Journal of Polytechnic, 23*(4), 1197–1204.

Delbecq, S., Budinger, M., Ochotorena, A., Reysset, A., & Defay, F. (2020). Efficient sizing and optimization of multirotor drones based on scaling laws and similarity models. *Aerospace Science and Technology, 102*, 105873.

Gaggero, S., Tani, G., Villa, D., Viviani, M., Ausonio, P., Travi, P., Bizzarri, G., & Serra, F. (2017). Efficient and multi-objective cavitating propeller optimization: An application to a high-speed craft. *Applied Ocean research, 64*, 31–57.

Gur, O., & Rosen, A. (2009). Optimization of propeller based propulsion system. *Journal of Aircraft, 46*(1), 95–106.

Kapsalis, S., Panagiotou, P., & Yakinthos, K. (2021). CFD-aided optimization of a tactical Blended-Wing-Body UAV platform using the Taguchi method. *Aerospace Science and Technology, 108*, 06395.

Lee, Y., Park, E.-T., Jeong, J., Shi, H., Kim, J., Kang, B.-S., & Song, W. (2020). Weight optimization of hydrogen storage vessels for quadcopter UAV using genetic algorithm. *International Journal of Hydrogen Energy, 45*, 33939–33947.

Magnussen, Ø., Ottestad, M., & Hovland, G. (2015). Multicopter design optimization and validation. *Modeling, Identication and Control, 36*(2), 67–79.

Mian, H. H., Wang, G., Zhou, H., & Wu, X. (2021). Optimization of thin electric propeller using physics-based surrogate model with space mapping. *Aerospace Science and Technology, 111*, 106563.

Zhang, H., Song, B., Li, F., & Xuan, J. (2021). Multidisciplinary design optimization of an electric propulsion system of a hybrid UAV considering wind disturbance rejection capability in the quadrotor mode. *Aerospace Science and Technology, 110*, 106372.

Concept Design and Analysis for a Fixed-Wing Unmanned Aerial Vehicle to Perform Surveillance and Mapping Missions

Osman Kumuk and Mustafa Ilbas

1 Introduction

A lot of recent research has been done on design of unmanned aerial vehicle (UAV) (Önal et al., 2019). They investigated the flight parameters of a designed unmanned aerial vehicle. As a result of the analysis, the optimum angles of attack for minimum power requirement and maximum range were obtained as 12° and 8°, respectively. In his work, Dereli designed and produced the unmanned aerial vehicle named Turaç, which has both horizontal and vertical take-off features (Dereli, 2014). Carbon-fiber composite material, which is widely used today, has been chosen as the material. He checked the stability and stability of the aircraft by making various static and fluid analyses. In her study, Gülbahar compared the analysis and test studies of the model created with the finite element model (CFD) in its design with experimental data (Gülbahar, 2015). As a result of the analyses, it has been seen that the unmanned aerial vehicle is sufficiently durable and suitable for flight. Park and his team investigated the static strength and buckling loads of the target drone wing under 5 g and − 1.5 g loads using the finite element method (Park et al., 2011). After the conceptual design and preliminary design stages, they analyzed the vehicle with computational fluid dynamics. It is quite difficult to create an aerodynamic design that can improve the persistence of the UAV. To obtain the preliminary design, an estimation of the aerodynamic coefficients was made together with the computational fluid dynamics analysis (Alan et al., 2019). They made a comparison and

O. Kumuk (✉)
Iskenderun Technical University, İskenderun, Hatay, Turkey
e-mail: osman.kumuk@iste.edu.tr

M. Ilbas
Department of Energy Systems Engineering, Faculty of Technology, Gazi University, Ankara, Turkey

T. H. Karakoc et al. (eds.), *New Achievements in Unmanned Systems*, Sustainable Aviation, https://doi.org/10.1007/978-3-031-29933-9_15

analysis of wing profiles, propulsion type, airframe design, and vehicle control surfaces. They compared a new design to the existing Universidad Militar Nueva Granada model (Rocha & Solaque, 2013).

2 Material and Method

During the airfoil selection phase, EH 2.0/10, FX 76-MP-120, NACA 4412, NACA 6412, and SD 7062 airfoils were selected as candidates. These profiles were determined as candidates by considering their suitable hump, easy production, and thickness to withstand wind force. The geometries of the candidate profiles are shown in Fig. 1.

Candidate airfoils were analyzed on the XFLR5 program and the lift coefficients, drag coefficients, and pitching moment coefficients of the two-dimensional profiles were obtained. Among the profiles, the EH 2.0/10 profile was chosen due to its low Reynolds numbers characteristic, sufficiently thick geometry, and suitable bearing coefficient. It is also preferable to have a wing profile that can be suitable for our design that does not have an elevator. Airfoils were analyzed in the XFLR5 program at −20 to 20° attack angles for five different profiles. Airfoil chart Cl, Cm, and Alpha are shown in Figs. 2 and 3, respectively.

The solid model has 724,002 mesh elements. Standard atmosphere conditions and free flow velocity of 20 m/s are given as input boundary conditions. The boundary condition for the control volume walls is set as frictionless. The k-ε turbulence model was used for the solution. Solid model isometric views and solid model meshing views are shown in Figs. 4 and 5. The designed airplane model parameters are given in Table 1.

EPS foam material was used for analysis and production. EPS foam material properties are given in Table 2.

A weighted estimate is measured during the design process (as seen in Table 3). In estimating the weight of the unmanned aerial vehicle, an EPS material with a density of 26.1 g/cm^3 is considered for future studies; therefore, conceptual and preliminary designs are measured.

Figures 6 and 7 display the velocity of air over the wing and the vortices formed on the air in the wing. Parameters were created to obtain the unmanned aerial vehicle performance for the simulation. The pressure on the wing is shown in Fig. 8. When the picture is examined, it is seen that the high pressure is in the wing edge and nose part. Figures 2 and 3 show the consequences of the lift produced by attack at all angles. An angle of attack shift is brought about by getting a lift coefficient of the simulated wing, as shown in this figure, and a stall angle of around 12° is shown. This angle should be kept from in-flight to reduce lift and strength. Model airplane test flight after manufacturing and assembly is shown in Fig. 9.

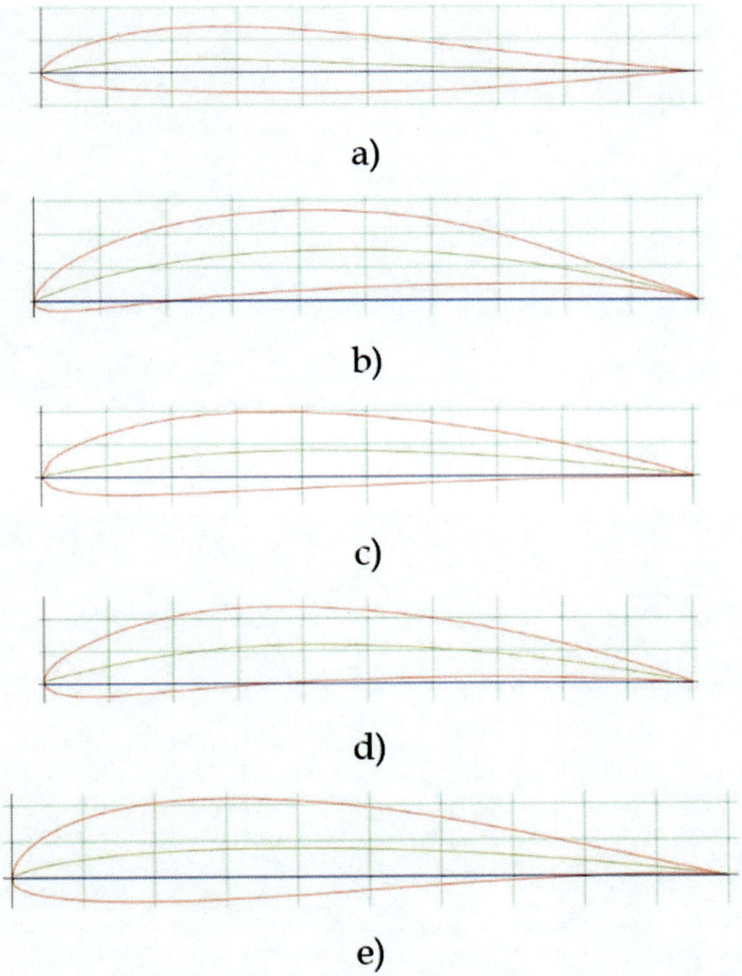

Fig. 1 (**a**) EH 2.0/10 (**b**) FX 76-MH-120 (**c**) NACA 4412 (**d**) NACA 6412 (**e**) SD 7062

3 Conclusion

Conceptual, preliminary, and aerodynamic operating conditions of an UAV designed under specific mission and flight circumstances have been studied. In fact, the concept design yielded a weight estimate that included all the necessary electric-electronic and structural elements for locating, observing, and mapping. The tail and fuselage are designed according to aerodynamic calculations to achieve the best glide angle that will provide static strength during the glide. So different wing

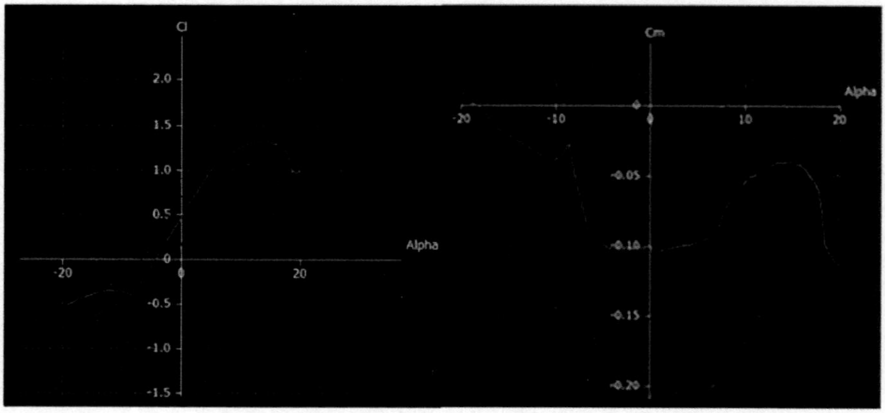

Fig. 2 Airfoil chart Cl/Alpha and Cm/Alpha

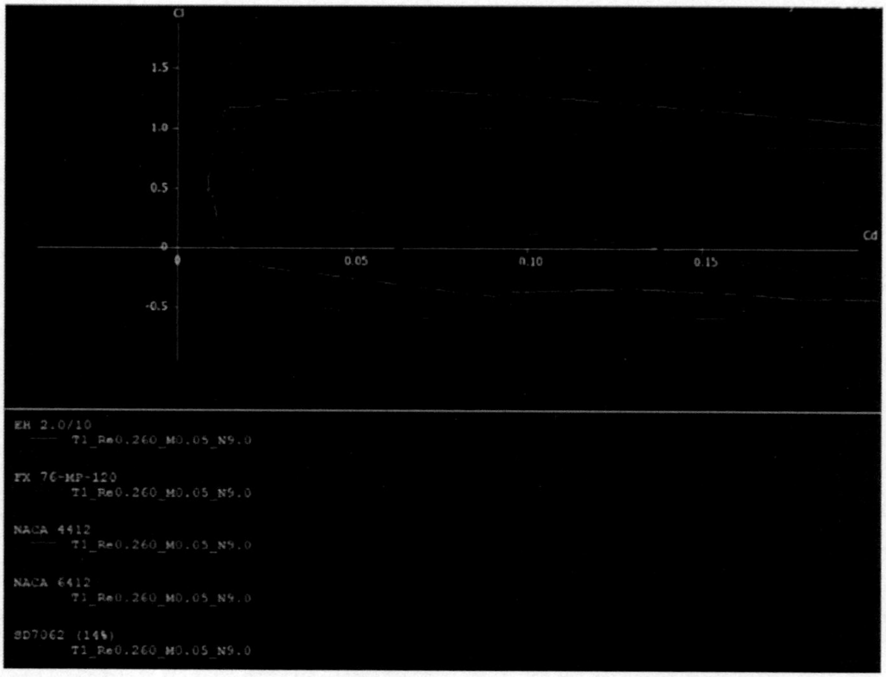

Fig. 3 Airfoil chart Cl/Cd

profiles were compared by thinking about the features and parameters of our model. An initial model based on the conceptual design was obtained and aerodynamic analysis was performed through CFD to analyze the lift generated by the unmanned aerial vehicle. This tool has proven valuable for checking design calculations to get an aircraft qualified of performing the flight task. The flight parameter required by

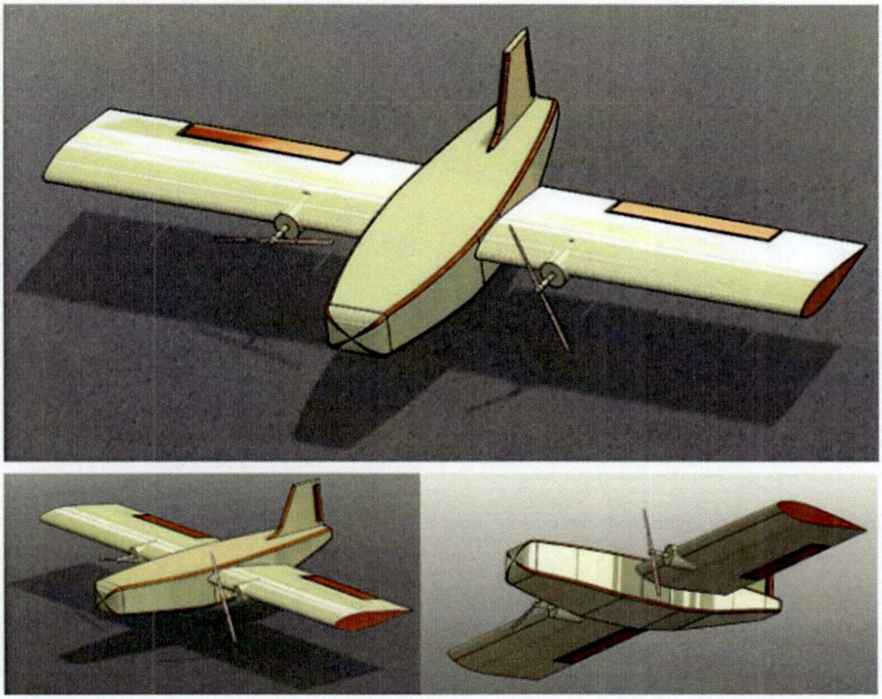

Fig. 4 Solid model isometric views

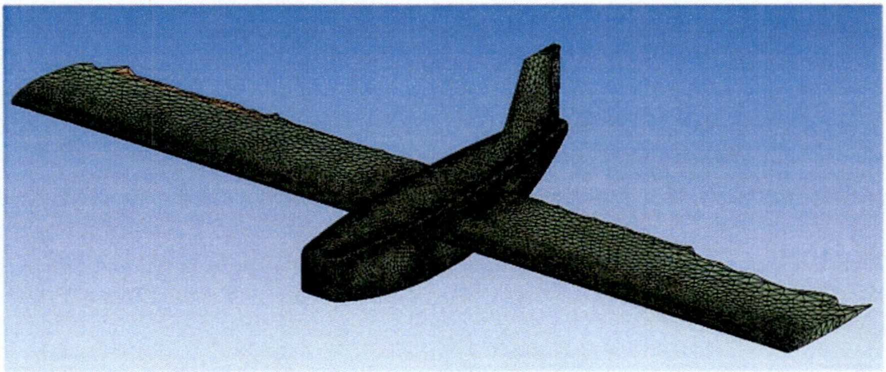

Fig. 5 Solid model meshing view

the UAV for stable flight and aerodynamic balance was determined as 12° angle of attack with the highest L/D ratio and a speed of 8.6 m/s. According to the conceptual model, the necessary equipment was produced and assembled, and the test flight was successfully carried out. The concept design of the designed and realized UAV gave positive and good results in aerodynamics, stability, and control studies.

Table 1 Model airplane dimensions

Item	Value[mm]
Wingspan	1150
Veter	194
Model width	115
Model length	565
Model height	95

Table 2 EPS foam material properties

Element	Estimated weight
Elastic modulus	2.21 Mpa
Density	26.1 kg/m3
Shear modulus	3.17 Mpa
Tensile strength	0.12 Mpa
Comprehensive strength	0.1 Mpa
Yield strength	0.18 Mpa

Table 3 Estimated unloaded weight of the model

Item	Quantity	Unit weight (gr)	Total weight (gr)
Electric engine	2	65	130
Servomotor	3	12	36
Battery	1	185	185
ESC	1	10	10
Flight control	1	49	49
Propeller	2	12	24
GPS	1	49	49
Camera	1	25	25
Approximate weight of the UAV	1	1570	1570
		Total	2078

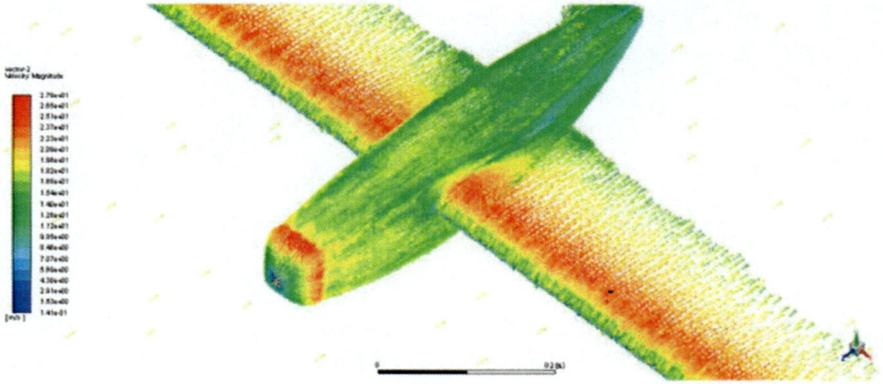

Fig. 6 Speed vectors on the model unmanned aerial vehicle

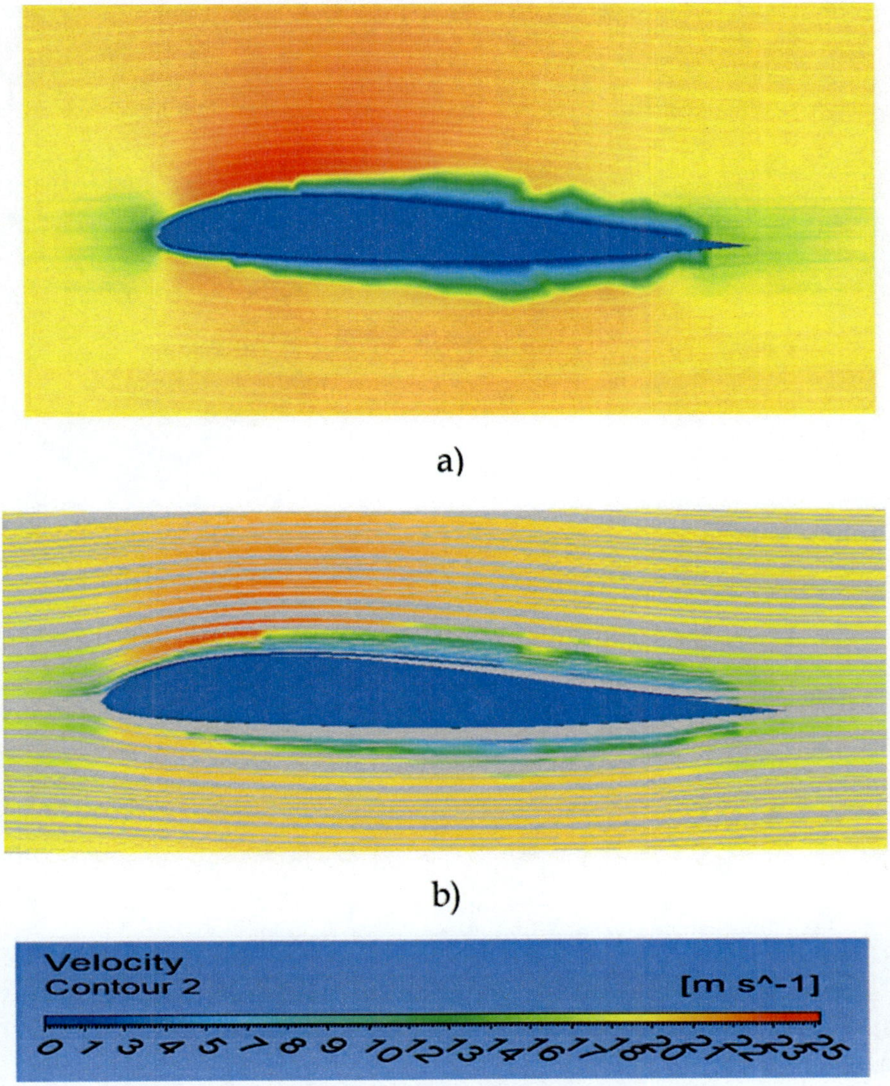

Fig. 7 (**a**) Air velocity over the wing contour view, (**b**) air velocity over the wing streamline view

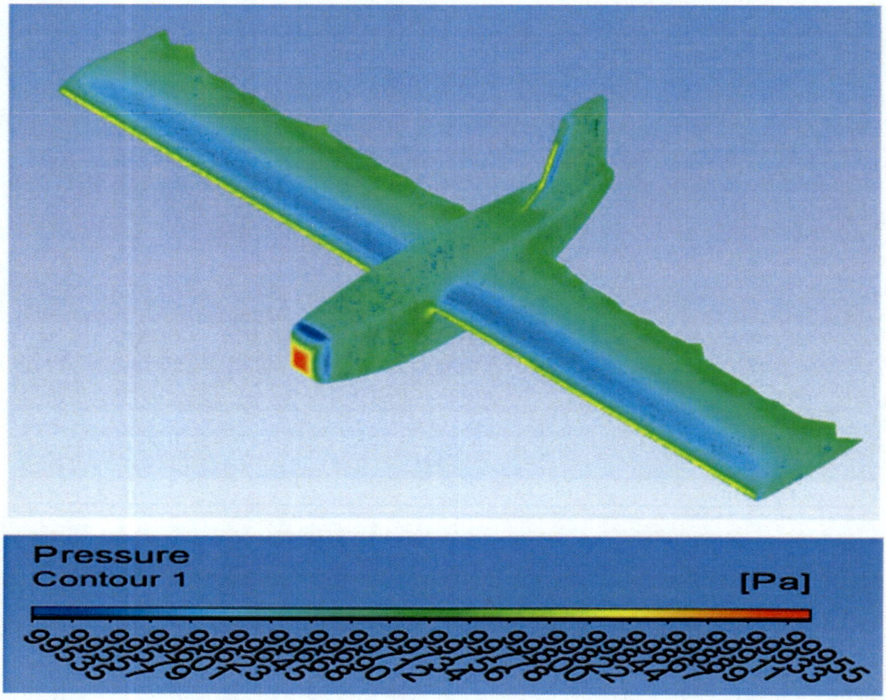

Fig. 8 Pressure on the model unmanned aerial vehicle

a)

Fig. 9 (**a**) UAV view manufactured and assembled, (**b**) test flights of the model UAV

b)

Fig. 9 (continued)

References

Alan, G. E. R., Omar, L. B., Luis, R. O., Patricia, Z. R., Luis, A. B., & Octavio, G. S. (2019). Conceptual design of an unmanned fixed-wing aerial vehicle based on alternative energy. *International Journal of Aerospace Engineering*, *2019*(1), 1–13.

Dereli, Y. (2014). *Turaç İnsansız Hava Aracının Yapısal Modelinin Hazırlanması Ve Analizlerinin Yapılması*. PhD thesis, İstanbul Teknik Üniversitesi Fen Bilimleri Enstitüsü, İstanbul.

Gülbahar, Ö. (2015). *Karayel İnsansız Hava Aracının Statik Ve Dinamik Analizleri*. PhD thesis, İstanbul Teknik Üniversitesi Fen Bilimleri Enstitüsü, İstanbul.

Önal, M., Çoban, S., Yapıcı, A., & Bilgiç, H. H. (2019). Determination of the optimum flight parameters for maximum range and minimum power requirement of a vertical take-off and landing UAV. *Journal of Aviation, 3*(2), 106–112.

Park, Y. B., Nguyen, K. H., Kweon, J. H., Choi, J. H., & Han, J. S. (2011). Structural analysis of a composite target-drone. *International Journal of Aeronautical and Space Sciences, 12*(1), 84–91.

Rocha, M. A., & Solaque, L. E. (2013). Concept design for an unmanned aerial vehicle that will perform exploration missions in Colombia. In *2nd IFAC workshop on research, education and development of unmanned aerial systems*, November 20–22, 2013, Compiegne, France.

Flow Patterns in Double Planar Synthetic Jets

Eva Muñoz and Soledad Le Clainche

Nomenclature

a_m	Amplitude of the mode m
D	Orifice diameter
D_p	Cavity diameter and depth
f	Frequency of the piston movement
L	Length of the domain
Re	Reynolds number
R_e	Exterior radius of the domain
St	Strouhal number
U	Peak of velocity in the cavity
u_p	Velocity of the piston
u_m	Mode given by the HODMD
x	Coordinate in the stream-wise direction
y	Coordinate in the span-wise direction

Greek Letters

δ_m	Growth rate of the mode
ν	Kinetic energy
ϕ	Phase shift between the jets, radians

E. Muñoz (✉) · S. Le Clainche
School of Aeronautics and Space Engineering of the Polytechnic University of Madrid, Madrid, Spain
e-mail: eva.munoz.salamanca@ulb.be; soledad.leclainche@upm.es

T. H. Karakoc et al. (eds.), *New Achievements in Unmanned Systems*, Sustainable Aviation, https://doi.org/10.1007/978-3-031-29933-9_16

Subscripts

i Index for the discretization in the coordinate x
j Index for the discretization in the coordinate y
k Index for the discretization in time
m Index for the number of mode

Superscripts

l Magnitude which is described in the mode

1 Introduction

During recent years, the investigation on alternative propulsion systems is increasingly important. In this context, synthetic jets come up as a new system of marine propulsion (Buren et al., 2018). These devices are also quite relevant in the field of aerospace engineering, especially for their application as a mechanism for active flow control of the boundary layer (Leschziner & Lardeau, 2011), reducing the drag and improving the aircraft efficiency. Hence, to fully understand the physical mechanisms driving the flow motion in synthetic jets is very relevant. This knowledge could facilitate the task of using synthetic jets to maximize the efficiency in aircrafts in a simple and systematic way.

Synthetic jets are devices that create vortex rings that move forward producing thrust, similar to the mechanism driving the flow motion of some marine animals such as squids or jellyfish. The jet stream is formed by a periodic movement composed by two different phases. In the injection phase, the fluid, which is contained in a cavity, is driven by a piston or a membrane to move downstream, passing through the jet nozzle. At this phase, the vortex ring, producing momentum, is generated downstream the jet nozzle. In the suction phase, the fluid returns to the initial cavity passing through the same orifice. A saddle point separates the fluid that continues moving downstream, from the fluid that returns to the cavity.

The main characteristic of synthetic jets is that they generate stream-wise momentum without the need of additional mass injection (zero-net-mass-flux ZNMF). That is thanks to the periodical movement of the membrane or piston, which forces the fluid to go into and leave the cavity through an orifice, the jet nozzle. This characteristic makes very attractive the use of synthetic jets for several industrial applications: i.e., flow control, enhancement of heat transfer, improving the fluid mixing properties, among others (Glezer & Amitay, 2002).

This work studies in detail the flow physics in two-planar synthetic jets, with the aim at exploring their future application as alternative propulsion mechanism and for flow control, to improve the efficiency in aircrafts. The research presented explores the application of data-driven methods, introduced as machine-learning tools, to identify the main physical mechanisms driving the flow motion in synthetic jets. This knowledge can be exploited in future research for flow control applications. Controlling the flow, it is possible to use synthetic jets in a systematic way, to maximize the system efficiency (i.e., improve the efficiency in aircrafts, among other applications).

2 Method

The two jet devices have been modeled performing numerical simulations in a two-dimensional domain and for an incompressible fluid. Figure 1 represents the computational domain, specifying in the upper part the dimensions of the domain and in the lower part the boundary conditions. The dimensions, non-dimensionalized with the diameter of the orifice, $D = 1$, are defined as: the cavity diameter and depth are $D_p = 5D$, and the length, L, and exterior radius, R_e, are $L = 300D$ and $R_e = 240D$. These two dimensions were chosen with a mesh independence study.

The boundary conditions are defined as follows (see more details in (Le Clainche, 2019)). (i) Inlet: $u_p = U\frac{1}{5}\sin(2\pi ft)$, with $U = 1$ and f as the forcing frequency, for the velocity and Neumann boundary conditions for the pressure. (ii) The walls defining the jet exit area are: non-slip condition. (iii) The outflow boundary is defined: with Dong's boundary condition (Dong et al., 2014).

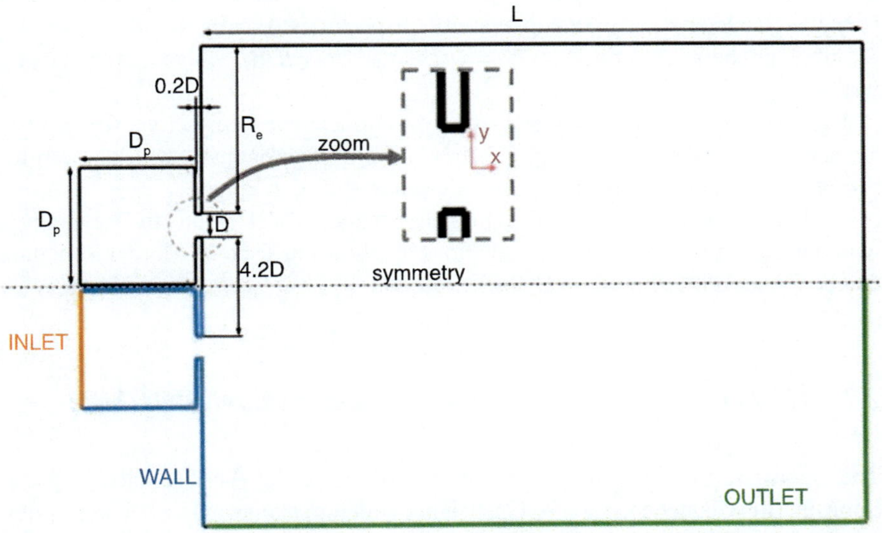

Fig. 1 Computational domain for the numerical simulations of two planar synthetic jets: dimensions and boundary conditions. (Figure not drawn to scale)

Table 1 Simulations carried out grouped by their type of simulation and their Reynolds number

Simulation Reynolds	1 jet	2 jets $\phi = 0$	$\phi = \pi/2$	$\phi = \pi$
10/15		x	x	
25			x	
50		x	x	x
100	x	x	x	x
150		x	x	
200	x	x		
500	x			
1000	x			

The simulations were performed using the numerical solver Nek5000, an open-source code whose spatial discretization is based on the spectral element method (SEM). For the temporal discretization, for the viscous terms we use an implicit backwards differentiation scheme and for the non-linear terms we use an explicit third-order extrapolation scheme. The simulations were carried out solving 1000 non-dimensional time units, which correspond to 30 cycles of the jet oscillations.

The mesh is composed of 3182 macro elements, each of them discretized using 19 Gauss–Lobatto–Legendre points, which represent a polynomial order of 18. In order to ensure the good resolution of the results, the number of grid points is larger in the area near and downstream the two jets. It is also remarkable that the polynomial order was defined after a mesh independence study.

The flow patterns of the synthetic jets vary depending on the Reynolds number and the Strouhal number (Carter & Soria, 2002), distinguishing four flow regimes. The definition of both dimensionless numbers requires characteristic values: D and U as the characteristic length and velocity, respectively. Finally, the Reynolds and Strouhal numbers are defined as $Re = U\,D/\nu$ and $St = fD/U$, where ν is the kinetic viscosity.

Four types of simulations were carried out: (a) a single jet is active; (b) two jets are active moving synchronously; (c) they move asynchronously, with phase shift, $\phi = \pi/2$ rad; and (d) asynchronously with phase shift $\phi = \pi$ rad.

In all the simulations the Strouhal number is fixed, $St = 0.03$, and the Reynolds is varied in each different types of simulations, as shown in Table 1. More information can be found in (Muñoz & Le Clainche, 2022).

2.1 Higher Order Dynamic Mode Decomposition (HODMD)

The algorithm presented in this section was developed by Le Clainche, Vega, & Soria in (Le Clainche et al., 2017). Higher order dynamic mode decomposition (HODMD) (Le Clainche & Vega, 2017) is a data-driven method suitable to study the flow physics in complex flows (Le Clainche et al., 2020). The method is introduced as an extension of the standard dynamic mode decomposition (DMD) presented by Schmid (Schmid, 2010).

For simplicity, the spatio-temporal data analyzed are organized in tensor form $v^l(x_i, y_j, t_k)$, where $l = 1, 2, 3$ denote the stream-wise and span-wise velocity component and the pressure, respectively. The discrete grid points defined in the two spatial directions and the time are defined as $x_i = (i - 1)\Delta x$, $y_j = (j - 1)\Delta y$ and $t_k = (k - 1)\Delta t$, being Δx, Δy, and Δt, the distance between two consecutive grid points of the time step between the snapshots analyzed. HODMD decomposes the data as a temporal expansion in M DMD modes $u_m(l, x_i, y_j)$ as

$$v^l(x_i, y_j, t_k) \cong \sum_{m=1}^{M} a_m u_m^l(x_i, y_j) e^{(\delta_m + i\omega_m)t_k}, \tag{1}$$

where M is the spectral complexity and is referred to the number of terms appearing in the expansion, and each mode u_m is defined by its amplitude, a_m, its growth rate, δ_m, which must be zero when the decomposition is realized in data temporally converged, and its frequency, ω_m.

3 Results and Discussion

The identification of flow patterns is one of the pillars of this work and it is carried out with an analysis of the vorticity, streamlines, and stream-wise velocity for each simulation of Table 1. Figure 2 shows the domain defined in $x\epsilon[-5.2,10]$ and $y\epsilon[-10,5]$, where $x = 0$ is the jet exit. The positive vorticity (counterclockwise

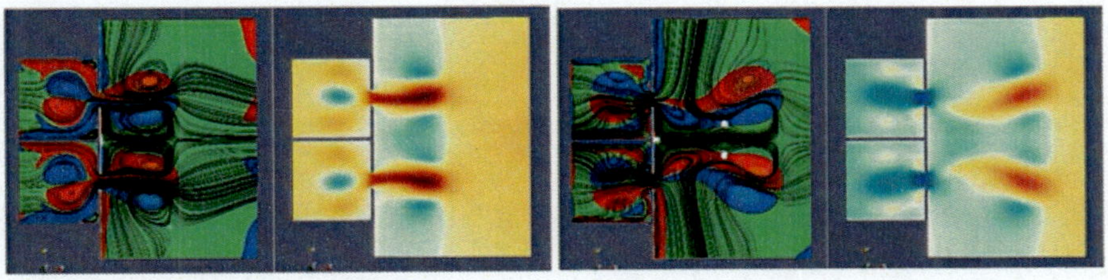

Snapshot extracted at $\varphi = 3\pi/4$ in the injection phase

b) Snapshot extracted at $\varphi = 5\pi/4$ in the suction phase

Scale of the figures a) and b)

Fig. 2 Two synchronous jets at Re = 100. Left: vorticity and streamlines, right: stream-wise velocity

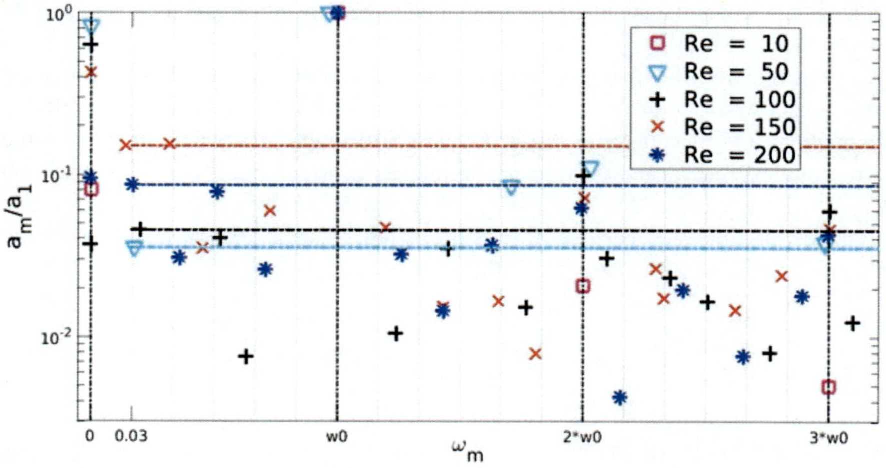

Fig. 3 Frequencies vs. Amplitudes of the DMD modes calculated at different Reynolds number in two synchronous jets

movements) is represented with warm colors and negative vorticity (clockwise movements) with cool colors, and the white dots represent the saddle points, characteristics of the suction phase.

Figure 2 illustrates the injection and suction phase of the two jets moving synchronously at Reynolds number Re $= 100$. During the injection phase, a vortex is expelled from each cavity; during the suction phase, the fluid is forced to enter the cavity while the vortices continue moving forward; these two movements are separated by saddle points.

As anticipated in the introduction, the flow behavior varies with the Reynolds number. This particular case (two synchronous jets at Reynolds number 100) is clearly symmetric. However, when the Reynolds number increases, it can be observed that the symmetry is broken, identifying a larger number of small size flow structures. A similar behavior is identified in the case of one jet and two asynchronous jets. More specifically, in this last case, the flow is never symmetric. These figures are not included for the sake of brevity, but will be presented at the time of the conference.

An analysis of the main flow structures and patterns driving the flow dynamics is carried out using HODMD, in order to find the origin of the symmetry breaking.

Figure 3 shows the modes calculated with this algorithm in the case of two synchronous jets. The horizontal axis corresponds to the frequency of each mode, and the vertical one to the amplitude of each mode, non-dimensionalized with one of the identified fundamental frequencies ω_0. In all the simulations, there are frequencies in common, the fundamental frequency ($\omega_0 = 2\pi \cdot 0.03 = 2\pi St$) and its harmonics ($\omega = 2\omega_0, 3\omega_0$, etc). The first one appears due to the periodical movement of the system (the fundamental frequency is the sixth harmonic of the forcing frequency $St = 0.03$). On the other hand, for Reynolds number higher or equal to 50, another relevant frequency appears: $St = 0.03$, the forcing frequency, and some cases its harmonics ($St = 0.06, 0.09$, etc.).

The forcing frequency appears with low amplitude for Reynolds numbers 50 and 100 and it increases in the cases where the symmetry is broken (Reynolds number 150 and 200). This behavior is also present in one jet simulations, suggesting that the symmetry breaking can be related with the forcing frequency, more details will be presented at the time of the conference.

4 Conclusion

Synthetic jets are devices with high potential for the aerospace industry, but solving realistic problem via numerical simulations need high resources. With the DMD modes, provided by the HODMD algorithm, it can be created a reduced order model that is able to predict the system dynamics, significantly reducing computational needs. Moreover, the method identifies the main patterns driving the flow. If the flow physics is known, then it is possible to control it, developing more efficient applications of synthetic jets for flow control of the boundary layer, improving the efficiency in aircraft.

Acknowledgments E.M.S. acknowledges the grant PID2020-114173RB-I00 funded by MCIN/AEI/10.13039/501100011033.

References

Buren, T., Floryan, D., & Smits, A. (2018). Bio-inspired underwater propulsors. *arXiv*, 1801.09714v2.

Carter, J. E., & Soria, J. (2002). The evolution of round zero-net-mass-flux jets. *Journal of Fluid Mechanics, 472*, 167–200.

Dong, S., Karniadakis, G. E., & Chryssostomidis, C. (2014). A robust and accurate outflow boundary condition for incompressible flow simulations on severely-truncated unbounded domains. *Journal of Computational Physics, 261*, 83105.

Glezer, A., & Amitay, M. (2002). Synthetic jets. *Annual Review of Fluid Mechanics, 34*, 503–529.

Le Clainche, S. (2019). Prediction of the optimal vortex in synthetic jets. *Energies, 12*(9), 1635.

Le Clainche, S., & Vega, J. M. (2017). Higher order dynamic mode decomposition. *SIAM Journal on Applied Dynamical Systems, 16*(2), 882–925.

Le Clainche, S., Vega, J. M., & Soria, J. (2017). Higher order dynamic mode decomposition for noisy experimental data: Flow structures on a zero-net-mass-flux jet. *Experimental Thermal and Fluid Science, 88*, 336–353.

Le Clainche, S., Izbassarov, D., Rosti, M., Brandt, L., & Tammisola, O. (2020). Coherent structures in the turbulent channel flow of an elastoviscoplastic fluid. *Journal of Fluid Mechanics, 888*. https://doi.org/10.1017/jfm.2020.31

Leschziner, M. A., & Lardeau, S. (2011). Simulation of slot and round synthetic jets in the context of boundary-layer separation control. *Philosophical Transactions of the Royal Society A, 369*, 1495–1512.

Muñoz, E., & Le Clainche, S. (2022). On the topology patterns and symmetry breaking in two planar synthetic jets. *Physics of Fluids, 34*, 024103. https://doi.org/10.1063/5.0080834

Schmid, P. J. (2010). Dynamic mode decomposition of numerical and experimental data. *Journal of Fluid Mechanics, 656*, 5–28.

Coordinated Path-Following for Multi-Agent Fixed-Wing Aircraft

Hugo S. Costa, Stephen Warwick, Paulo Oliveira, and Afzal Suleman

1 Introduction

Unmanned aerial vehicles (UAVs) have found increasing use in academic, military, and practical applications, such as reconnaissance and surveillance. Their restricted capabilities in terms of power, sensing, communication, and computation have generated interest in the use of teams of UAVs to improve the capability, capacity, and flexibility of the unmanned aerial system (UAS). Advantages include the ability to parallelize individual tasks, increase the tolerance of the whole system to sensor and hardware failures, and the possibility of giving different capabilities to each UAV (Kaminer et al., 2017; Xargay et al., 2013). A large subset of these missions requires that the UAVs coordinate their motion in relative and/or absolute time (Kaminer et al., 2017). Examples of these cooperative mission scenarios are sequential auto-landing and coordinated ground-target suppression for multiple UAVs (Xargay et al., 2013).

Ihle et al. (2007) developed a passivity approach for coordinate path-following of general non-linear systems by combining a passive path-following control algorithm for each UAV with a passive synchronization algorithm. Ghabcheloo et al. (2009) studied the coordinated path-following under communication losses and time delays and derived sufficient conditions for the stability of the system. Xargay et al. (2013) extends and applies this last study to address the problem of steering a fleet of UAVs along desired paths while meeting spatial and temporal constraints. The missions

H. S. Costa (✉) · P. Oliveira
Instituto Superior Técnico, Lisbon, Portugal
e-mail: hugosequeiracosta@tecnico.ulisboa.pt; paulo.j.oliveira@tecnico.ulisboa.pt

S. Warwick · A. Suleman
Department of Mechanical Engineering, University of Victoria, Victoria, BC, Canada
e-mail: swarwick@uvic.ca; suleman@uvic.ca

T. H. Karakoc et al. (eds.), *New Achievements in Unmanned Systems*, Sustainable Aviation, https://doi.org/10.1007/978-3-031-29933-9_17

considered require that the UAVs follow collision-free paths and that all vehicles arrive at the same time to their final destinations.

The objective of the work described in this chapter is to fly two or three UAVs in a coordinated flight testing campaign. To achieve this, the problem is reformulated as one of coordinated path-following. This problem is decomposed into two. First, two path-following algorithms are presented that solve a path-following problem by commanding the attitude of the vehicle. Second, a synchronization algorithm that synchronizes the vehicle's along-path parameters according to given coordination maps by commanding their speeds is specialized to each of the path-following algorithms. Preliminary simulations of these algorithms are presented.

2 Method

The coordinated path-following problem addressed in this chapter assumes that for each UAV a desired path has been generated, with a corresponding desired speed profile. In the next section, the mathematical model for each vehicle is presented, followed by the description of the two path-following algorithms and the coordination algorithm.

2.1 Mathematical Model

It is assumed that the autopilot is capable of maintaining each aircraft at the desired altitude set-point and that each aircraft follows a coordinated turn. Under these assumptions, the kinematic model for the aircraft will be taken as

$$\dot{\boldsymbol{p}} = \boldsymbol{R}(\psi) V \boldsymbol{e}_1 + \boldsymbol{w}, \tag{1}$$

$$\dot{\psi} = \frac{g}{V} \tan \phi, \tag{2}$$

where $\boldsymbol{p} = (x, y)^T$ is the north–east position of the aircraft in a local inertial reference frame, ψ is the heading of the aircraft,

$$\boldsymbol{R}(\psi) = \begin{pmatrix} \cos\psi & -\sin\psi \\ \sin\psi & \cos\psi \end{pmatrix}$$

is the rotation matrix from the direction of the speed of the aircraft to the local inertial reference frame, V is the airspeed of the aircraft, $\boldsymbol{e}_1 = (1, 0)^T$, $\boldsymbol{w} \in \mathbb{R}^2$ is the wind speed in the local inertial reference frame, g is the local acceleration of gravity, and ϕ

is the roll angle of the aircraft. The airspeed and the roll angle are considered to be the inputs, since it is assumed that the autopilot controls these aircraft variables on a much faster time-scale than the path following algorithms that will be presented.

The roll angle input is transformed to the turn rate command by

$$\omega^c = \frac{g}{V} \tan\phi.$$

In addition, no wind will be assumed. These considerations lead to the model that will be considered for the path-following algorithms to be presented:

$$\dot{\boldsymbol{p}} = \boldsymbol{R}(\psi) V \boldsymbol{e}_1$$

$$\dot{\psi} = \omega^c, \tag{3}$$

subject to the input constraints

$$0 < V_{\min} \leq V \leq V_{\max} \tag{4}$$

$$|\ \omega^c\ | \leq \omega_{\max}\ (V). \tag{5}$$

These constraints are imposed by the flight characteristics of fixed-wing UAVs. In the first constraint, the lower bound is due to the existence of a minimum airspeed under which the aircraft is not capable of maintaining level flight, stall speed, and the maximum bound is due to the aircraft's maximum thrust capability. The second constraint is imposed by the maximum roll angle that is to be allowed.

2.2 *Linearised Path-Following Controller*

This controller works by approximating the desired path for each aircraft by a concatenation of straight-line segments, whose end-points will be denoted by $\{\boldsymbol{w}_0, \boldsymbol{w}_1, \ldots, \boldsymbol{w}_N\}$. The transition between the different segments is implemented using the half-plane switching method (Beard & McLain, 2012). In this method, the path controller switches to the next segment when it has crossed the bisecting plane between this and the current segment. This switching condition is satisfied when the UAV's position is within the half-plane:

$$\mathcal{H} = \left\{\boldsymbol{p} \in \mathbb{R}^2 : (\boldsymbol{p} - \boldsymbol{w}_i)^T \boldsymbol{n}_i \geq 0\right\},$$

where $\overline{\boldsymbol{w}_{i-1}\boldsymbol{w}_i}$ is the current segment, $\overline{\boldsymbol{w}_i\boldsymbol{w}_{i+1}}$ is the next segment,

$$\boldsymbol{n}_i = \boldsymbol{q}_{i-1} + \boldsymbol{q}_i$$

is normal to the half-plane that points to the interior, and

$$\boldsymbol{q}_i = \frac{\boldsymbol{w}_{i+1} - \boldsymbol{w}_i}{||\boldsymbol{w}_{i+1} - \boldsymbol{w}_i||}$$

are the unit vectors of each segment.

The controller is designed to steer the UAV to the current segment being followed. The objective is to reduce the lateral distance of the UAV to the path to zero. For this purpose, we denote the lateral distance to the path by δ, considering it to be positive when the UAV is to the right of the path, negative otherwise. As such, its dynamics are described by

$$\dot{\delta} = V \sin \tilde{\psi},$$

where $\tilde{\psi} = \psi - \Psi_0 \in]-\pi, \pi]$ is the heading relative to the path's direction. Linearizing, for a given airspeed V, this last equation, combining it with Eq. (3), and introducing the integral state

$$\dot{\delta}_I = \delta + \delta_{aw},$$

where δ_{aw} is an anti-wind-up discharge rate, the following linearized system is obtained:

$$\begin{pmatrix} \dot{\delta}_I \\ \dot{\delta} \\ \dot{\tilde{\psi}} \end{pmatrix} = \begin{pmatrix} 0 & 1 & 0 \\ 0 & 0 & V \\ 0 & 0 & 0 \end{pmatrix} \begin{pmatrix} \delta_I \\ \delta \\ \tilde{\psi} \end{pmatrix} + \begin{pmatrix} 0 \\ 0 \\ 1 \end{pmatrix} \omega^c + \begin{pmatrix} 1 \\ 0 \\ 0 \end{pmatrix} \delta_{aw}.$$

The following PID control law is implemented:

$$\omega_\ell^c = -k_{\delta_I}\delta_I - k_\delta \delta - k_{\tilde{\psi}}\tilde{\psi},$$

but to satisfy the constraint in the turn-rate Eq. (5), this command is saturated:

$$\omega^c = sat_{-\omega_{\max}}^{\omega_{\max}} \omega_\ell^c,$$

and the back-calculation anti-wind-up solution is implemented by discharging the integral state by the amount

$$\delta_{aw} = k_{\delta_I,\text{aw}}(\omega_\ell^c - \omega^c),$$

where $k_{\delta_I,\text{aw}}$ is a gain.

2.3 Vector-Field Curved Path-Following Controller

In this subsection, a non-linear curved path-following controller is considered. Let $\boldsymbol{q} : I \subset \mathbb{R} \to \mathbb{R}^2$ be the parametrization of the path of a given UAV, parametrized by the path-length variable s. Define the along and cross-track errors as

$$\begin{pmatrix} e_s \\ e_d \end{pmatrix} = R^T\left(\psi_q(s)\right)(\boldsymbol{p} - \boldsymbol{q}(s)),$$

where $\psi_q(s)$ is the course angle of the path for the given point parametrized by s. To make the UAV follow its assigned path by converging the along and cross-track errors to zero, the following updating law for the path parameter and control law for the turn rate (Zhao et al., 2020) are implemented:

$$\begin{aligned} \dot{s} &= k_s e_s + V \cos \tilde{\psi} \\ \omega^c &= -k_\omega(\tilde{\psi} - \tilde{\psi}_d(e_d)) + \kappa(s)\dot{s} + \dot{\tilde{\psi}}_d(e_d), \end{aligned}$$

where

$$\tilde{\psi}_d(e_d) = -\tilde{\psi}^\infty \tanh(k_d e_d)$$

is a desired course differentiable vector field, $\tilde{\psi}^\infty \in \left[0, \frac{\pi}{2}\right]$, $\tilde{\psi} = \psi - \psi_q(s)$, $\kappa(s)$ is the curvature of the path at the point parametrized by s, and k_s, k_d, and k_ω are positive control gains.

2.4 Coordination Algorithm

The desired path length of the UAV along its assigned path for time t_d is given by (Kaminer et al., 2017)

$$\ell_{d,i}(t_d) = \int_0^{t_d} v_{d,i}(\tau) d\tau,$$

where $v_{d,i}(\tau)$ is the desired speed of UAV i at time τ. Since the desired speed of each fixed-wing UAV is strictly positive (Eq. 4), this is a strictly increasing function of the desired time t_d, so it is invertible (Kaminer et al., 2017). This inverse function η gives the desired time that the aircraft should be given the along path parameter, and shall henceforth be referred to as the UAV's virtual time.

For the first path-following algorithm, it is assumed that the speed profile is constant along each path segment. As such,

$$\eta(s) = \frac{\Delta\ell_{i,j}}{v_{d,i,j}} + \sum_{j\in\text{previous segments}} \frac{L_{i,j}}{v_{d,i,j}},$$

where $L_{i,\,j}$ is the total length of path segment j of aircraft i, $v_{d,\,i,\,j}$ is the desired speed at path segment j of aircraft i, and $\Delta\ell_{i,\,j}$ is the length that aircraft has covered in the current path segment, given by

$$\Delta\ell_{i,j} = (\cos\,\Psi_0,\ \sin\,\Psi_0)(\mathbf{p} - \mathbf{w}_{i-1}).$$

For the second path-following algorithm, no restriction is placed on the virtual time beyond the ones for the general case.

The dynamics of each UAV's virtual time are given by

$$\dot{\xi}_i = \frac{\dot{s}_i}{v_{d,i}(\xi_i)}.$$

As a first step, this equation is feedback-linearized. In the case of the first path-following algorithm, the input velocity is transformed by

$$V^c{}_{\ell,i} = v_{d,i}\, u_{\text{coord},i}$$

For the second one, the transformation is

$$V_i = \frac{-k_s\, e_{s,i} + v_{d,i}\, u_{\text{coord},i}}{\cos\,\tilde{\psi}_i}.$$

By these transformations, single integrator dynamics are achieved for each of the virtual time coordination states:

$$\dot{\xi}_i = u_{\text{coord},i} \tag{6}$$

The coordination control law implemented for each leader i is (Kaminer et al., 2017)

$$u_{\text{coord},i} = -k_P \sum_{j\in\mathcal{N}_i(t)} (\xi_i - \xi_j) + 1, \tag{7}$$

while the one implemented for the followers is (Kaminer et al., 2017)

$$u_{\text{coord},i} = -k_P \textstyle\sum_{j\in\mathcal{N}_i(t)} (\xi_i - \xi_j) + \chi_{I,i},$$

$$\dot{\chi}_{I,i} = -k_I \sum_{j \in \mathcal{N}_i(t)} (\xi_i - \xi_j) + k_{\text{coord,aw}}(V - V_{\text{unsat}}), \chi_{I,i}(0) = 1,$$

where $\mathcal{N}_i(t)$ is the set of neighbours of i at time t, and k_P, $\chi_{I,\ i}$, and $k_{\text{coord, aw}}$ are positive control gains.

It is assumed that the communication between each UAV is bidirectional and that the information is transmitted continuously with no delays, each vehicle only exchanges its coordination state $\xi_i(t)$ with its neighbours, and the communication graph is connected in an integral sense, i.e., it satisfies the condition

$$\frac{1}{N}\frac{1}{T}\int_t^{t+T} Q\,L(t)Q^T\,d\tau \geq \mu \mathbb{I}_{N-1}, \quad \text{for all } t \geq 0$$

where $L(t) \in \mathbb{R}^{N \times N}$ is the Laplacian of the communications graph, Q an $(N-1) \times N$ matrix such that $Q1_N = 0$, and $QQ^T = \mathbb{I}_{N-1}$, 1_N which is the vector in $\mathbb{R}^n$ whose components are all 1, and the parameters $T, \mu > 0$ represent the level of connectivity of the communications graph (Xargay et al., 2013).

To ensure the algorithms satisfy the airspeed constraint in Eq. (4), the speed commands are saturated between these two bounds.

3 Results and Discussion

In this section, simple simulations are presented for each of the algorithms. The simulations were performed in Simulink, using as the model for each aircraft the kinematic model described by Eqs. (1) and (2). The parameters used for g, $V_{\min}$, and $V_{\max}$ are 9.81 m s^{-2}, 15 ms^{-1}, and 44 ms^{-1}, respectively.

The chosen communication topology is shown in Fig. 1. In this topology, each aircraft implements its own virtual leader, with each aircraft communicating its actual virtual time only with its own virtual leader and the virtual leaders communicating their virtual times between themselves. This addition of virtual leaders with "uncertainty-free dynamics" improves coordination robustness in the presence of disturbances like winds and gusts (Kaminer et al., 2017; Xargay et al., 2012).

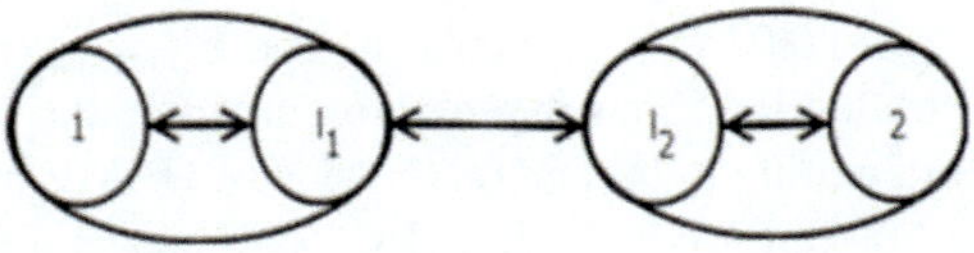

Fig. 1 Communication topology. 1 and 2 represent the physical vehicles, while l_1 and l_2 are the virtual leaders

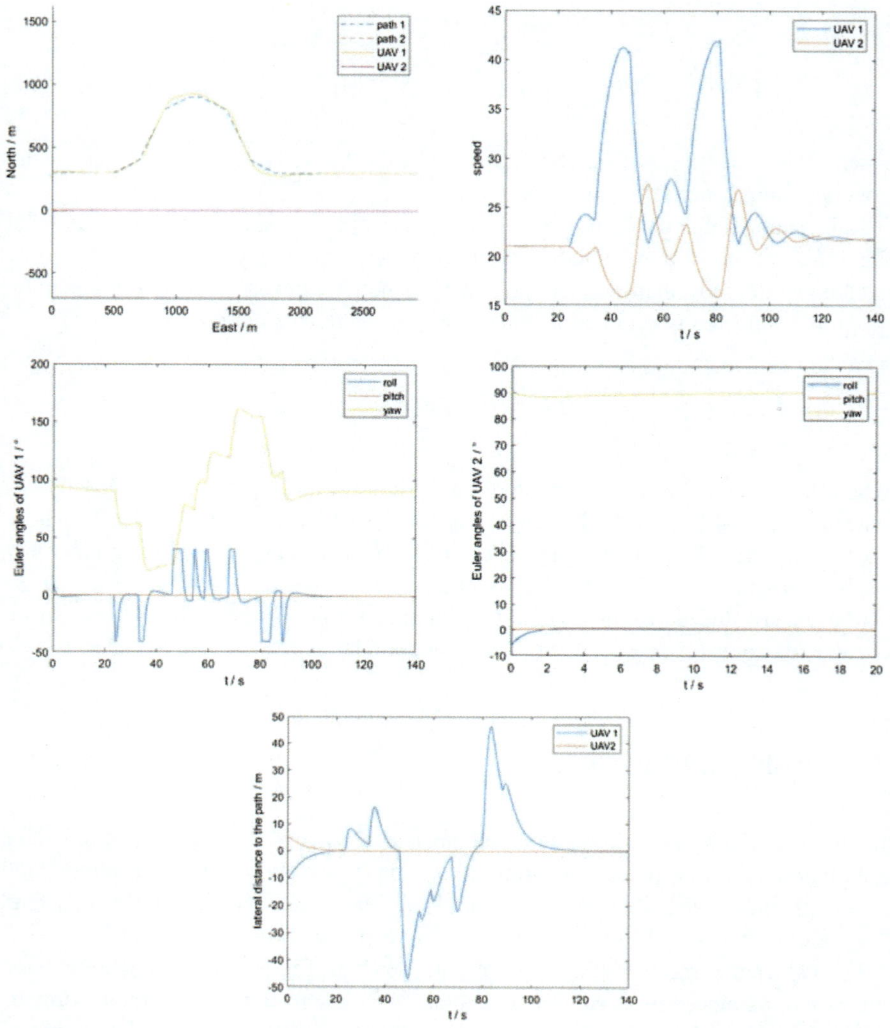

Fig. 2 Algorithm 1 simulation for a polygonal detour (trajectories, airspeed, attitudes, and lateral distance)

The results of the simulation of the first algorithm are shown in Fig. 2. The parameters used in the path controller are 0.01 m^{-1} s^{-1}, 1.4 s^{-1}, 7× 10^{-5} m^{-1}, and 20 m for k_δ, k_{δ_I}, $k_{\tilde{\psi}}$, and $k_{\delta_I,\text{aw}}$, respectively. The parameters used in the coordination are 0.9 s^{-1}, 0.03 s^{-2}, and 0.02 m^{-1} s^2, for k_P, k_I, and $k_{\text{coord, aw}}$. In this scenario, the mission is designed so that UAV 1 is required to make a detour (e.g., to circumvent an obstacle), while attempting to remain on the side of UAV 2, who continues on a straight line. It may be seen in Fig. 2 that UAV 1 has several spikes in its lateral distance, due to the discontinuity introduced by the chosen segment switching method. This, in turn, leads to significant variations in the speed command. Nonetheless, the vehicle is able to converge to its path, even though the roll set-point is

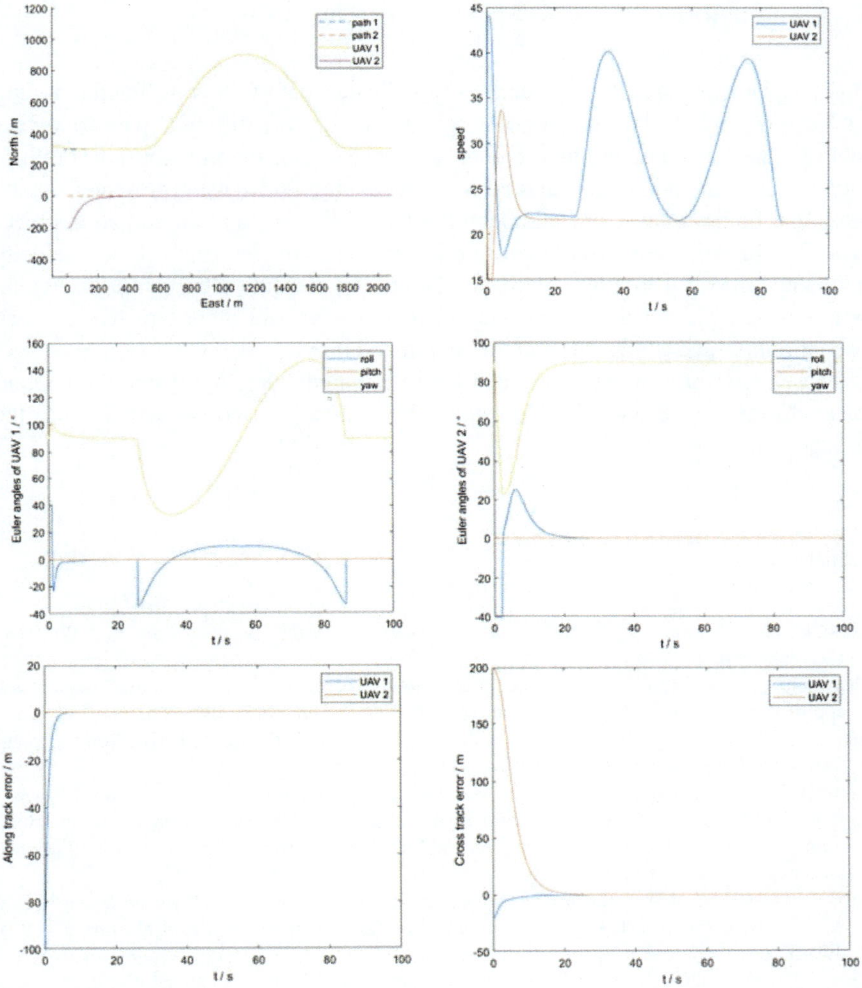

Fig. 3 Algorithm 2 simulation for a cubic spline detour path (trajectories, airspeed, attitudes, along-track, and cross-track errors)

always saturated upon each switch, and the vehicles finish the simulation with vehicle 1 behind 2 by only 2 meters. These effects introduced by the discontinuity of the transition may be mitigated by changing the switching algorithm to a smoother one such as a fillet transition (Beard & McLain, 2012).

Simulation results for the second algorithm are shown in Fig. 3. The parameters used in the path controller are 1 s^{-1}, 10 s^{-1}, 0.01 m, and 70 ° for k_s, k_ω, χ^∞, and k_d. The parameters used in the coordination are 2 s^{-1}, 0.1 s^{-2}, and 2 m^{-1} s^2, for k_P, k_I, and $k_{\text{coord, aw}}$. The mission is identical to the former, but the detour is described by two cubic spline segments. UAV 1 starts behind UAV 2, but it may be seen that they are able to coordinate and finish the simulation alongside each other with similar speeds, while also converging to their respective paths. Compared to the former algorithm, UAV 1 tracks its path more precisely. In addition, the speed command and roll set-point are smoother.

4 Conclusion

This chapter addressed the problem of coordinated path-following for fixed-wing UAVs for a fixed altitude set-point by presenting two different path-following control algorithms and combining them with a synchronization algorithm. Preliminary results simulating the operations of these algorithms were presented which show that both are able to accomplish the path-following and coordination objectives. The linearization-based algorithm is shown to produce spikes in the velocity command and roll set-point commands due to the discontinuities in the switching between the different path segments, whereas the second algorithm has no such discontinuities and produces smoother command inputs.

These algorithms are currently being implemented using an off-board computer for controlling fixed-wing UAVs using a PX4 autopilot and simulated using the Gazebo simulator.

References

Beard, R. W., & McLain, T. W. (2012). *Small unmanned aircraft: Theory and practice*. Princeton University Press.

Ghabcheloo, R., et al. (2009). Coordinated path-following in the presence of communication losses and time delays. *SIAM Journal on Control and Optimization, 48*(1), 234–265.

Ihle, I. F., Arcak, M., & Fossen, T. I. (2007). Passivity-based designs for synchronized path-following. *International Journal of Green Energy, 43*(9), 1508–1518.

Kaminer, I., et al. (2017). *Time-critical cooperative control of autonomous air vehicles*. Elsevier.

Xargay, E. et al. (2012). Convergence of a PI coordination protocol in networks with switching topology and quantized measurements. In *2012 IEEE 51st conference on decision and control*, December 10, 2012. Champaign, IL.

Xargay, E., et al. (2013). Time-critical cooperative path following of multiple unmanned aerial vehicles over time-varying networks. *Journal of Guidance, Control, and Dynamics, 36*(2), 499–516.

Zhao, S., et al. (2020). Integrating vector field approach and input-to-state stability curved path following for unmanned aerial vehicles. *IEEE Transactions on Systems, Man, and Cybernetics: Systems, 50*(8), 2897–2904.

Onboard Trajectory Coordination of Multiple Unmanned Air Vehicles

James Sease, Stephen Warwick, and Afzal Suleman

1 Introduction

In recent years, the increased use of unmanned air vehicles (UAVs) in aviation has created new challenges and opportunities for improvement in the operation of multi-agent flight systems. A solution for multiple cooperative agent guidance within an urban air mobility environment has been proposed (Yang et al., 2020).

There are numerous potential benefits for agents if they are able to coordinate their flight trajectories. These benefits include more efficient use of airspace, improved energy efficiency, and more effective mission completion. For example, when aircraft fly in formation they are able to gain fuel efficiently by taking specific positions in formation. The fuels savings have been measured as high as 18% (Vachon et al., 2003).

Using communication and path coordination, UAVs have the potential to plan their paths more efficiently with respect to each other. It could also allow aircraft to safely fly closer together in a congested airspace. Depending on the mission, proper coordination could significantly improve their mission effectiveness. As an example, in a search and rescue operation, multiple high-altitude aircraft could cover an area quickly and call in low-altitude UAVs to provide confirmation on any findings needing further verification.

J. Sease (✉) · S. Warwick · A. Suleman
Department of Mechanical Engineering, Centre for Aerospace Research, University of Victoria, Sidney, BC, Canada
e-mail: jsease@uvic.ca; swarwick@uvic.ca; suleman@uvic.ca

T. H. Karakoc et al. (eds.), *New Achievements in Unmanned Systems*, Sustainable Aviation, https://doi.org/10.1007/978-3-031-29933-9_18

2 Background

There are many different ways to coordinate a group of agents within a system. Two such methods are virtual leader and flocking.

Virtual leader uses the same coordination scheme as a leader follower setup, but replaces the leader with a virtual agent. This has previously been simulated in combination with the artificial potential field method (Zhang et al., 2018). Because when the virtual leader is simulated, its movement can be predicted. It is also at no risk of a physical collision with other agents if something goes wrong. Because the leader is virtual, all effects that the environment plays on the behavior of the leader must specifically be programmed into the simulation. For example, if the formation were flying into a strong headwind, the virtual leader may need to be programmed to accommodate a change to its flight speed so as to maintain the optimal airspeed for the UAVs it is leading.

Flocking is a system made up of a number of different decision-making algorithms that can be combined in different ways (Reynolds, 1987). The core algorithms are cohesion, which steers agents toward other agents Fig. 1a, and alignment, which causes the agents to steer in the same direction that all other agents are traveling, as illustrated in Fig. 1b, and separation, which steers the agents away from other agents that start to get to close (Fig. 1c). Proper balancing between these algorithms can lead to agents acting similar to a flock of birds or school of fish.

Many existing multi-AAV systems are actively controlled from a ground station. This allows for the UAV movement to be optimized and verified by a powerful computer, with human oversight. The ground system directs the aircraft through a radio link. It may not always be possible to provide a stable ground link to a group of UAVs operating far from the base station. As an example, UAVs being used for search and rescue in a mountainous environment may not be able to maintain communication links to a ground station at all times.

3 Theoretical Design

A new multi-UAV coordination system was developed and tested using a combination of Virtual Leader and Flocking to coordinate the aircraft. The coordination software was designed to run on an onboard computer, mounted to each of the UAVs in the system. The aircrafts were able to communicate information with each other using a mesh radio architecture.

The virtual leader of the formation is simulated onboard of one of the aircrafts. This aircraft sends the coordination information consisting of the leader's latitude, longitude, altitude, and its velocity to all cooperative agents.

Each aircraft tracks its target position with respect to the virtual leader. The formation used in this study was a delta formation. The formation was originally configured so that the aircraft would position themselves 5 m to the left, front, and

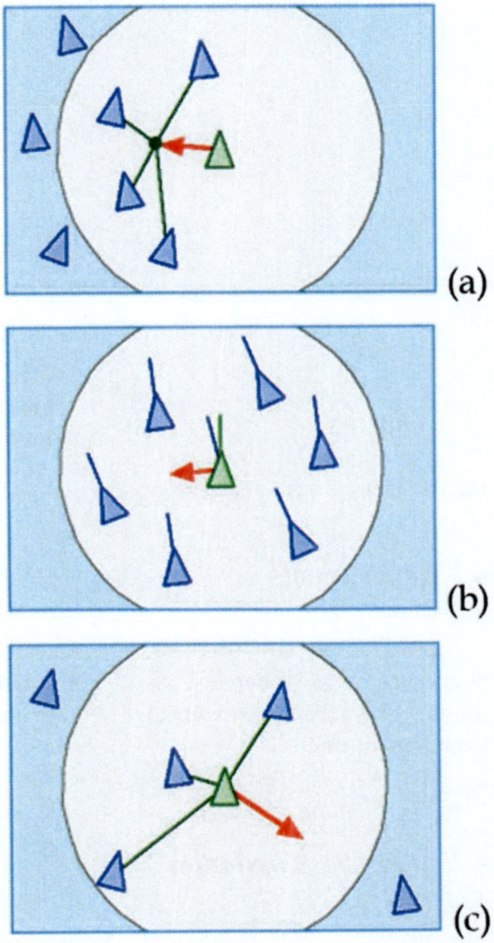

Fig. 1 Flocking algorithms: (**a**) cohesion, (**b**) alignment, (**c**) avoidance. (Reynolds, 2001)

right of the virtual leader Fig. 2. This was increased during flight testing to 10 m to provide a greater safety margin to mitigate against the risk of aerial collisions.

The follower aircraft were controlled with velocity commands fed into a previously tuned PX4 multi-rotor controller. For safety proposes, the multi-rotors where limited to a relatively low maximum speed.

In order to reduce risk and prevent collisions, some elements of flocking were added. Each aircraft shares its location with the rest of the fleet. If the distance to another aircraft is too small, an aircraft will take evasive maneuvers to increase the distance to reduce the risk of collision.

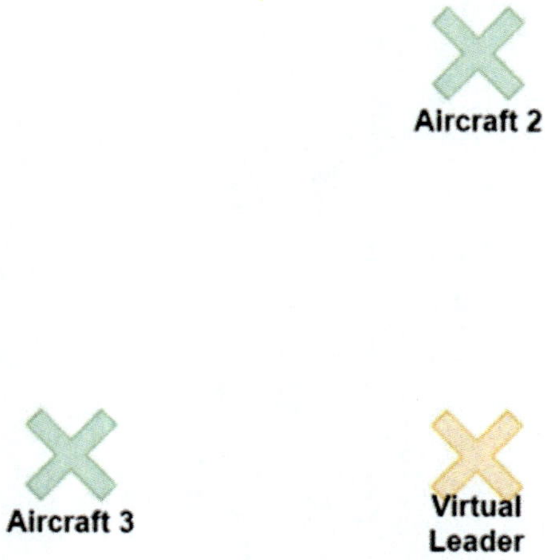

Fig. 2 Aircraft theoretical formation

4 Experiments

The proposed system was tested in simulation using Gazebo configured with a representative flight dynamics model of a multirotor aircraft. The simulations were followed by a flight test campaign. Following the flight tests, some improvements were implemented.

4.1 Gazebo Simulation

The system was built using ROS. The simulation was performed using Gazebo and the PX4 autopilot firmware software-in-the-loop platform. When simulated in Gazebo, an unexpected communication delay, consistent with hardware tests, was found. This contributed to a transport delay of 0.3 s to all the position and velocity feedback used in each of the aircraft.

The aircraft logged their state information for analysis. The logs are correlated with system timestamps. The time offsets between the aircrafts clocks was also recorded. Figure 3 shows the position errors of each of the three aircraft simulated as part of the formation. The formation was moving with constant speed to provide a strait and level flight test.

The data in Fig. 3 shows that the aircraft is able to maintain a relatively constant position error at approximately 0.75 m from their target location.

A second configuration was tested where the formation was required to track through a coordinated turn providing a heading change for formation position error evaluation. The position errors of this test are shown in Fig. 4. The executed turn was

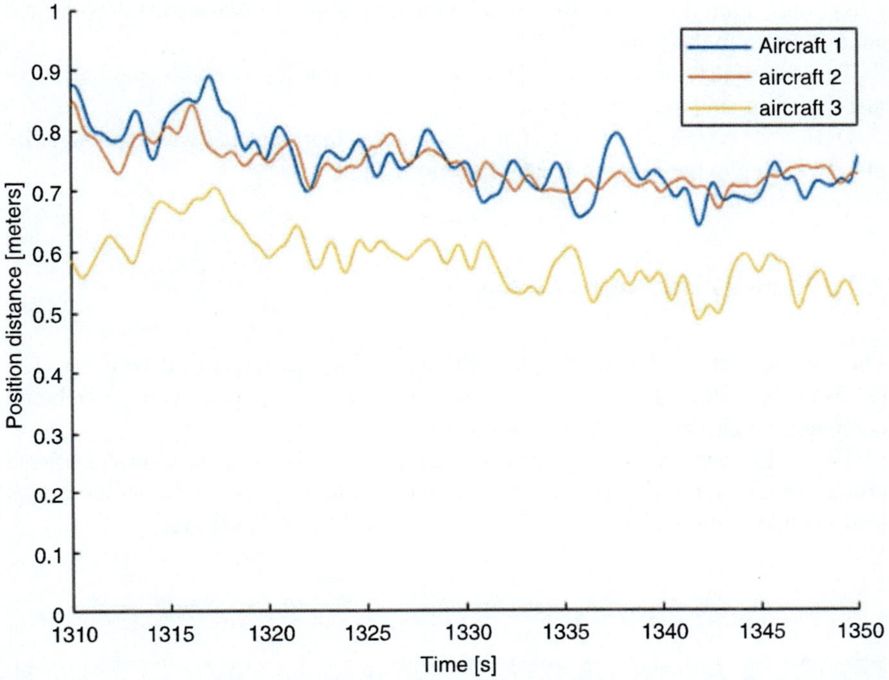

Fig. 3 Gazebo simulation, formation position error for straight and level flight

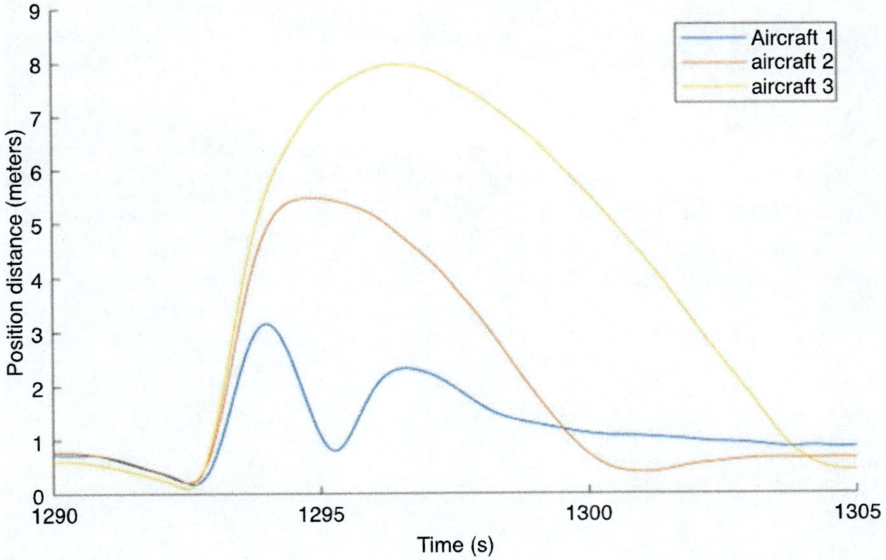

Fig. 4 Gazebo simulation, formation position error tracking through coordinated turn

a right turn, aircraft 1 was on the inside of the turn, aircraft 3 was on the outside, and aircraft 2 was in the front.

This shows that the maximum error for aircraft 1 was just over 3 m, aircraft 2 was just over 5 m, and aircraft 3 was at 8 m.

During the coordinated turn test, it was observed that aircraft 1 was required to fly in reverse due to the rotational rate of the formation.

4.2 Real-World Flight Testing

The system was deployed onto a multi-agent UAV test platform (Fig. 5). The platform consisted of three multi-rotor UAVs, each equipped with an onboard computer which runs the coordination algorithms.

The virtual center was simulated on-board on one of the aircrafts. Due to a voltage problem with aircraft 2, it could not be operated. Aircraft 1 was set to run the virtual leader while flying and aircraft 3 flew as a standard agent in the system.

Fig. 5 Multi-agent test platform

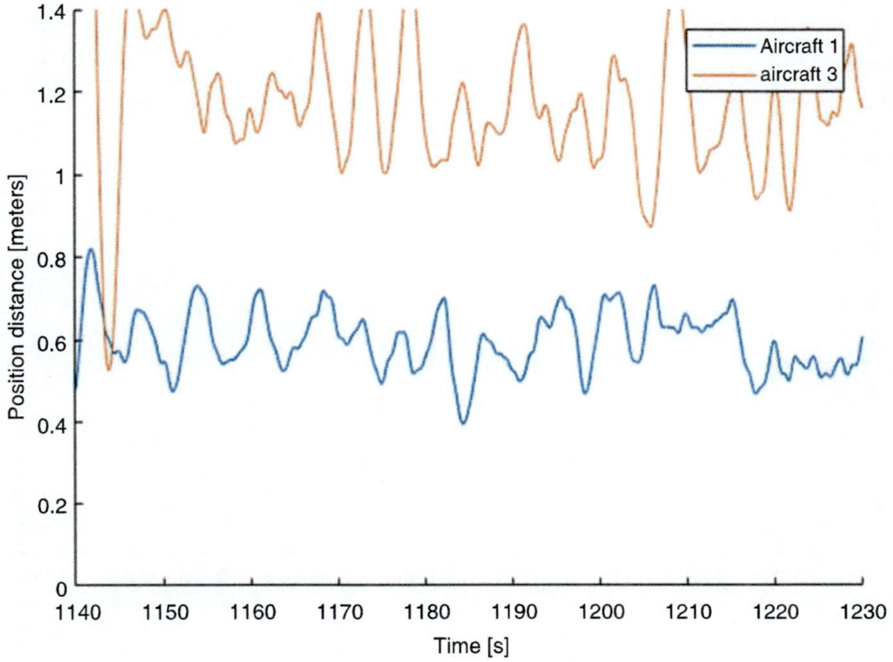

Fig. 6 Real-world flight test, formation position error for straight and level flight

Figure 6 shows the results of the straight and level flight test. It can be seen that aircraft 1 performed similarly to the simulated performance, while aircraft 3 performed noticeably worse, with a position error of around 1.2 m.

Figure 7 shows the formation position error while the virtual leader performed a coordinated turn. As with the straight flight error, the real-world performed quite similarly to the simulated environment. The position errors of both aircraft where slightly larger in the real-world than in the simulation, about 1 m and 1.5 m higher maximum errors for aircraft 1 and 3, respectively. These real-world errors are approximately 18–33% larger than in simulation.

The spacing between aircraft 1 and aircraft 3 is expected to be 20 m, Fig. 8 shows that during the coordinated turn, they maintained this distance with a peek error of 0.6 m.

4.3 Parameter Modification and Re-simulation

Because the aircraft spacing was increased from 5 m to 10 m, the speeds required for each aircraft to maintain position during the turn exceeded the previously set safety limit. In the case of aircraft 3, it was observed that its position in the formation was traveling more than twice its safety limit.

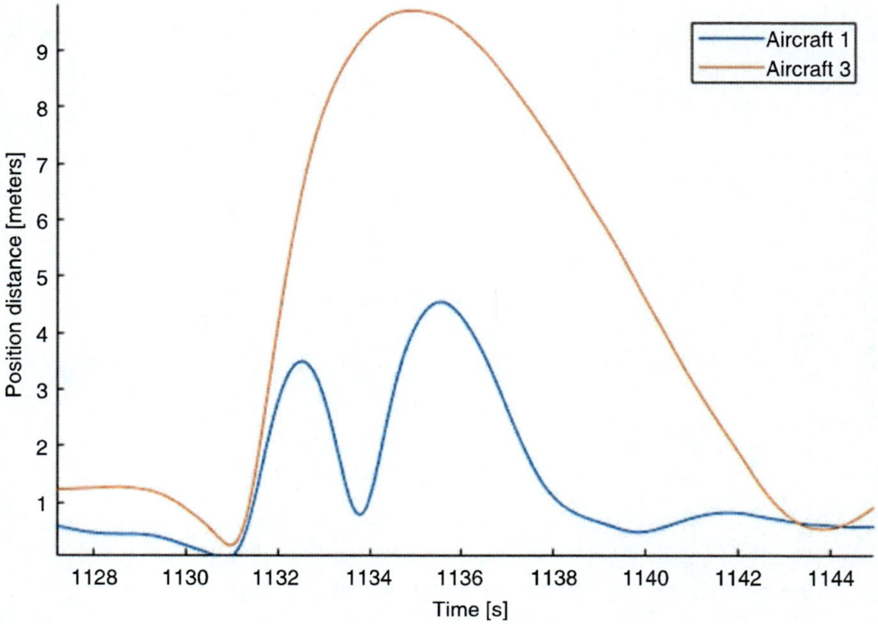

Fig. 7 Real-world flight test, formation position error tracking through coordinated turn

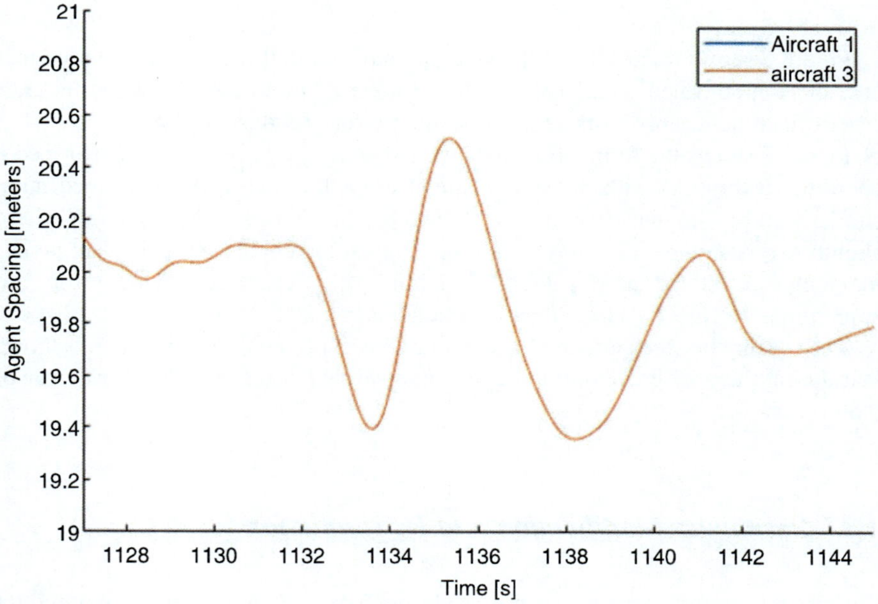

Fig. 8 Real-world flight test, agent separation through coordinated turn

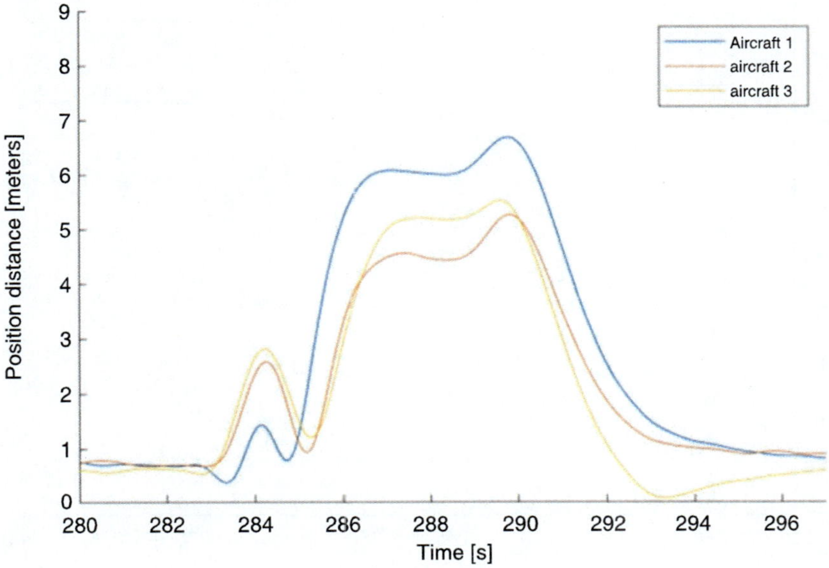

Fig. 9 Re-simulation, formation position error tracking through coordinated turn

Due to the findings in the simulation and in the flight test, modifications to tuning were applied to provide immediate improvement to performance. The maximum turning rate of the formation center was decreased, which ensures that the maximum required flight speed would be significantly reduced. In addition, the maximum allowed speed of all aircraft were increased.

The level flight matched with the earlier simulations. Figure 9 shows the coordinated turn errors of the system after these changes. These figures show that the straight and level error was not changed, but all three aircrafts coordinated turn errors are now comparable.

The maximum error of aircraft 3 was significantly reduced, aircraft 2 was reduced by a small factor, and aircraft 1 drastically increased.

The distance between aircraft 2 and each of the other aircrafts should be 14.14 m, which will be the smallest distance for each aircraft. Figure 10 shows that the aircraft achieves this with a peak error of 0.2 m during the coordinated turn operation.

5 Results and Discussion

The flight testing data shows that the aircraft are able to maintain a position within a formation with an error of approximately 1.2 m in straight and level flight. The coordinated turn error peaks at 9 m. Improvements made and verified in simulation are expected to transfer to the real world, reducing the peak error in coordinated turns to less than 7 m.

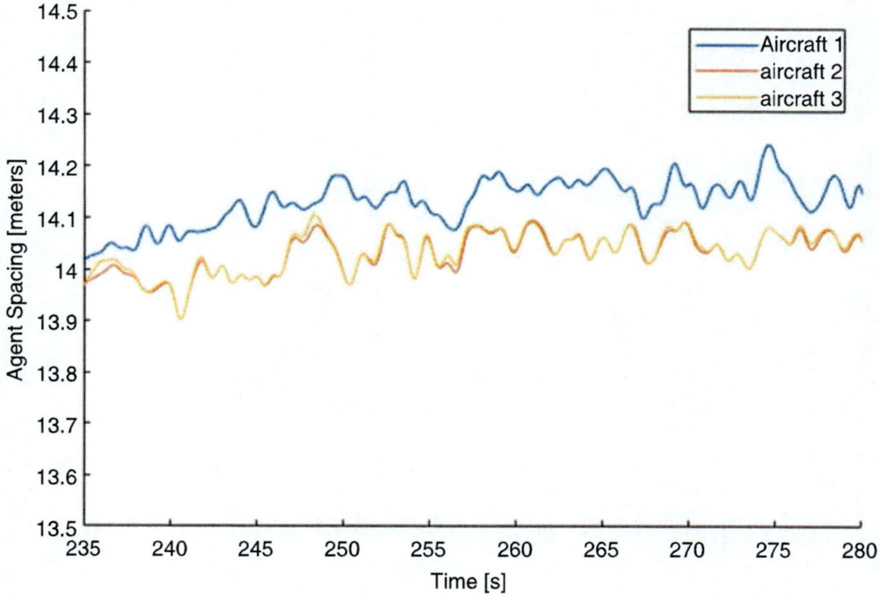

Fig. 10 Re-simulation, agent separation through coordinated turn

Given the spacing between the aircraft, the position errors represented a 6% straight and level flight position error. The coordinated turn error was recorded as up to a 95%, but it should be easily improvable to 75%. The aircraft spacing was normally close to the expected distances and any significant variations were larger distances between aircraft not smaller distances.

6 Conclusions

The virtual leader follower aircraft control was shown to be successful. The spacing data shows that fleet safety was never compromised by the errors that were present.

The comparison of the initial simulation with the flight test shows that the simulator is a good representation of an aircraft that has the fleet coordination system onboard of the aircraft. The effect of the radio link on the system performance should be modeled to improve fidelity.

This system provides a foundation that can be used for further multi-agent development or can be used to allow multiple aircraft to be operated in formation flight.

References

Reynolds, C. W. (1987). Flocks, herds, and schools: A distributed behavioral model, in computer graphics. In *SIGGRAPH '87 conference proceedings*.

Reynolds, C. (2001, September 6). *Boids: Background and update*. Accessed 14 Aug 2020. http://www.red3d.com/cwr/boids/

Vachon, M., Ray, R., Walsh, K., & Ennix, K. (2003). *F/A-18 performance benefits measured during the autonomous formationi flight project*. https://doi.org/10.2514/6.2002-4491

Yang, X., Deng, L., Wei, P., Li, H., & Liu, J. (2020). *Multi-agent autonomous operations in urban air mobility with communication constraints*. https://doi.org/10.2514/6.2020-1839.

Zhang, J., Yan, J., & Zhang, P. (2018). Fixed-wing UAV formation control design with collision avoidance based on an improved artificial potential field. *IEEE Access, 6*, 78342–78351. https://doi.org/10.1109/ACCESS.2018.2885003

In-Flight Nonlinear System Identification for UAS Adaptive Control

Sean Bazzocchi and Afzal Suleman

Nomenclature

UAS	Unmanned aerial system
CFD	Computational fluid dynamics
SID	System identification
MIAC	Model identification adaptive controller
SINDy	Sparse identification of nonlinear dynamics
u	Linear rate on x body axis, m/s
w	Linear rate on x body axis, m/s
q	Pitch angular rate, rad/s
u	Command
X	States
T	Training time, s
k	Harmonic multiplier
t	Time, s
R^2	R-Squared fitness value
θ	Pitch angle, rad
Θ	Library of candidate nonlinear functions
Ξ	Sparse coefficient matrix
ξ	Sparse coefficient vector
η	Noise magnitude
$_c$	Canard
$_t$	Throttle
$_j$	Input index
$_k$	Harmonic index

S. Bazzocchi (✉) · A. Suleman
University of Victoria (UVic), Center for Aerospace Research (CfAR), Victoria, BC, Canada
e-mail: seanbazzocchi@uvic.ca; suleman@uvic.ca

T. H. Karakoc et al. (eds.), *New Achievements in Unmanned Systems*, Sustainable Aviation, https://doi.org/10.1007/978-3-031-29933-9_19

^{T}	Transpose
^	Estimated noisy state

1 Introduction

The conventional design and tuning method of an autopilot system for an UAS is an expensive process. This procedure usually starts with ground-based analysis, such as CFD or wind tunnel, in order to generate the data needed to develop a simulation model of the vehicle. After defining control laws, the model and controller can then be tested in SITL simulations with the goal of tuning control parameters for the desired performance. The controller is then deployed into hardware and the vehicle flight tested while recording all observable states from the sensors. This data is then processed to correct the initial simulation model allowing for greater precision in a second tuning iteration. This process needs to be repeated for all prototypes and when a system goes through any modification that alters the dynamics of the vehicle.

The concept of adaptive controllers is to eliminate this lengthy procedure and automate the tuning process. But rapid prototyping and the ability to easily overcome changes in system dynamics are not the only motivations that are driving this research field. There are several other advantages including:

- Detection and mitigation of failures.
- Ability to cope with morphing aerodynamic properties.
- Dynamic flight envelope correction.
- Exploration of nonlinear dynamics for complex aircraft configuration.
- Fast generation of a time varying system model.

Some of the well-known agencies and companies in the aerospace industries are pursuing these objectives, including the US Air Force with project RESTORE, Boeing with project RACE, and NASA with the L2F project (Heim et al., 2020).

This study first presents an effective architecture to achieve the stated adaptive control benefits and second validates a system identification method in the context of a UAS platform developed at the University of Victoria Center for Aerospace Research (UVic CfAR).

2 Method

The selected control architecture is categorized as a Model Identification Adaptive Controller (MIAC).

From the block diagram displayed in Fig. 1, the high-level structure of the MIAC is similar to a conventional autopilot architecture. The main difference lies in the presence of the System Identification Process which collects data from the controller

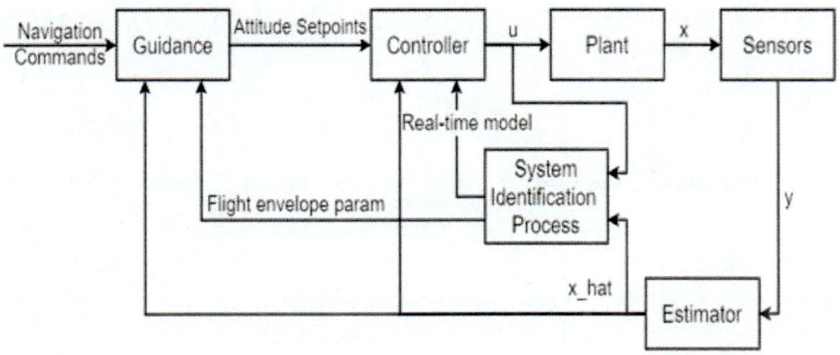

Fig. 1 Model identification adaptive controller architecture

output and state estimator with the task of generating a near real-time model of the vehicle. This information is then fed to the model-based controller which becomes 'adaptive' by virtue of the updating dynamic model. It is also possible to extrapolate model performance parameters from the identified system which can be used to tune the guidance and navigation controller.

One of the main advantages of this type of architecture compared to other data-driven controllers is the human readability. This property allows for formal validation of the controller. In other words, it is possible to validate the control behaviour without testing every single condition, something that is not possible with a Neural Network type of architecture for example (Van Wesel & Goodloe, 2017). Furthermore, the generation of a real-time SID model allows for rapid investigation of the system performance and analysis of any changes to the dynamics during flight.

2.1 System Identification Method

The selection of the system identification method is crucial since the MIAC is highly dependent on the capability of the SID process to effectively identify the vehicle dynamics. The method employed in this research is the Sparse Identification of Nonlinear Dynamics, or SINDy for short (Brunton et al., 2016). This sparse regression technique was theorized by Brunton and Proctor in 2016 and in the last 4 years has been applied to several dynamic systems (Kaiser et al., 2018) and adapted to include control identification (Brunton et al., 2016), detection of abrupt changes (Quade et al., 2018), physical constraints, and model structure selection.

The scheme shown in Fig. 2 illustrates the high-level SID process for the longitudinal dynamics (u, w, q, θ) of the Eusphyra aircraft designed at UVic CfAR.

During flight, the actuator commands (u_c and u_t) generated by the controller are occasionally perturbed by unique, zero-mean, multi-frequency signals (δ_c and δ_t). This practice ensures that the vehicle dynamics are effectively excited to allow for comprehensive system identification. In this study, the nonlinear flight dynamic model was generated using CFD analysis and coded in Simulink to calculate the

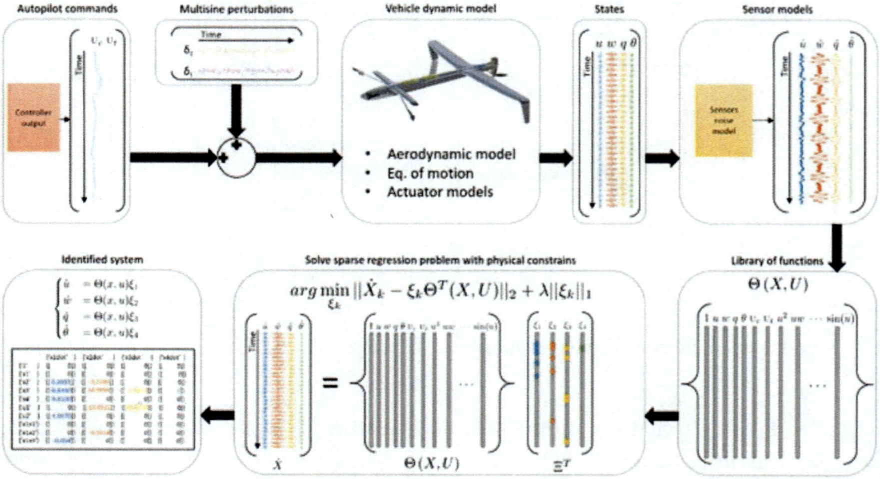

Fig. 2 System identification process

system response (X). The true states are then fed into the sensors and estimator models, which apply noise to the response depending on the signals' standard deviation. The estimated states are then used to generate the matrix Θ which holds the pool of candidate nonlinear equations describing the system. In this case, it contains constant, polynomial, and trigonometric terms. Because only a few terms of the proposed library of functions will be active in each differential equation, a sparse regression problem is formulated to determine the vector of coefficients forming the matrix Ξ. Each column ξ_i is a sparse vector of coefficients determining which terms of Θ are active in the corresponding differential equation:

$$\dot{x}_i = f_i(x) = \Theta\left(x^T\right)\xi_i \tag{1}$$

2.2 Command Perturbation

One of the most important components in the system identification is the correct perturbation of the available actuators. This not only serves the purpose of identifying control effectiveness but also excites the system dynamics across a defined spectrum of frequencies. The amplitude of the perturbations must also be correctly set to ensure that the signal-to-noise ratio remains above the usable threshold.

The perturbation signals used in this project are orthogonal, phase-optimized, multisine inputs (Grauer, 2018). The principal benefit of this type of input is that the orthogonal nature allows simultaneous perturbations without correlating responses. These inputs are also effective in generating the steady-state response data needed to estimate frequency responses.

Defining the duration of the excitation T is the first step towards the design of a multisine perturbation. This also delimits the fundamental frequency $1/T$ and by extension the harmonic multiples k/T. Each actuator u_j has a unique subset of multipliers K_j which define the frequency range spanned during excitation. The multisine signal is then generated as a sum of sinusoids:

$$u_j(t) = \sum_{k \in K_j} a_k \sin\left(\frac{2\pi k}{T} t + \phi_k\right) \tag{2}$$

The amplitudes a_k are calculated according to the desired power spectra for each input and the phase angles ϕ_k are optimized to reduce the peak value of the multisine signal in order to keep the response near the set flight condition.

3 Results and Discussion

The simplest way of validating the proposed method is to overlay the response of the identified model with the response of the true dynamic system.

The graph in Fig. 3 displays the simulation results of a model trained with only 5 s of dynamic data and a noise-to-signal ratio of 0.01. In this short timeframe, the SID process is capable of generating a model with 88% goodness of fit averaged across

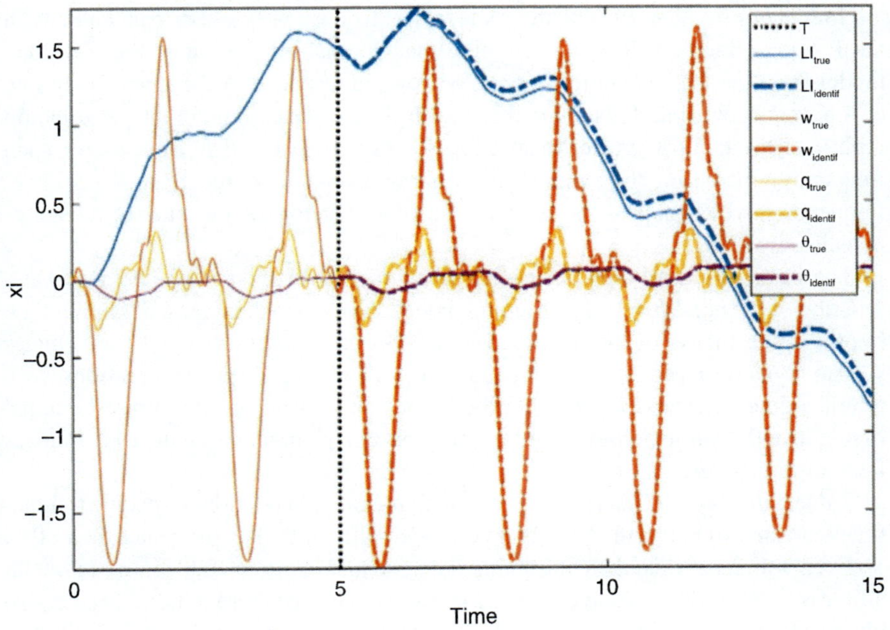

Fig. 3 Validation of the identified system

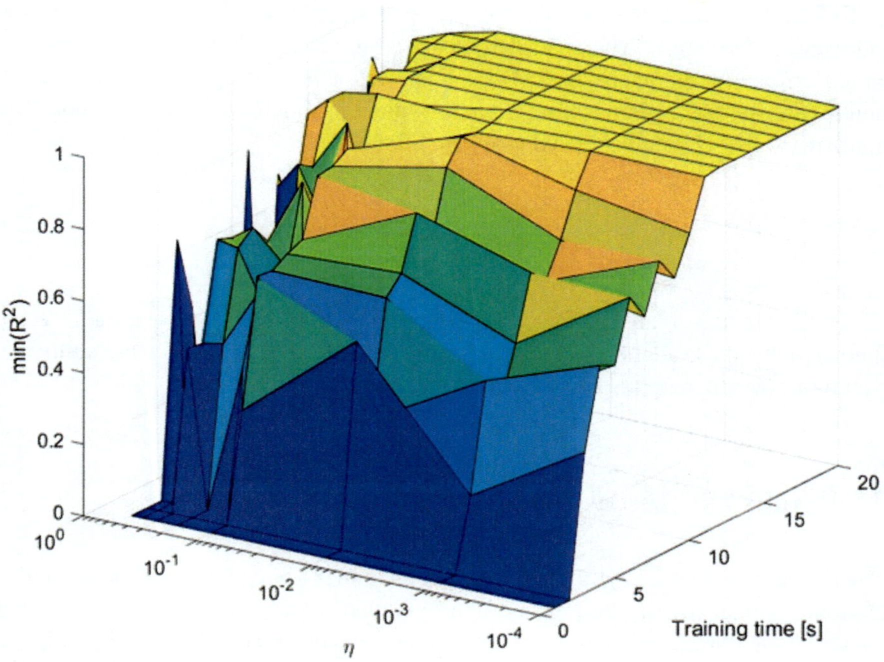

Fig. 4 Fitness level in function of noise and training time

the four states (u, w, q, θ). The fitness level in this chapter has been calculated with the R-Square method. It is worth noting that the main deviation of the identified model from the true dynamics is observed on the response of the linear body rate u. A primary contributing factor to this error is the frequency limit of the perturbation input (set to equal or greater than 2/T, allowing at least two full cycles for each frequency). Increasing the training time T allows excitation of the dynamics at lower frequencies which appear to improve the identification of the slower dynamics associated with u.

The surface in Fig. 4 reports the minimum fitness level of the identified model as a function of the noise level η and training time T. As expected, increased training time improves the fitness of the identified model. In this particular study, the vehicles system dynamics can be correctly identified in all flight-tested conditions with training times greater than 9 s. On the other hand, increasing the noise-to-signal ratio above 0.5 while holding training time constant significantly reduces the fitness level of the model.

Table 1 displays the maximum noise-to-signal ratio, in relation to specific training times, for the SID algorithm to obtain a model with an R-Square greater than 80%.

Even with large increases of training data, the maximum tolerable noise to obtain a fitness level of 80% averages around $\eta = 0.64$. This asymptotic trend can be also observed in Fig. 5. The yellow-coloured area defines the region in which it is possible to correctly identify the studied system. The curve that outlines this region

Table 1 Required training time for extreme noise-to-signal ratios

H	T [s]	R2 [%]
0.01	5	88
0.22	10	99
0.35	20	98
0.42	30	98
0.53	50	94
0.61	80	86
0.63	120	83
0.63	160	81

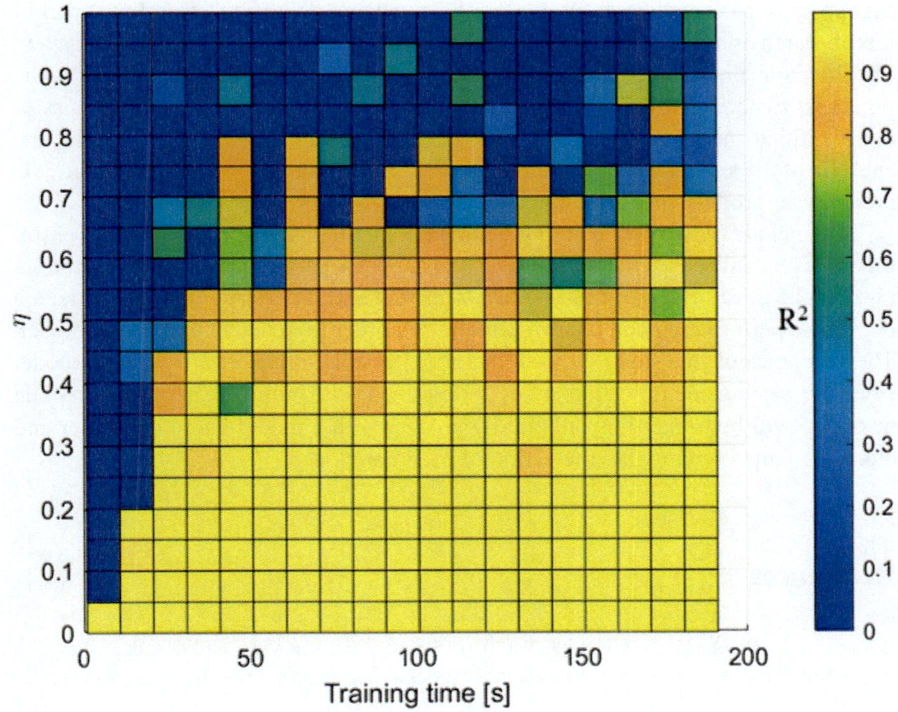

Fig. 5 Fitness-level distribution for high level of noise and increased training time

flattens for T > 60 s; therefore, for higher levels of noise increasing the training time does not show any substantial improvement in the model fitness.

It is important to mention that if the training time is increased, the harmonic multipliers K_j should also increase in value to ensure that the higher frequencies are still being excited. Failing to do so results in degradation of the fitness level.

4 Conclusion

The innovation in vehicle configurations and the increase of onboard computational power are contributing to the growing interest towards adaptive controllers and data-driven models.

For the objectives of this project, model identification adaptive controller (MIAC) stands as the preeminent architecture, allowing for exploration of a machine learning technique while maintaining a certain level of verification for the safety of flight.

In this chapter, the high-level MIAC structure was presented, highlighting the benefits of having a human readable controller. This method heavily relies on the SID process to correctly identify the vehicle dynamics, therefore a significant part of this research is focused on the study of an effective technique for online system identification processing. SINDy method was proven to successfully identify the nonlinear dynamics of the flight test vehicle. The results demonstrate the fitness of the identified model in several conditions of noise and training time. Lastly, the analysis of the simulation data established the limit conditions within which the SID process can produce an accurate model.

To conclude, it is worth pointing out the conflicting nature of this type of control theory. The controller's main objective is to suppress perturbations and track the set point in the most actuator-efficient way, while the SID process requires some degree of vehicle perturbation and control actuation in order to generate a dynamic model. The outcomes of this study indicate that it is possible to generate a suitable model from the system identification process described with minimal perturbation. The next step will be to pair this online SID process with a model-based controller and close the loop studying the interaction of each component.

References

Brunton, S. L., Proctor, J. L., & Kutz, J. N. (2016). Sparse identification of nonlinear dynamics with control (SINDYc). *IFAC-PapersOnLine*. Elsevier B.V., *49*(18), 710–715. https://doi.org/10.1016/j.ifacol.2016.10.249

Brunton, S. L., et al. (2016). Discovering governing equations from data by sparse identification of nonlinear dynamical systems. *Proceedings of the National Academy of Sciences of the United States of America, 113*(15), 3932–3937. https://doi.org/10.1073/pnas.1517384113

Grauer, J. (2018, November). Dynamic Modeling usingOutput-Error Parameter Estimationbased on Frequency ResponsesEstimated with Multisine Inputs. NASA/TM–2018–220108

Heim, E. et al. (2020). *NASA's learn-to-fly project overview*. Available at: https://ntrs.nasa.gov/search.jsp?R=2019002721

Kaiser, E., Kutz, J. N., & Brunton, S. L. (2018). Sparse identification of nonlinear dynamics for model predictive control in the low-data limit. *Proceedings of the Royal Society A: Mathematical, Physical and Engineering Sciences, 474*(2219). https://doi.org/10.1098/rspa.2018.0335

Quade, M., et al. (2018). Sparse identification of nonlinear dynamics for rapid model recovery. *Chaos, 28*(6), 1–10. https://doi.org/10.1063/1.5027470

Van Wesel, P., & Goodloe, A. E. (2017, June). *Challenges in the verification of reinforcement learning algorithms NASA STI program ... in profile* (pp. 2017–219628). Available at: https://ntrs.nasa.gov/archive/nasa/casi.ntrs.nasa.gov/20170007190.pdf

Drone Simulation, Mapping and Navigation via ROS

Demet Canpolat Tosun

1 Introduction

Robotic simulators play an important role in robot development operations as they provide fast results, efficiency analysis and the opportunity to try without loss of cost. Gazebo is one of the most effective 3D simulators, providing accurate and efficient simulation of robots in real environments with a wide variety of sensors and visualization tools.

ROS is a platform that enables the development of robotic applications in software and hardware terms. Gazebo can be used in integration with the ROS platform by means of some Gazebo plugins. This provides great convenience in creating robot models, customizing existing models, testing robots with different sensors, and developing algorithms. It is also possible to use these platforms for aerial vehicles.

The ease of combining hardware and software provided by ROS has contributed to the adoption of it by many researchers and the increase in the user and developer community in the last decade.

By ensuring autonomy in aerial vehicles and increasing the level of autonomy, its use in many areas such as agriculture, delivery, photography has also increased. Drones can be used efficiently in agricultural use both in terms of fertilization and crop control (Puri et al., 2017). Drones are also used to provide medical assistance. In Rwanda, a company named Zipline uses drones to deliver blood to patients in need, in a much shorter time than normal delivery (Ackerman & Strickland, 2018).

Drones used for search and rescue operations in natural disasters (Mishra et al., 2020) are also vital.

D. C. Tosun (✉)
Eskişehir Technical University, Eskişehir, Turkey
e-mail: demetcanpolat@eskisehir.edu.tr

T. H. Karakoc et al. (eds.), *New Achievements in Unmanned Systems*, Sustainable Aviation, https://doi.org/10.1007/978-3-031-29933-9_20

In addition to these applications, there are new studies on human–drone interaction (HDI) (Tezza & Andujar, 2019).

It is possible to interact with the drone through speech, gaze, gesture (Mashood et al., 2015), touch (Abtahi et al., 2017), and even a brain–computer interface (LaFleur et al., 2013) with methods called natural user interface (Peshkova et al., 2017).

This study focused on ROS-based mapping and navigation of a drone model in the Gazebo environment. At the first stage, the model of the drone, the sensors to be used and the environment to be operated are determined. In the second stage, the environment mapping of the drone is provided. Finally, the navigation of the drone is provided with various algorithms.

2 Robot Operating System (ROS)

Robot Operating System (ROS Wiki: Documentation, 2021) is a platform that enables robot and robot systems to be controlled and communicated on a computer. Although ROS has an operating system in its name, it is not actually an operating system; it needs an operating system to run. The main objective of ROS is to establish a standard for robotic applications. Applying this standard in any robot saves time for developers, which allows the developer to focus on the actual problem.

ROS offers powerful tools with its modular structure and data communication network.

In ROS, it is possible to split the code into reusable code blocks. Communication between all these code blocks or between subprogrammes can be achieved through structures called ROS topics.

ROS also offers users diversity in terms of software language, recognizes different software languages and establishes communication between different software languages without any problems.

3 Gazebo

Gazebo (Gazebo Simulator, 2021) is an open source multi-robot simulator developed for both indoor and outdoor environments. It has the ability to simulate multiple robots, sensors, and objects in a three-dimensional environment.

Gazebo is divided into several libraries:

- Physics updates the physical state of the simulation.
- Scene visualizes the simulation situation.
- Sensor generates sensor data.
- Communication manages interprocess communication.

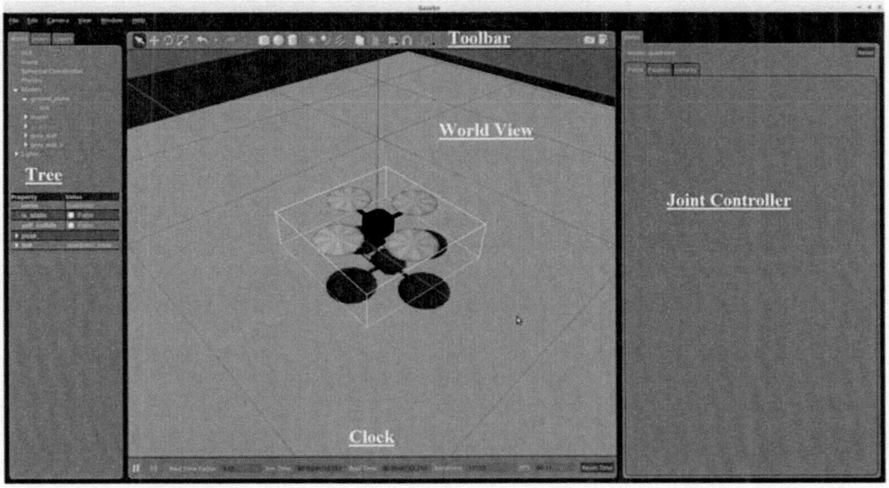

Fig. 1 Graphical user interface

- Graphical user interface (GUI) provides visualization and manipulation of the simulation.

These libraries are used by two main processes:

- Server (gzserver) runs physics update-loops and generates sensor data.
- Client (gzclient) provides visualization of user interaction and simulation.

The graphical user interface provides the display of the simulation shown in Fig. 1 and the interaction between the user and the simulation through its five components.

Tree: This list provides hierarchical display of models in the scene. It is also used to add a new model to the scene.

Toolbar: Using this toolbar, it is possible to zoom in and out of the scene, move a model on the scene, rotate it, change the camera angle and position, add cubes, cylinders, spheres and light sources.

World View: It is the area where the created environment and models are displayed. Models can be added, removed or manipulated in this area.

Clock: It is used to start and stop the simulation. It also displays how slow the simulation is running in real time or relative to real time.

Joint Controller: This component is used to apply force, speed, or change the angle of the joints of the models in the scene.

In addition to creating realistic models and environments, Gazebo has become a very useful tool for developers with packages and plugins that allow it to work integrated with ROS.

4 Methodology

The study consists of three parts. In the first stage, the drone model is selected and the appropriate Gazebo environment is created. In the next stage, the drone is moved manually and mapping is ensured by scanning its environment with the laser distance sensor on it. In the last stage, it is ensured to reach a selected location according to the map.

4.1 Environment Setup and Drone Configuration

The drone model used in the study is shown in Fig. 2. Some design features of the drone model have been changed by editing URDF files. In the next step, a laser distance sensor is placed on the drone model. Features such as range, scan angle, sampling rate and resolution of the sensor are set at the desired level in the created world file.

After creating the drone model with the needed sensors, a simple world environment shown in Fig. 3 is created to implement autonomous navigation.

4.2 Mapping

The Gmapping algorithm (Grisetti et al., 2007) is used in the mapping phase. The algorithm creates a 2D map of the environment with the data from the laser distance sensor and the sensors that give the instant position of the drone.

During the mapping process, the slam_gmapping node runs on the ROS platform. This node subscribes to the /scan topic, which contains data from the laser distance

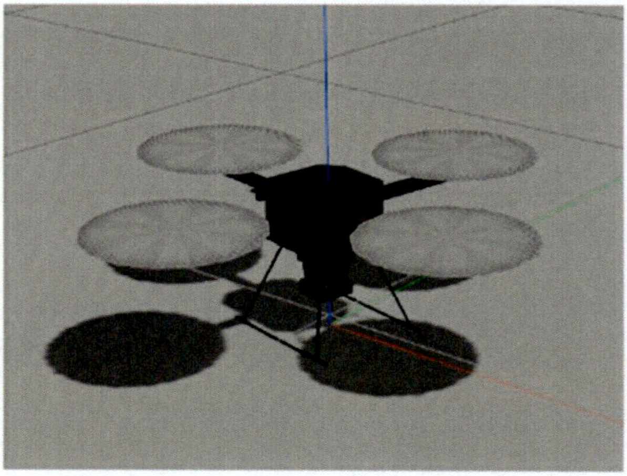

Fig. 2 Drone model

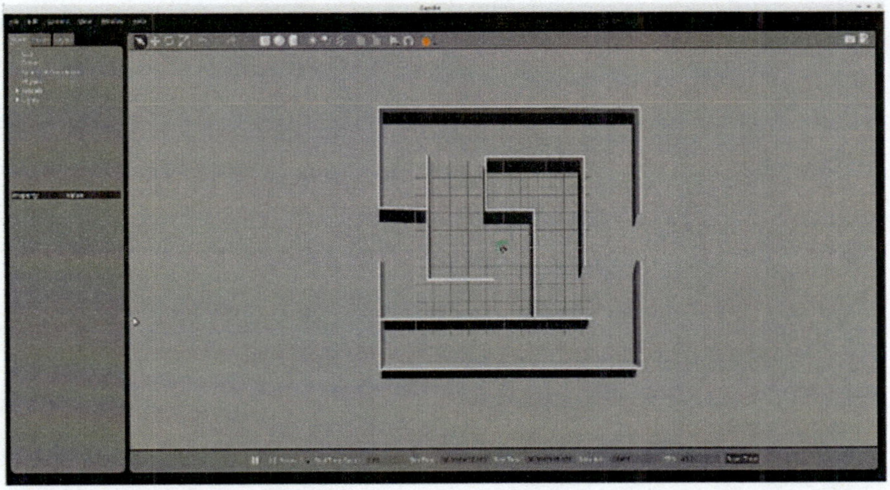

Fig. 3 Simple indoor environment

Table 1 Published/subscribed topics for mapping

Topic name	Message type	Function
/scan	sensor_msgs/ LaserScan	Provides laser data
/tf	tf/tfMessage	Provides transformation from the map frame to the odometry frame
/map	nav_msgs/ OccupancyGrid	Provides occupancy grid map

sensor, and to the /tf topic, which provides transformation between coordinate frames. The /slam_gmapping node publishes the data it receives from these two topics to which it subscribes with a /map topic. The /map topic provides a continuously updated occupancy grid map.

The topics and their functions used in mapping are given in Table 1.

The drone was controlled manually while mapping. It has been moved so that it can detect obstacles and safe areas around it with the laser distance sensor.

The images obtained in the Gazebo environment and the Rviz visualization tool while the drone is hovering for mapping purposes are shown in Fig. 4.

The situation in which the environment map is completed is as in Fig. 5.

4.3 *Navigation*

At this stage, according to the map obtained in the previous section, it is ensured that the drone reaches a target determined autonomously.

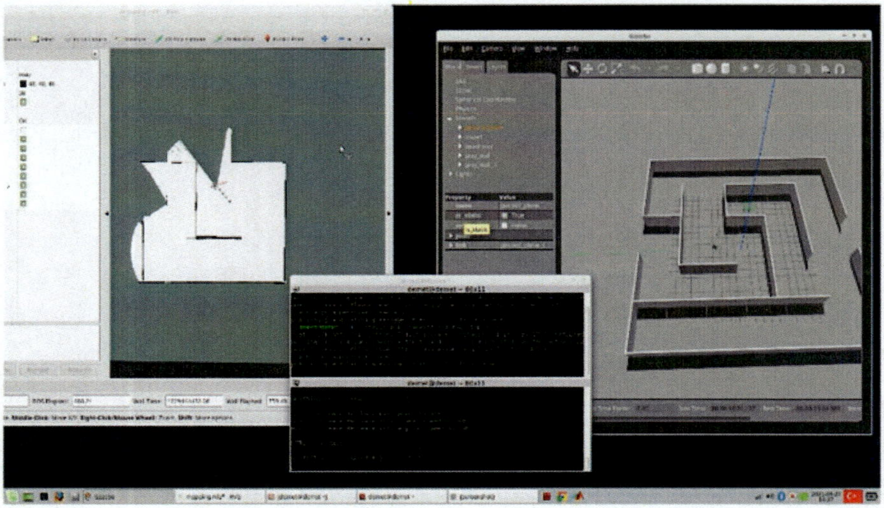

Fig. 4 Mapping of the environment

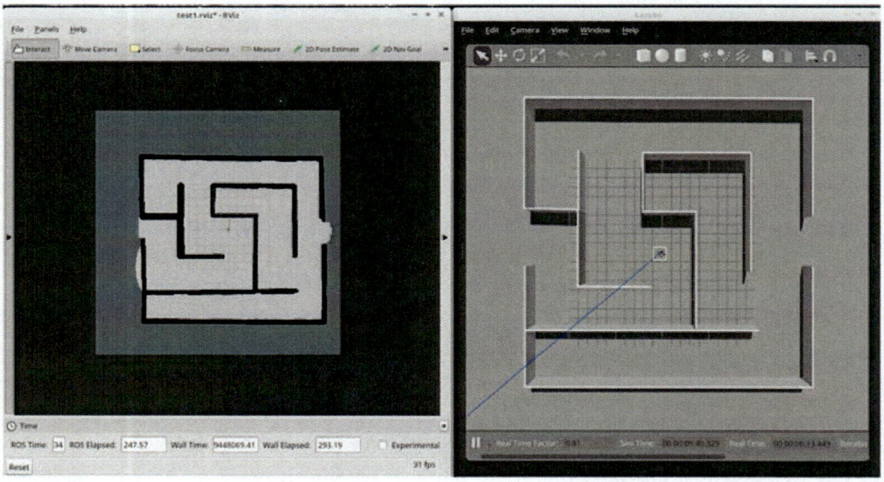

Fig. 5 Completed map of the environment

Active ROS nodes and subjects during navigation are shown in Fig. 6.

The drone autonomously travels to a selected point on the map with the Rviz visualization tool as shown in Fig. 7.

Figures 8 and 9 also show path plans created for different destinations.

In these tests, the time it takes for the drone to reach its target depends on its velocity and the length of the path.

The average velocity of the drone at different distances and the time it takes to travel these distances are given in Table 2.

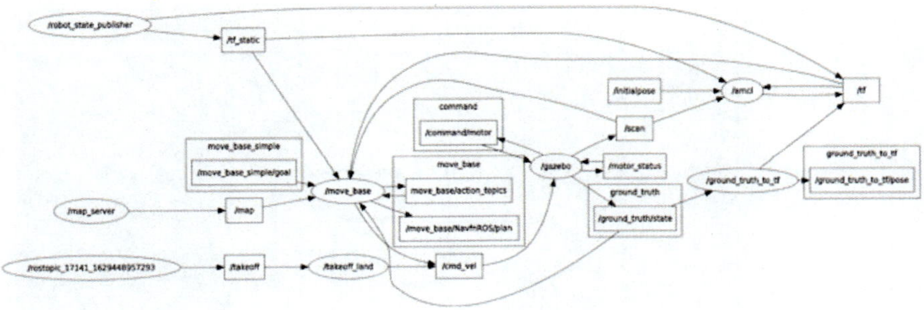

Fig. 6 Active ROS nodes/topics during navigation

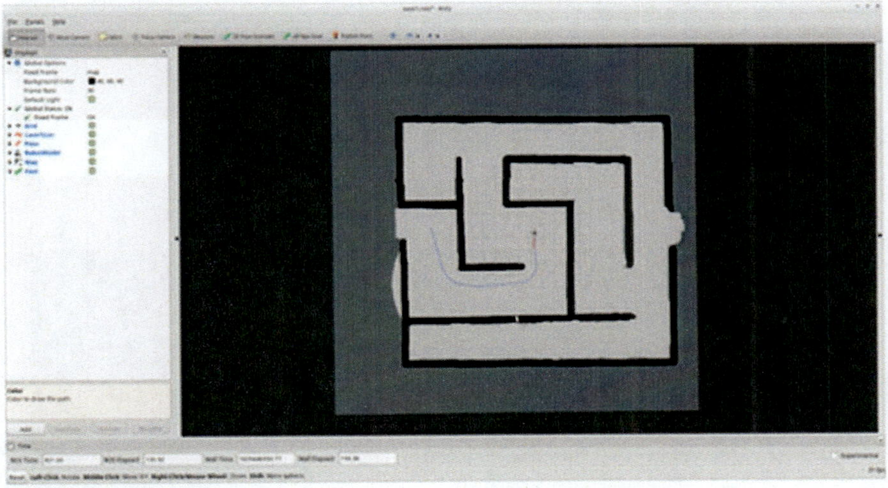

Fig. 7 Navigation test -1-

The autonomous navigation of the drone is provided by the move_base node. The move_base node contains the generated map, planners, and files for localizing the drone. Regions on the map are represented as local and global cost maps. Local and global cost maps show safe or collision-prone areas for the drone. On the other hand, planners use two different methods to create the shortest path for the drone to reach its destination.

One of them is Dijstra Algorithm (Dijsktra, 1959), a global planning method, and the other is Dynamic Window Approach (DWA), a local planning method.

Dijkstra's Algorithm (Fox et al., 1997) is one of the graph search methods that forms the basis of the A* algorithm. The algorithm allows to calculate the shortest path between a node and all other nodes in the graph.

The dynamic window approach is a velocity-based local path planning algorithm. The dynamic window is the set of velocities that can be reached in a short time interval. The planner selects the velocity to ensure the movement of the vehicle in the

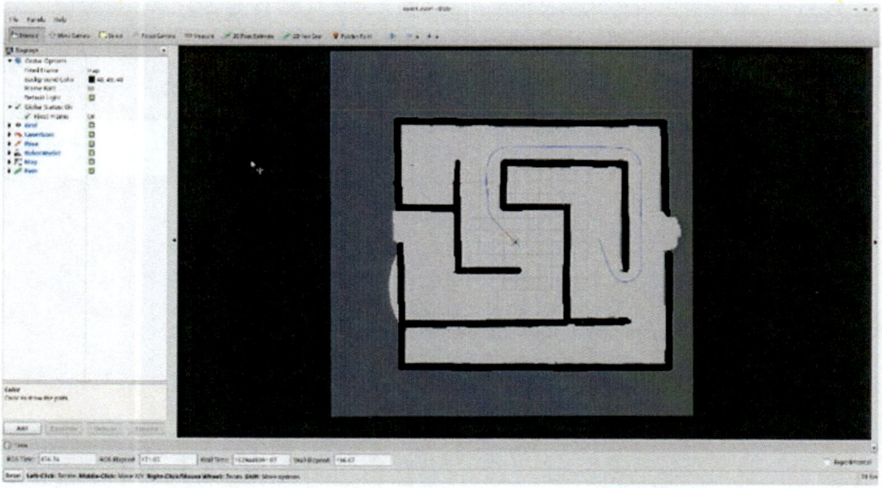

Fig. 8 Navigation test -2-

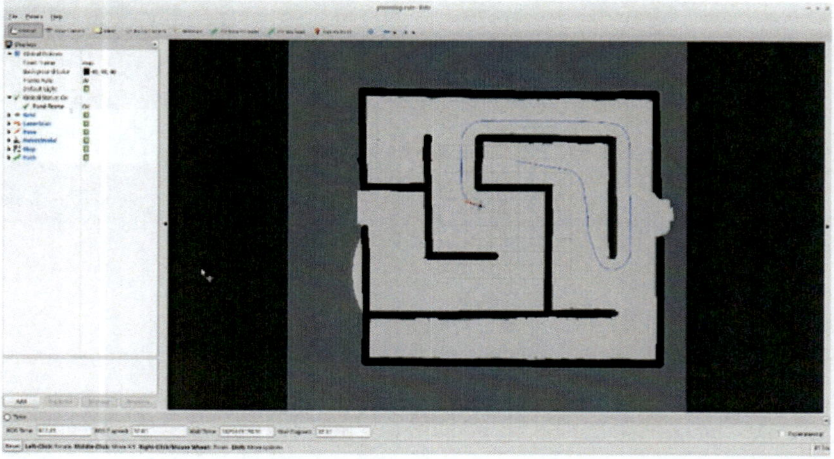

Fig. 9 Navigation test -3-

Table 2 Navigation test results for different paths

Path distance (m)	Average velocity (m/s)	Elapsed time (s)
26.45	0.256	103
35.27	0.249	141
37.53	0.235	159
41.80	0.345	121
51.26	0.251	204

safe zone. The optimum velocity is chosen by maximizing an objective function that considers the progress towards the target, the velocity of the vehicle and the distance from any obstacle on the path.

5 Results and Discussion

The use of the ROS platform for aerial vehicles is quite new. Due to its dynamics, the drone cannot move in every trajectory and speed that is suitable for mobile robots. The results obtained are suitable in terms of the path and velocities that the drone can travel.

Algorithms that provide navigation have produced fast and safe path plans. From the drone point of view, it is accessible.

It can be seen as a disadvantage that the global method used needs a map, but in order for the path to be optimal, it is necessary to use a global method. In addition to the global method, the use of the DWA method also ensures that the drone avoided local obstacles.

6 Conclusion

The ease of communication between hardware and software provided by the ROS platform also makes real-world applications possible. It is very advantageous to use the ROS platform, which is frequently used in the field of robotics, for aerial vehicles with more limited test stages in terms of time and cost.

In this study, point-to-point navigation is provided for a drone in a pre-mapped environment. The navigation problem is one of the most fundamental problems in robotics. Based on the solution of this problem, it can be applied to a real-world problem such as providing security in public space or indoor environments, emergencies, delivery.

Navigation methods used in this study require a map of the environment. The study can be developed using methods that do not require maps, where the drone does not know the environment.

References

Abtahi, P., Zhao, D. Y., Jane, L. E., & Landay, J. A. (2017). Drone near me: Exploring touch-based human-drone interaction. *The Proceedings of the ACM on Interactive, Mobile, Wearable and Ubiquitous Technologies, 1*(3), 34.

Ackerman, E., & Strickland, E. (2018). Medical delivery drones take flight in East Africa. *IEEE Spectrum, 55*, 34–35.

Dijkstra, E. W. (1959). A note on two problems in connexion with graphs. *Numerische Mathematik, 1*, 269–271.

Fox, D., Burgard, W., & Thrun, S. (1997). The dynamic window approach to collision avoidance. *IEEE Robotics and Automation Magazine, 4*(1), 23–33.

Gazebo Simulator. (2021). http://gazebosim.org/. Accessed 25 Aug 2021.

Grisetti, G., Stachniss, C., & Burgard, W. (2007). Improved techniques for grid mapping with rao-blackwellized particle filters. *IEEE Transactions on Robotics, 23*, 34–46.

LaFleur, K., Cassady, K., Doud, A., Shades, K., Rogin, E., & He, B. (2013). Quadcopter control in three-dimensional space using a noninvasive motor imagery-based brain-computer interface. *Journal of Neural Engineering, 10*(4), 046003.

Mashood, A., Noura, H., Jawhar, I., & Mohamed, N. (2015, May). A gesture based kinect for quadrotor control. In *Proceedings of the international conference on information and communication technology research (ICTRC)* (pp. 298–301).

Mishra, B., Garg, D., Narang, P., & Mishra, V. (2020). Drone-surveillance for search and rescue in natural disaster. *Computer Communications, 156*, 1–10.

Peshkova, E., Hitz, M., & Kaufmann, B. (2017). Natural interaction techniques for an unmanned aerial vehicle system. *IEEE Pervasive Computing, 16*(1), 34–42.

Puri, V., Nayyar, A., & Raja, L. (2017). Agriculture drones: A modern breakthrough in precision agriculture. *Journal of Statistics & Management Systems, 20*, 507–518.

ROS Wiki: Documentation. (2021). http://wiki.ros.org/Documentation. Accessed 15 July 2021.

Tezza, D., & Andujar, M. (2019). The state-of-the-art of human–drone interaction: A survey. *IEEE Access, 7*, 167438–167454.

Hierarchical Adaptive Fault-Tolerant Model Predictive Control of a Quadrotor

Binyan Xu, Yang Shi, and Afzal Suleman

1 Introduction

With the expansion of unmanned aerial vehicle (UAV) technology, using quadrotor UAV as a remote-sensing platform on persistent surveillance becomes a reasonable idea. Quadrotor with the potential applications for maritime domain sensing should be able to conduct long-endurance and large-area-coverage flight without ground support. Therefore, the development of a performance-optimized and resilient quadrotor control system is required. Many nonlinear control methods have been reported in the open literature (Voos, 2009; Wang et al., 2020; Zhou et al., 2019). Nevertheless, these controllers are developed without actuator faults, which may compromise system stability in the absence of support from ground stations.

The combination of fault-tolerant control and model predictive control (MPC) may be a straightforward perspective to achieve optimal control subject to unexpected faults and control constraints. Some representative results on fault-tolerant MPC include passive methods utilizing the inherent robustness of MPC (Kufoalor & Johansen, 2013) and active methods based on online estimation (Yang & Maciejowski, 2015; Izadi et al., 2011) and control reconfiguration (Gopinathan et al. 1998). However, many of the existing results are known to be computationally time-consuming, thus not applicable to real-time flight control requiring fast processing speeds.

In this chapter, a hierarchical control design for a quadrotor subject to actuator faults is proposed. For the higher level position control, we developed a novel adaptive fault-tolerant MPC scheme by combining adaptive online parameter estimation with the Lyapunov-based MPC framework. The proposed design has the following advantages:

B. Xu · Y. Shi · A. Suleman (✉)
Department of Mechanical Engineering, University of Victoria, Victoria, Canada
e-mail: binyanx@uvic.ca; yshi@uvic.ca; suleman@uvic.ca

T. H. Karakoc et al. (eds.), *New Achievements in Unmanned Systems*, Sustainable Aviation, https://doi.org/10.1007/978-3-031-29933-9_21

- The adaptive fault-tolerant MPC algorithm is simple, less time-consuming, and it exhibits satisfactory fault compensation performance. A comprehensive closed-loop analysis is provided considering both the tracking and the parameter estimation errors.
- The proposed design provides a more practical way to apply MPC to real-time flight control. The hierarchical control structure enables the MPC to work at a slower time scale, thus having a longer period to solve the constrained optimization problem between two sampling instants. The designed fast-responding lower level attitude controller ensures the rapid response of the control system.
- In most of the existing results on quadrotor control, solving the nonlinear transcendental function is required by the higher level position control to extract the desired attitude from the newly defined control vector. Our work applied MPC, thereby avoiding this tedious and time-consuming procedure.

The rest of this chapter is organized as follows. In Sect. 2, quadrotor dynamics model is presented, and actuator faults are modeled by addictive uncertainties. The hierarchical control architecture developed based on the time-scale separation of quadrotor dynamics is introduced. In Sects. 3 and 4, the lower level attitude control and the higher level position control are developed, respectively. Section 5 gives the simulation results. Lastly, in Sect. 6, conclusions are given.

2 Problem Statement

2.1 Quadrotor Dynamics

The quadrotor is a six-degree-of-freedom rigid body moving in three-dimensional space. Its position is denoted by the Euclidean coordinates as $\boldsymbol{\zeta} = [x, y, z]^T$, and its orientation is denoted by the Euler angles as $\boldsymbol{\eta} = [\phi, \theta, \psi]^T$. The translational and rotational motions of the quadrotor are represented by the following differential equations:

$$\ddot{\zeta} = \begin{bmatrix} \ddot{x} \\ \ddot{y} \\ \ddot{z} \end{bmatrix} = \begin{bmatrix} (\cos\phi \sin\theta \cos\psi + \sin\phi \sin\psi)T/m \\ (\cos\phi \sin\theta \sin\psi - \sin\phi \cos\psi)T/m \\ \cos\phi \cos\theta \;\; T/m - g \end{bmatrix} \tag{1}$$

$$\ddot{\eta} = \begin{bmatrix} \ddot{\varphi} \\ \ddot{\theta} \\ \ddot{\psi} \end{bmatrix} = \begin{bmatrix} \dot{\theta}\dot{\psi}(I_{yy} - I_{zz})/I_{xx} + \tau_{\phi}/I_{xx} \\ \dot{\phi}\dot{\psi}(I_{zz} - I_{xx})/I_{yy} + \tau_{\theta}/I_{yy} \\ \dot{\phi}\dot{\theta}(I_{xx} - I_{yy})/I_{zz} + \tau_{\psi}/I_{zz} \end{bmatrix} \tag{2}$$

where m is the mass of the quadrotor, g is gravity acceleration, $I_{xx,\ yy,\ zz}$ are the moments of inertia. The control inputs of the quadrotor dynamics are the total thrust force T and the applied torques $\tau_{\phi,\ \theta,\ \psi}$, both coming from the thrust forces of the four rotors.

2.2 *Actuator Faults*

As the actuators of the quadrotor control system, the rotors have the possibility of malfunction during the flight potentially due to a low voltage supply or blade defection, causing a reduction in the rotor's thrust. To describe the impact of the actuator faults, we define σ_ϕ, σ_θ, σ_ψ as fault parameters denoting the loss of thrust or torques. Then, we can consider the actuator faults as additive uncertainty, as follows:

$$\begin{bmatrix} T \\ \tau_\phi \\ \tau_\theta \\ \tau_\psi \end{bmatrix} = \begin{bmatrix} u_T \\ u_\phi \\ u_\theta \\ u_\psi \end{bmatrix} + \begin{bmatrix} \sigma_T \\ \sigma_\phi \\ \sigma_\theta \\ \sigma_\psi \end{bmatrix} \tag{3}$$

where u_T, u_ϕ, u_θ, u_ψ are the commanded force and torques.

2.3 *Hierarchical Control and Time-Scale Separation*

Considering that the quadrotor dynamics is an under-actuated system with six state variables and only four control inputs, it is difficult to control all of the states simultaneously. The solution is to adopt a hierarchical control structure. As shown in Fig. 1, the whole system is composed of a lower level attitude control loop and a higher level position control loop. The position coordinates x, y, z are the state variables to be controlled, and the Euler angles ϕ,θ, ψ are the internal states linking the two control loops. The position controller receives the desired position and generates the desired attitude angles, while the attitude controller is developed to track the desired attitude angles.

The hierarchical control strategy is developed in the context of a time-scale separation between the two control loops. For the quadrotor control system, the overall stability is guaranteed only when the time scale of the rotational dynamics is much smaller than that of the translational dynamics. In the next two sections, we design the attitude control and position control for quadrotor's rotational dynamics and translational dynamics separately, where the time-scale separation is achieved by tuning the control gains jointly.

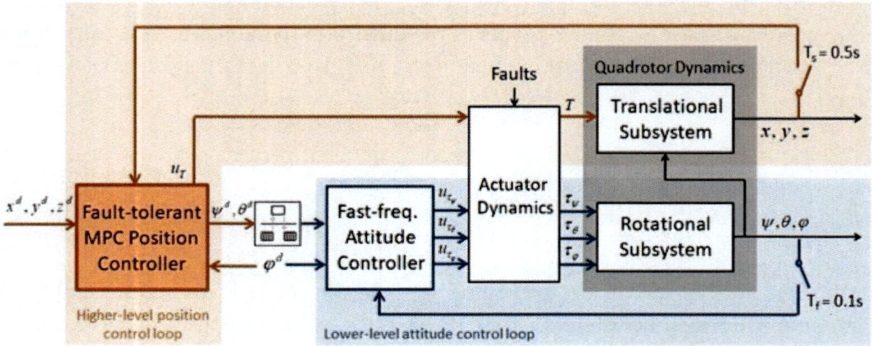

Fig. 1 Hierarchical control architecture

3 Lower Level Attitude Control

The attitude control objective is to design the torque control command leading to stabilized rotational dynamics.

3.1 Attitude Tracking Error Dynamics

Given the desired attitude angles, the attitude tracking errors are defined as $\widetilde{\phi} = \phi - \phi^d, \widetilde{\theta} = \theta - \theta^d, \widetilde{\psi} = \psi - \psi^d$. Note that the desired attitude angles generated by the slow-responding position control can be assumed to remain unchanged in the fast time-scale of the attitude control loop, namely $\dot{\phi}^d, \ddot{\phi}^d, \dot{\theta}^d, \ddot{\theta}^d, \dot{\psi}^d, \ddot{\psi}^d = 0$. Then, defining $x_a = \left[\widetilde{\phi}, \dot{\widetilde{\phi}}, \widetilde{\theta}, \dot{\widetilde{\theta}}, \widetilde{\psi}, \dot{\widetilde{\psi}}\right]^T$ as the state vector, and $\mathbf{u}_a = [u_\phi, u_\theta, u_\psi]^T$ as the control vector, we obtain the following attitude tracking error dynamics subject to actuator faults:

$$\dot{x}_a = A_a x_a + B_a\left(f_a(x_a) + I^{-1} u_a + \sigma_a - \ddot{\eta}^d\right) \tag{4}$$

with

$$A_a = \begin{bmatrix} 0 & 1 & 0 & 0 & 0 & 0 \\ 0 & 0 & 0 & 0 & 0 & 0 \\ 0 & 0 & 0 & 1 & 0 & 0 \\ 0 & 0 & 0 & 0 & 0 & 0 \\ 0 & 0 & 0 & 0 & 0 & 1 \\ 0 & 0 & 0 & 0 & 0 & 0 \end{bmatrix}, B_a = \begin{bmatrix} 0 & 0 & 0 \\ 1 & 0 & 0 \\ 0 & 0 & 0 \\ 0 & 1 & 0 \\ 0 & 0 & 0 \\ 0 & 0 & 1 \end{bmatrix}, I = \begin{bmatrix} I_{xx} & 0 & 0 \\ 0 & I_{yy} & 0 \\ 0 & 0 & I_{zz} \end{bmatrix},$$

$$f_a(\mathbf{x}_a) = \begin{bmatrix} \dot{\theta}\dot{\psi}(I_{yy} - I_{zz})/I_{xx} \\ \dot{\phi}\dot{\psi}(I_{zz} - I_{xx})/I_{yy} \\ \dot{\phi}\dot{\theta}(I_{xx} - I_{yy})/I_{zz} \end{bmatrix}, \sigma_a = \begin{bmatrix} \sigma_\phi/I_{xx} \\ \sigma_\theta/I_{yy} \\ \sigma_\psi/I_{zz} \end{bmatrix}, \ddot{\eta}^d = \begin{bmatrix} \ddot{\phi}^d \\ \ddot{\theta}^d \\ \ddot{\psi}^d \end{bmatrix}.$$

3.2 Feedback Linearization Control Law

Applying feedback linearization, the control command vector for the torques is devised as

$$u_a = -\mathrm{I}\left(K_a x_a + f_a(x_a) - \ddot{\eta}^d\right) \tag{5}$$

where

$$K_a = \begin{bmatrix} k_{a0} & k_{a1} & 0 & 0 & 0 & 0 \\ 0 & 0 & k_{a0} & k_{a1} & 0 & 0 \\ 0 & 0 & 0 & 0 & k_{a0} & k_{a1} \end{bmatrix}$$

with k_{a0} and k_{a1} being positive control gains. In order to achieve the time-scale decomposition, it is necessary to choose relatively large attitude control gains.

3.3 Stability Analysis

Submitting the control law (5) into the open-loop dynamics (4) yields the closed-loop dynamics of the attitude tracking error:

$$\begin{bmatrix} \ddot{\tilde{\phi}} \\ \ddot{\tilde{\theta}} \\ \ddot{\tilde{\psi}} \end{bmatrix} + k_{a1} \begin{bmatrix} \dot{\tilde{\phi}} \\ \dot{\tilde{\theta}} \\ \dot{\tilde{\psi}} \end{bmatrix} + k_{a0} \begin{bmatrix} \tilde{\phi} \\ \tilde{\theta} \\ \tilde{\psi} \end{bmatrix} - \begin{bmatrix} \sigma_\phi / I_{xx} \\ \sigma_\theta / I_{yy} \\ \sigma_\psi / I_{zz} \end{bmatrix} = \begin{bmatrix} 0 \\ 0 \\ 0 \end{bmatrix} \tag{6}$$

which is a stable second-order system perturbed by the faults parameters.

It is shown by the closed-loop tracking error dynamics (6) that the attitude angles converge to the steady states as $\phi \rightarrow \phi^s$, $\theta \rightarrow \theta^s$, $\psi \rightarrow \psi^s$ with $\phi^s = \phi^d + \sigma_\phi/(I_{xx}k_{a0})$, $\phi^s = \phi^d + \sigma_\phi/(I_{xx}k_{a0})$, $\psi^s = \psi^d + \sigma_\psi/(I_{zz}k_{a0})$. The exact values of the steady-state errors are unknown due to the lack of fault information. However, it can be assumed that they are sufficiently small with sufficiently large k_{a0}, thus being not likely to affect the balancing of the quadrotor to a large extend. In the later position control design, these unknown errors can be regarded as uncertainties added to the control inputs and then be compensated by the designed fault-tolerant control.

4 Higher Level Position Control

The position control objective is to determine the force control command u_T and the desired attitude angles ϕ^d, θ^d, ψ^d in order to track a desired position trajectory. The desired position trajectory is assumed to be twice differentiable at least.

The Lyapunov-based MPC framework proposed in (Mhaskar et al., 2005, 2006) is adopted to design the position controller. Figure 2. shows the system architecture, composed of three blocks: (i) a model predictive controller that generates the control commands by solving a constrained optimization problem at each sampling instant; (ii) an auxiliary controller that is used to construct a Lyapunov function decay constraint for the MPC problem to ensure the stability; (iii) an adaptive online

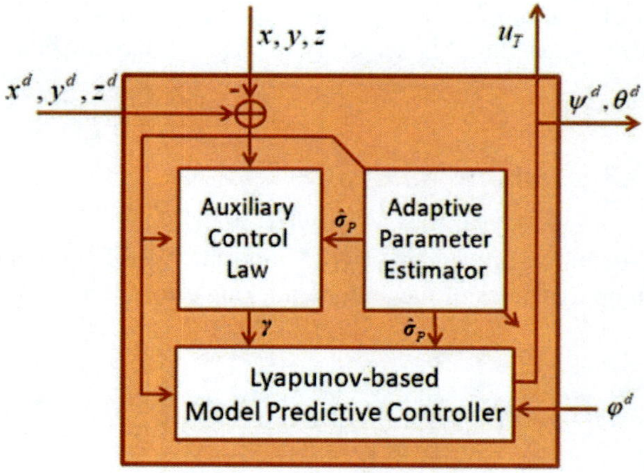

Fig. 2 Fault-tolerant MPC position controller (zoom in)

parameter estimator that estimates the unknown fault parameters and then passes the fault information to the auxiliary controller and the model predictive controller.

4.1 Position Tracking Error Dynamics

The fast-responding rotational dynamics can be regarded as a static block reaching its steady-state instantly, so we have $\phi = \phi^s$, $\theta = \theta^s$, $\psi = \psi^s$. Given the desired position, the position tracking errors are defined as $x_p = \left[\tilde{x}, \dot{\tilde{x}}, \tilde{y}, \dot{\tilde{y}}, \tilde{z}, \dot{\tilde{z}}\right]^T$. Defining $\mathbf{u}_p = [\phi^d, \theta^d, \psi^d, u_T]^T$ as the control vector, we obtain the following position tracking error dynamics subject to the actuator faults:

$$\dot{x}_p = A_p x_p + B_p\left(f_p(u_p) + \sigma_p - \ddot{\xi}^d\right) \tag{7}$$

with

$$A_p = \begin{bmatrix} 0 & 1 & 0 & 0 & 0 & 0 \\ 0 & 0 & 0 & 0 & 0 & 0 \\ 0 & 0 & 0 & 1 & 0 & 0 \\ 0 & 0 & 0 & 0 & 0 & 0 \\ 0 & 0 & 0 & 0 & 0 & 1 \\ 0 & 0 & 0 & 0 & 0 & 0 \end{bmatrix}, B_p = \begin{bmatrix} 0 & 0 & 0 \\ 1 & 0 & 0 \\ 0 & 0 & 0 \\ 0 & 1 & 0 \\ 0 & 0 & 0 \\ 0 & 0 & 1 \end{bmatrix}, \ddot{\xi}^d = \left[\ddot{x}^d \;\; \ddot{y}^d \;\; \ddot{z}^d\right],$$

$$\sigma_p = f_p(u_T + \sigma_T, \phi^s, \theta^s, \psi^s) - f_p\left(u_T, \phi^d, \theta^d, \psi^d\right)$$

$$f_p(\mathbf{u}_p) = \begin{bmatrix} \left(\cos\phi^d \sin\theta^d \cos\psi^d + \sin\phi^d \sin\psi^d\right) u_T/m \\ \left(\cos\phi^d \sin\theta^d \sin\psi^d - \sin\phi^d \cos\psi^d\right) u_T/m \\ \cos\phi^d \cos\theta^d \;\; u_T/m - g \end{bmatrix},$$

where $\boldsymbol{\sigma}_p$ is a lumped uncertain parameter, which is also bounded since the sine and cosine terms always remain in $[-1,1]$ and u_T is constrained in a compact set.

4.2 Auxiliary Control Law

We first define a new control vector as

$$\gamma \triangleq \boldsymbol{f}_p(\boldsymbol{u}_p) \tag{8}$$

Then, the auxiliary control law is developed based on the proportional-derivative formula as

$$\gamma = h(\mathrm{x}_p, \widehat{\sigma}_p) = -K_p x_p + \ddot{\zeta}^d - \widehat{\sigma}_p \tag{9}$$

where

$$K_p = \begin{bmatrix} k_{p0} & k_{p1} & 0 & 0 & 0 & 0 \\ 0 & 0 & k_{p0} & k_{p1} & 0 & 0 \\ 0 & 0 & 0 & 0 & k_{p0} & k_{p1} \end{bmatrix}$$

with k_{p0} and k_{p1} being positive control gains and $\widehat{\boldsymbol{\sigma}}_p$ denotes the estimate of the unknown $\boldsymbol{\sigma}_p$ obtained from the adaptive parameter estimator designed below.

4.3 Adaptive Parameter Estimation

The adaptive updating law for estimating the unknown parameter online is designed as

$$\dot{\widehat{\sigma}}_p = k_\sigma B_p^T P_p x_p \tag{10}$$

with k_σ being a positive constant and the solution of the Lyapunov equation:

$$(A_p - B_p K_p)^T P_p + P_p (A_p - B_p K_p) = -Q_p \tag{11}$$

4.4 Fault-Tolerant Lyapunov-Based Model Predictive Control

For the MPC design, the control at t is obtained by solving online a finite horizon optimal control problem, formulated based on the current state measurement $\mathbf{x}_p(t)$, the current estimated parameter $\widehat{\sigma}_p(t)$, and the auxiliary control law $\mathbf{h}(\cdot)$:

$$\min_{\overline{u}_p(\cdot)} \int_t^{t+T_p} \left(\|\overline{x}_p(s;x_p(t),\widehat{\sigma}_p(t),t)\|_{\overline{Q}}^2 + \|\overline{u}_p(s)\|_{\overline{R}}^2 \right) ds \tag{12}$$

subject to

$$\dot{\overline{x}}_p(s) = A_p\overline{x}_p(s) + B_p\left(f_p(\overline{u}_p(s)) + \overline{\widehat{\sigma}}_p(s) - \ddot{\zeta}^d\right), \qquad s \in [t, t+T_p) \tag{13}$$

$$\dot{\overline{\widehat{\sigma}}}_p(s) = k_\sigma B_p^T P_p \overline{x}_p(s), \qquad s \in [t, t+T_p) \tag{14}$$

$$\overline{x}_p(s) = x_p(t), \qquad s \in [t, t+\Delta) \tag{15}$$

$$\overline{\widehat{\sigma}}_p(s) = \widehat{\sigma}_p(t), \qquad s \in [t, t+\Delta) \tag{16}$$

$$\overline{u}_p(s) \in \mathbb{U}, \qquad s \in [t, t+T_p) \tag{17}$$

$$\begin{aligned} &2\overline{x}_p^T(s)P_p\left(A_p\overline{x}_p(s) + B_p\left(f_p(\overline{u}_p(s)) + \overline{\widehat{\sigma}}_p(s) - \ddot{\zeta}^d\right)\right) \\ &\le 2\overline{x}_p^T(s)P_p\left(A_p\overline{x}_p(s) + B_p\left(h\left(\overline{x}_p(s),\overline{\widehat{\sigma}}_p(s)\right) + \overline{\widehat{\sigma}}_p(s) - \ddot{\zeta}^d\right)\right), s \in [tt+\Delta), \end{aligned} \tag{18}$$

where $\overline{Q}$ and $\overline{R}$ are positive-definite, symmetric weighting matrices, Δ is the sampling period, T_p is the prediction horizon, $\overline{x}_p\left(\cdot;x_p(t),\widehat{\sigma}_p(t),t\right)$ is the predicted trajectory of the prediction model (13) driven by $\overline{u}_p(\cdot)$ with the initial state defined as (15). $\overline{\widehat{\sigma}}_p(s)$ is the predicted parameter estimate driven by (14) with the initial value defined as (16). $\mathbb{U}$ represents the admissible input constraint. The constraint of (18) ensures that over the current sampling period $[t, t+\Delta)$, the control computed by the MPC enforces a decay in the value of the Lyapunov function by at least the rate achieved by the auxiliary control law $\mathbf{h}(\cdot)$.

Note that the above model predictive controller is implemented in a receding horizon and sample-and-hold fashion. The optimization problem is solved repeatedly at each sampling instant $t = N\Delta$ and the optimal control profile is applied over $[N\Delta, (N+1)\Delta)$ until the next measurement is available at $(N+1)\Delta$. Thus, the applied closed-loop position control is defined as

$$u_p(s) \triangleq \overline{u}_p^*\left(s;x_p(N\Delta),\widehat{\sigma}_p(N\Delta),N\Delta,N\Delta+T_p\right), s \in [N\Delta,(N+1)\Delta). \tag{19}$$

4.5 Stability Analysis

To proof the closed-loop stability under the control of the developed fault-tolerant MPC, first define a positive definite Lyapunov function as

$$V_p = \|x_p\|_{P_p}^2 + 1/k_\sigma \|\widetilde{\sigma}_p\|^2 \tag{20}$$

where $\widetilde{\boldsymbol{\sigma}}_p = \boldsymbol{\sigma}_p - \widehat{\boldsymbol{\sigma}}_p$ denotes the estimation error. Differentiating V_p along the trajectory of the open loop trajectory of (7) and substituting the adaptive updating law (10) gives

$$\dot{V}_p = 2x_p^T P_p \left(A_p x_p + B_p \left(f_p(u_p) + \widehat{\sigma}_p - \ddot{\xi}^d\right)\right) \tag{21}$$

The feasibility of the MPC at each sampling instant is guaranteed since the auxiliary control law $\mathbf{h}(\cdot)$ can be a feasible solution. Recalling the constraint (18), we have that for $s \in [t, t + \Delta)$, there always exists an optimal control $\mathbf{u}_p$ leading to a decay of the Lyapunov function by at least the rate achieved by the auxiliary control (9)

$$\begin{aligned} \dot{V}_p(x_p(s)) &= 2x_p^T(s) P_p \left(A_p x_p(s) + B_p \left(f_p\left(\bar{u}_p^*(s)\right) + \hat{\sigma}_p(s) - \ddot{\xi}^d\right)\right) \\ &\leq x_p^T(s) P_p (A_p x_p(s) + B_p(h(x_p(s)) + \hat{\sigma}_p(s) - \ddot{\xi}^d)) \\ &= -\|x_p(s)\|_{Q_p}^2 \end{aligned} \tag{22}$$

Then, integrating (22) yields the non-increasing property of the Lyapunov function. That is, for $0 \leqslant t_1 \leqslant t_2 \leqslant \infty$, there is

$$V_p(x_p(t_1)) - V_p(x_p(t_2)) \leq -\int_{t_1}^{t_2} \|x_p(t)\|_{Q_p}^2 dt \tag{23}$$

We can then conclude the Lyapunov stability. First, the boundedness of $V_p(\mathbf{x}_p(t))$, $\mathbf{x}_p$ and $\widetilde{\boldsymbol{\sigma}}_p$ is proven. Because control $\mathbf{u}_p$ is bounded, we can then conclude from (7) the boundedness of $\dot{\mathbf{x}}_p$. Moreover, taking inductions of (23) gives

$$0 \leq V_p(\infty) \leq V_p(0) - \int_0^\infty \|x_p(\mathrm{t})\|_{Q_p}^2 dt \tag{24}$$

which implies that $\int_0^\infty \|x_p(\mathrm{t})\|_{Q_p}^2 dt$ exists. Thus, the tracking error $\mathbf{x}_p(t)$ is continuous, bounded, and integrable. Using Barbalat's lemma, we proved the asymptotic convergence that $\mathbf{x}_p(t) \rightarrow 0$ as $t \rightarrow 0$.

5 Simulation Results

To substantiate the effectiveness of the control design, simulations have been taken on a quadrotor with $l = 0.30486\text{m}$, $m = 2.618\text{kg}$, $g = 9.81\text{m/s}^2$, $I_{xx,\ yy} = 0.043467\text{kg} \cdot \text{m}^2$, $I_{zz} = 0.063267\text{kg} \cdot \text{m}^2$. A low voltage fault is simulated by reducing the rotor's lift force by 0.3N at the 25th second.

The initial values are selected as $\widehat{\sigma}_p(0) = [0, 0, 0]^T x_p(0) = [9, 0, 0.1, 0, 0, 0]^T$, $x_a(0) = [0.5, 0, 0.5, 0.0, 0]^T$. The control gains are chosen as $k_{a0} = 25$, $k_{a1} = 5$, $k_{p0} = 0.5$, $k_{p1} = 1$, Then the resulting time scale of the rotational dynamics is five times bigger than that of the translational dynamics. Thus, we choose the sampling periods as $\Delta_a = 0.1\text{s}$ and $\Delta = 0.5\text{s}$. The adaptive gain $k_\sigma = 0.0001$. In the Lyapunov function, the positive-definite matrix $\mathbf{Q}_p =$ diag ([10^3, 1, 10^3, 1, 10^3, 1]). In the MPC formulation, the prediction horizon $T_p = 2.5\text{s}$, and the weighting matrices are selected as $\overline{Q} = \text{diag}\left(\left[10^3, 1, 10^3, 1, 10^3, 10^3\right]\right), \overline{R} = \text{diag}\left(\left[10^{-3}, 10^5, 10^5\right]\right)$.

The simulation results are shown in Figs. 3, 4, 5, 6, 7 and 8. Figures 3 and 4 show that the quadrotor tracks the desired position trajectory. The tracking performance deteriorates when the fault happens but then quickly recovers, which implies the effectiveness of the fault-tolerant design. Figure 5 shows that the attitude of the quadrotor is regulated. Figure 6. is the adaptation process of the parameter estimator.

Fig. 3 Position trajectory tracking

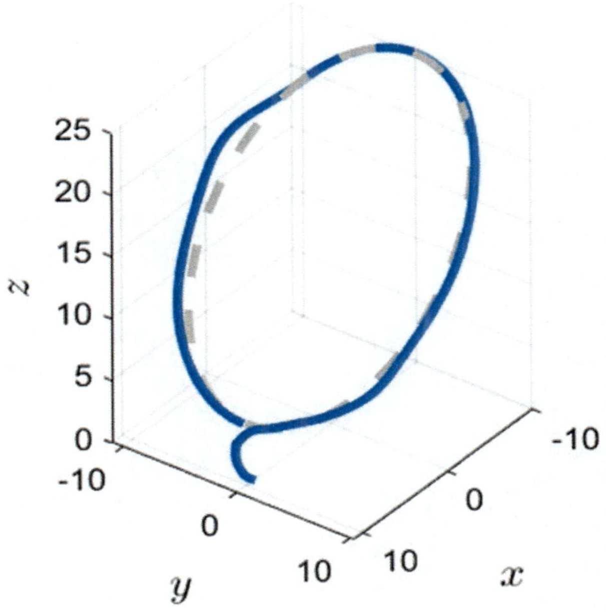

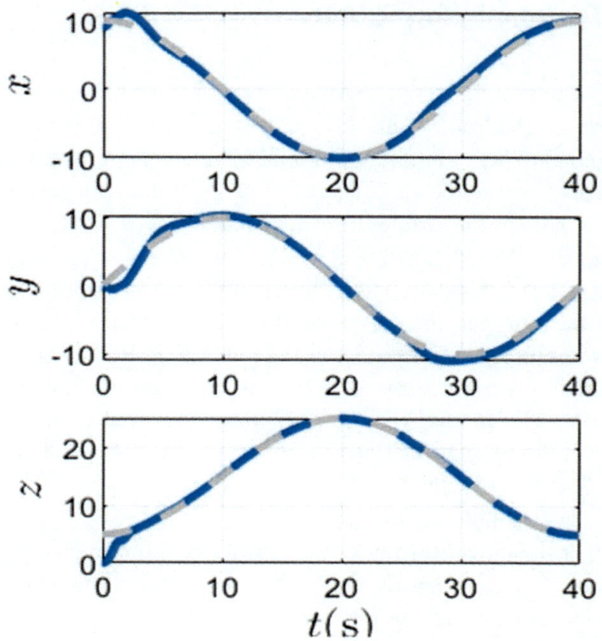

Fig. 4 Position profile

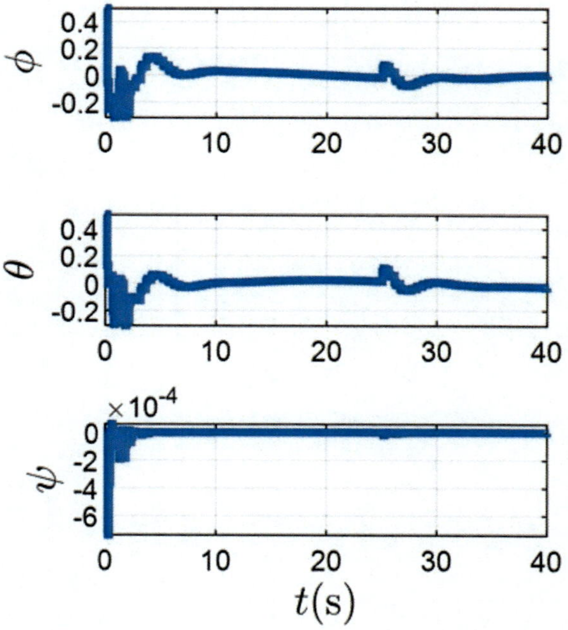

Fig. 5 Attitude profile

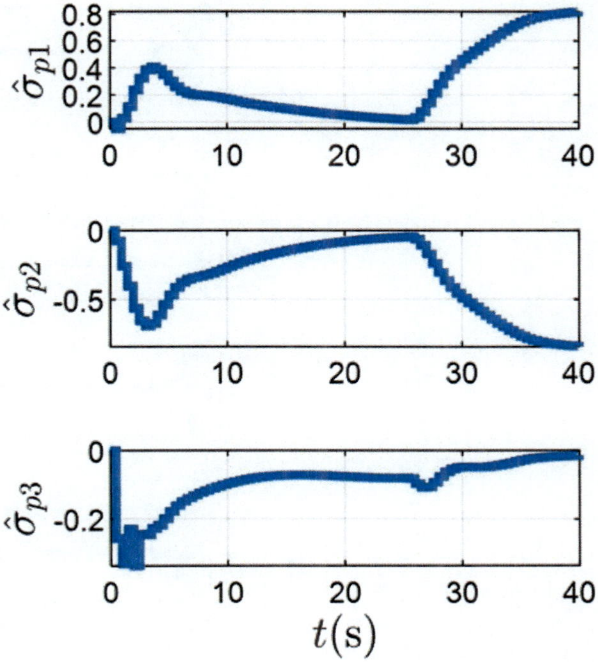

Fig. 6 Parameter adaption

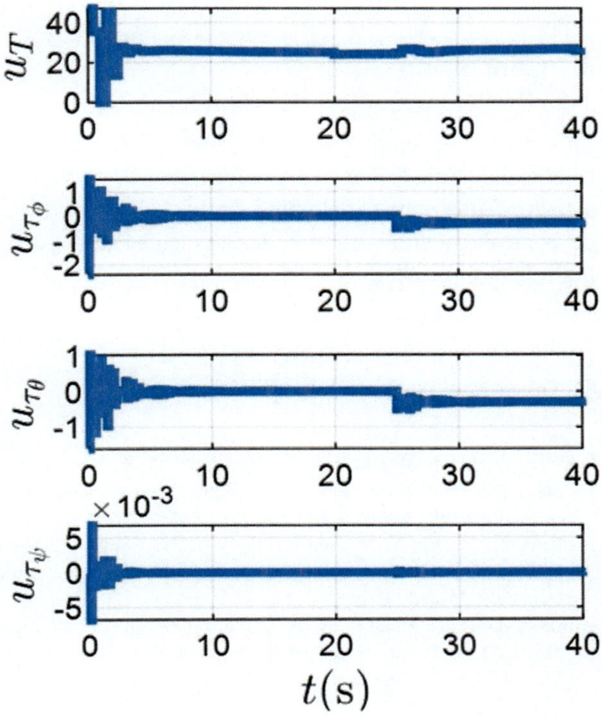

Fig. 7 Force and torques

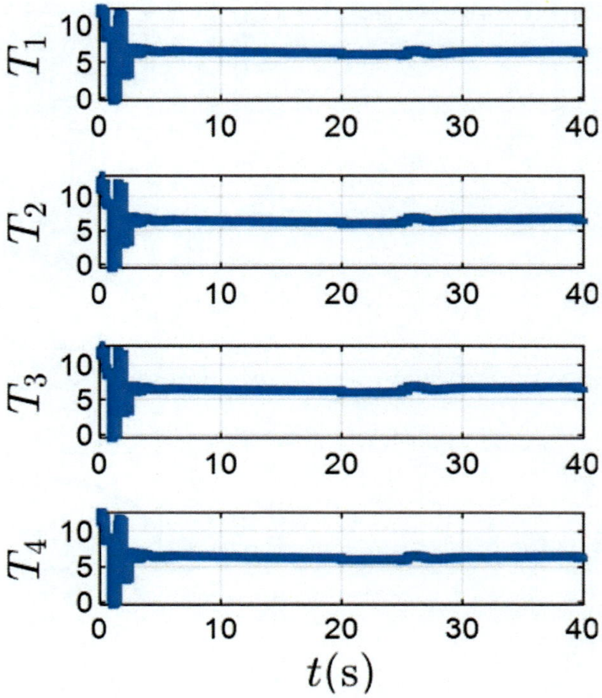

Fig. 8 Rotors' thrust

6 Conclusion

This chapter developed a novel adaptive fault-tolerant MPC for trajectory tracking control of a quadrotor. The design can adaptively compensate for unexpected actuator faults, thus maintaining the tracking performance. The proposed hierarchical control structure also provides a practical strategy to apply MPC to the real-time flight control problem.

References

Gopinathan, M., Boskovic, J.D., Mehra, R. K., & Rago, C. (1998, December). A multiple model predictive scheme for fault-tolerant flight control design. In *IEEE conference on decision and control*, Florida, USA.

Izadi, H. A., Zhang, Y., & Gordon, B. W. (2011, August 28–September 2). Fault tolerant model predictive control of quad-rotor helicopters with actuator fault estimation. In *IFAC World Congress*, Milano, Italy.

Kufoalor, D. K., & Johansen, T. A. (2013, June 17–19). Reconfigurable fault-tolerant flight control based on nonlinear model predictive control. In *American control conference*, Washington, DC., USA.

Mhaskar, P., El-Farra, N. H., & Christofides, P. D. (2005). Predictive control of switched nonlinear systems with scheduled mode transitions. *IEEE Transactions on Automatic Control, 50*(11), 1670–1680.

Mhaskar, P., El-Farra, N. H., & Christofides, P. D. (2006). Stabilization of nonlinear systems with state and control constraints using Lyapunov-based predictive control. *Systems & Control Letters, 55*(8), 650–659.

Voos, H. (2009, April). Nonlinear control of a quadrotor micro-UAV using feedback-linearization. In *IEEE international conference on mechatronics*, Málaga, Spain.

Wang, K., Hua, C., Chen, J., & Cai, M. (2020). Dual-loop integral sliding mode control for robust trajectory tracking of a quadrotor. *International Journal of Systems Science, 51*(2), 203–216.

Yang, X., & Maciejowski, J. (2015). Fault-tolerant control using Gaussian processes and model predictive control. *International Journal of Applied Mathematics and Computer Science, 25*, 133–148.

Zhou, Z., Wang, H., Hu, Z., Wang, Y., & Wang, H. (2019). A multi-time-scale finite time controller for the quadrotor UAVs with uncertainties. *Journal of Intelligent & Robotic Systems, 94*(2), 521–533.

Higher Order Dynamic Mode Decomposition to Model Reacting Flows

Adrián Corrochano, Giuseppe D'Alessio, Alessandro Parente, and Soledad Le Clainche

1 Introduction

Halting climate change is one of the greatest challenges facing the modern world. Improving the efficiency in combustion systems in aircraft, i.e., reducing the amount of fuel consumption, reducing the polluting fumes released from turbulent combustion or replacing fossil fuels by synthetic fuels are some of the strategies that could be addressed to reduce air pollution.

Modelling reacting flows is a highly complicated task, due to (i) the high complexity of the underlying physical problem and (ii) the need of using large computational resources, generally required to solve realistic problems. The alternative is to develop and use reduced order models (ROMs). ROMs model the main flow dynamics with relatively high accuracy at a reduced computational cost (in terms of memory and computational time).

A novel strategy has been carried out in this chapter to study the main dynamics of an axisymmetric, time varying, non-premixed laminar co-flow flame. The method identifies the main patterns describing the flow physics. Such patterns can be used to reconstruct the original solution with high accuracy, resulting in a ROM that can be used to model the problem studied. For such aim, we use higher order dynamic mode decomposition (HODMD) (Le Clainche & Vega, 2017), a data-driven method, suitable for the analysis of complex flows (see more details in Le Clainche et al., 2020a, b). Invoking the previous knowledge and based on the robustness of the

A. Corrochano (✉) · S. Le Clainche
School of Aerospace Engineering, Universidad Politécnica de Madrid, Madrid, Spain
e-mail: adrian.corrochanoc@upm.es; soledad.leclainche@upm.es

G. D'Alessio · A. Parente
École polytechnique de Bruxelles, Aero-Thermo-Mechanics Laboratory, Université Libre de Bruxelles, Bruxelles, Belgium
e-mail: Giuseppe.DAlessio@ulb.be; Alessandro.Parente@ulb.be

T. H. Karakoc et al. (eds.), *New Achievements in Unmanned Systems*, Sustainable Aviation, https://doi.org/10.1007/978-3-031-29933-9_22

method presented in complex flows (i.e. multiscale, noisy, turbulent, et cetera), either numerically or experimentally (Vega & Le Clainche, 2020), a pioneering study of a combustion database has been carried out, showing that this algorithm is also suitable for this complex field of knowledge.

The chapter is organized as follows. In Sect. 2, the algorithm used in this work is introduced and the database where this method is going to be applied is presented in Sect. 3. The main results and the conclusions are explained Sects. 4 and 5, respectively.

2 Method

Higher order dynamic mode decomposition (HODMD) (Le Clainche & Vega, 2017) is a data-driven algorithm used to analyse complex nonlinear dynamical systems. This algorithm is an extension of the classical dynamic mode decomposition (DMD) (Schmid, 2010), which is suitable for the analysis of complex flows, such as turbulent flows (Le Clainche et al., 2020a, b), noisy experiments (Le Clainche et al., 2017, 2018), transitional flows (Le Clainche & Vega, 2017), etc., to develop ROMs (Le Clainche & Ferrer, 2018; Le Clainche et al., 2020a, b) or identify flow patterns and coherent structures (Le Clainche & Vega, 2018).

HODMD decomposes the original data v_k, collected at time instant t_k, which are the snapshots, as a Fourier expansion of M DMD modes u_m in the following way (in discrete form):

$$v_k \simeq \sum_{m=1}^{M} a_m \boldsymbol{u}_m e^{(\delta_m + i\omega_m)t_k}, k = 1, \ldots, K, \quad (1)$$

where a_m, δ_m, ω_m are their amplitudes, growth rates and frequencies, respectively. The temporal dimension can be referred to as K, which are the number of snapshots analysed by the method, and the number of modes M is called the spectral complexity. The number of DMD modes M depends on a tunable tolerance ε_1, which can be adjusted to the desired accuracy to reconstruct the original data.

The algorithm can be summarised in two steps. Initially, the data are collected into a snapshot tensor **A**. The first dimension of the tensor represents the variables studied. The next two or three dimensions are the spatial dimensions, depending on whether the problem is bi or tri dimensional. The last dimension will be the temporal dimension. The first step applies a higher order singular value decomposition (HOSVD) to the tensor, retaining the large size scales and reducing the noise and the spatial redundancies. A second tunable tolerance ε_2 will be used to define the HOSVD modes retained. In the second step, a DMD-like algorithm is applied to the previously processed data. This algorithm is called the DMD-d algorithm. Similarly to the sliding window process from the power spectral density (PSD) analysis, the DMD-d algorithm uses d time-delayed snapshots from the snapshot matrix. The parameter d is tunable, as well as the tolerance ε_1 mentioned before, which will be applied now to define the number of DMD modes retained in the expansion Eq. 1.

Finally, the original data can be reconstructed as in Eq. (1), obtaining a new tensor A^{DMD}. The error made in the calculations can be evaluated measuring the root mean square error, as follows

$$\text{RMSE} = \sqrt{\frac{|\mathbf{A} - \mathbf{A}^{\text{DMD}}|^2}{|\mathbf{A}|^2}}. \tag{2}$$

More details about the algorithm used in this chapter can be found in (Le Clainche et al., 2017). The Matlab codes are presented in (Vega & Le Clainche, 2020).

HODMD will be applied to identify the main frequencies leading the non-linear solution of a numerical simulation modelling a turbulent combustion. The high complexity of the data makes HODMD a suitable algorithm to analyse the data.

3 Numerical Simulation

Numerical simulations have been carried out using the numerical open source solver *laminarSMOKE*, a CFD solver based on *OpenFOAM* and conceived to simulate laminar flows by means of a detailed kinetic mechanism (Cuoci et al., 2013). This solver uses finite volumes as spatial discretization. Simulations model an axisymmetric, time varying, non-premixed laminar co-flow flame: the fuel consists of 65% of methane and 35% of nitrogen on molar basis, while the oxidizer is regular air. Both streams are fed at ambient temperature and atmospheric pressure. The transient behaviour is induced by a sinusoidal perturbation to the parabolic velocity profile of the fuel stream, with frequency f and amplitude A. Thus, the fuel velocity profile is as follows:

$$v(r,t) = v_{\max} * \left(1 - r^2/R^2\right) * (1 + \text{A} * \sin(2 * \pi * f * t)), \tag{3}$$

where r is the radial coordinate, R is the internal radius, t is the time, $v_{\max}$ is the velocity equal to 70 cm/s, $f = 10$ Hz and $\text{A} = 0.25$. The co-flow air is instead injected at constant velocity $v = 35$ cm/s. The current simulation arrives to ~0.12 s, and the time step is $\Delta t = 0.0001$ s. For the HODMD analysis, 1 of each 10 snapshots have been used, so the Δt between snapshots will be 0.001 s. A thorough description of the numerical setup can be found in (D'Alessio et al., 2020).

The kinetic mechanism used for the simulation is the *POLIMI_C1C3_HT_1412* (Ranzi et al., 2012), it accounts for 82 species and 1698 reactions.

4 Results

In this section, the algorithm mentioned before will be used to analyse the database corresponding to the numerical simulation explained above. The data base is a fourth order tensor, formed by 83 variables (temperature and 82 different species); two spatial dimensions composed by a 150 × 150 grid; and 117 snapshots equidistant in time. The order of magnitude of the variables ranges from 10^2 (temperature) to 10^{-11} (RALD3G, the smallest radical computed). To avoid that the small variables could be removed, the mean has been removed and all variables have been scaled with its range, as in Parente & Sutherland, 2013. This scaling method performs better for most of the main species, which contain the main flow dynamics.

The tolerances used in this case are $\varepsilon_1 = \varepsilon_2 = 10^{-3}$. This tolerance gives a good reconstruction in all the cases studied. Parameter d is calibrated performing several analyses using $d = 10$, 12 and 15. Figure 1 shows the spectrum for all the cases studied. It shows a clearly periodic solution, with a dominant frequency (with highest amplitude) of $\omega_p = 60$ and its harmonics, which are the modes with lower amplitude and frequency $\omega_n = n * \omega_p$, being $n \in \mathbb{N} > 1$.

The case with $d = 12$ is selected as a representative case for studying the solution of this problem.

Figure 2 shows the error made in the reconstruction using the previous DMD modes in the DMD expansion Eq. (1). As expected, the best reconstructed variables are the main ones, i.e. the air components (N_2 and O_2), the fuel components (CH_4), the main products of the combustion (CO_2 and H_2O) and the temperature. These variables have a reconstruction error lower than 1% except the methane that has an error of 1.2%. Moreover, the nitrogen has a reconstruction error below 10^{-3}. On the other hand, the worst reconstructed variables are some of the smallest radicals. Their maximum value ranges from 10^{-5} to 10^{-11}. RALD3B presents the largest reconstruction error with 62%, while the order of magnitude of its maximum value is 10^{-9}.

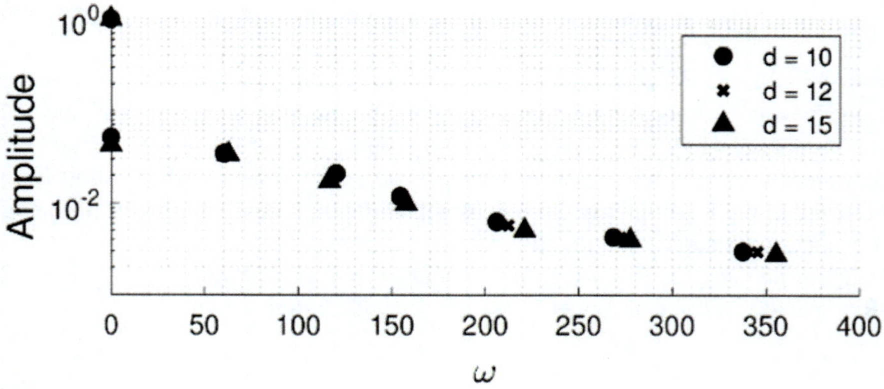

Fig. 1 Frequencies vs. amplitudes representing the DMD modes in the combustion database

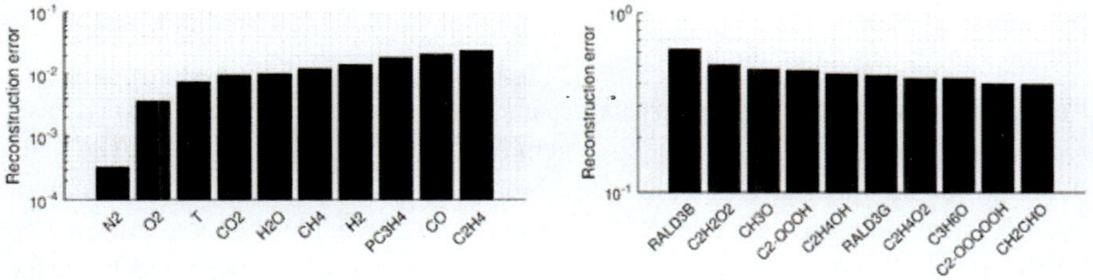

Fig. 2 Reconstruction error for the 10 best reconstructed variables (left) and the 10 worst reconstructed variables (right) using the DMD modes presented in Fig. 1 and the DMD expansion Eq. (1)

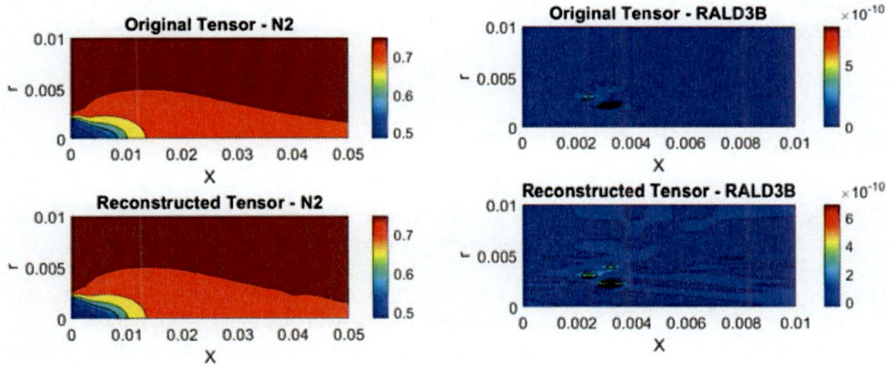

Fig. 3 Representative snapshot of the best reconstructed variable (N2). Right: Representative snapshot of the worst reconstructed variable (RALD3B)

Finally, a representative snapshot has been plotted with the best reconstructed variable and the worst one in Fig. 3. Nitrogen is very accurately reconstructed, as it shows the reconstruction error (below 10^{-3}). On the other hand, the RALD3B variable shows that, despite having high reconstruction error, the main structures can be identified, as seen in the reconstructed snapshot. This result suggests that the high error value could be connected to the fact that in most of the domain, the variable has a constant zero value. While the original data shows an order of magnitude of 10^{-23} on the zero part, on the reconstructed tensor reveals an order of magnitude of 10^{-12} on the zero part. This is between two and three orders of magnitude below the main structures. Therefore, with this method, the main structures of radical variables can be captured, although the whole snapshot may exhibit a large reconstruction error.

5 Conclusion

This work studies the main flow dynamics and the possibility of using HODMD as a reduced order model, to reconstruct transient data in turbulent combustion. The main dynamics of the flow have been found to be periodic, with a principal frequency of $\omega_p = 60$. The main species and the temperature are the variables better approximated by the method, while the worst reconstructed variables are represented by some radical species. Nevertheless, the method is also able to identify the main patterns in these poorly reconstructed variables, suggesting once more that HODMD is suitable to model turbulent combustion flows.

Acknowledgments A.C. and S.L.C. acknowledge the grant PID2020-114173RB-I00 funded by MCIN/AEI/ 10.13039/501100011033.

References

Cuoci, A., Frassoldati, A., Faravelli, T., & Ranzi, E. (2013). Numerical modeling of laminar flames with detailed kinetics based on the operator-splitting method. *Energy and Fuels, 27*(12), 7730–7753.

D'Alessio, G., Parente, A., Stagni, A., & Cuoci, A. (2020). Adaptive chemistry via pre-partitioning of composition space and mechanism reduction. *Combustion and Flame, 211*, 68–82.

Le Clainche, S., & Ferrer, E. (2018). A reduced order model to predict transient flows around straight bladed vertical Axis wind turbines. *Energies, 11*, 566.

Le Clainche, S., & Vega, J. M. (2017). Higher order dynamic mode decomposition. *SIAM Journal on Applied Dynamical Systems, 16*, 882–925.

Le Clainche, S., & Vega, J. M. (2018). Analyzing nonlinear dynamics via data-driven dynamic mode decomposition-like methods. *Complexity, 2018*, 6920783.

Le Clainche, S., Vega, J. M., & Soria, J. (2017). Higher order dynamic mode decomposition for noisy experimental data: Flow structures on a zero-net-mass-flux jet. *Experimental Thermal and Fluid Science, 88*, 336–353.

Le Clainche, S., Moreno, R., Taylor, P., & Vega, J. M. (2018). New robust method to study flight flutter testing. *Journal of Aircraft, 56*, 1–8.

Le Clainche, S., Izbassarov, D., Rosti, M., Brandt, L., & Tammisola, O. (2020a). Coherent structures in the turbulent channel flow of an elastoviscoplastic fluid. *Journal of Fluid Mechanics, 888*.

Le Clainche, S., Rosti, M., & Brandt, L. (2020b). Flow structures and shear-stress predictions in the turbulent channel flow over an anisotropic porous wall. *Journal of Physics: Conference Series, 1522*, 012016.

Parente, A., & Sutherland, J. C. (2013). Principal component analysis of turbulent combustion data: Data pre-processing and manifold sensitivity. *Combustion and Flame, 160*(2), 340–350.

Ranzi, E., Frassoldati, A., Grana, R., Cuoci, A., Faravelli, T., Kelley, A. P., & Law, C. K. (2012). Hierarchical and comparative kinetic modeling of laminar flame speeds of hydrocarbon and oxygenated fuels. *Progress in Energy and Combustion Science, 38*(4), 468–501.

Schmid, P. (2010). Dynamic mode decomposition of numerical and experimental data. *Journal of Fluid Mechanics, 656*, 5–28.

Vega, J. M., & Le Clainche, S. (2020). *Higher order dynamic mode decomposition and its applications*. Elsevier.

Fault-Tolerant Estimation of Relative Motion of Satellites in Cluster

Tuncay Yunus Erkec and Chingiz Hajiyev

Nomenclature

GPS	Global Positioning System
EKF	Extended Kalman Filter
AKF	Adaptive Kalman Filter

1 Introduction

In space missions, both coverage and functionality limitations can be overcome by constellation architecture. The use of satellites as a cluster means the coordinated movement of several spacecraft to achieve a common goal. The formation flight can also be used to distribute the payload on two spacecraft, depending on mission objectives (Erkec & Hajiyev, 2021).

Although the methods proposed in the literature for the relative navigation problem of satellites differ conceptually or applied, they are similar in terms of constraints such as intense computational load, line-of-sight vector separation, coordinate system transformations, and sensor noise (Erkec & Hajiyev, 2019; Sever & Hajiyev, 2020).

T. Y. Erkec (✉)
Turkish National Defence University, Hezarfen Aeronautics and Space Technologies Institute, Istanbul, Turkey
e-mail: terkec@hho.msu.edu.tr

C. Hajiyev
Istanbul Technical University, Istanbul, Turkey
e-mail: cingiz@itu.edu.tr

T. H. Karakoc et al. (eds.), *New Achievements in Unmanned Systems*, Sustainable Aviation, https://doi.org/10.1007/978-3-031-29933-9_23

Within the scope of the "fault-tolerant Kalman filter," errors were detected in the EKF-1 and EKF-2 filters, and GPS errors were eliminated with two different versions:

- First version: During the use of four GPS satellite data, by detecting the data corruption in one GPS satellite that occurs at any time, estimating the target and follower satellite positions and relative state vectors with the EKF system with three GPS satellites, with the reconfigured EKFs.
- Second version: During the use of four GPS satellite data, the relative situation using the target and follower satellite positions determined by AKFs as measurement, using the data of all of them, including the data of the fourth GPS satellite, which also contains the corrupted data, as well as the detection of data corruption in one GPS satellite that occurs at any time. Estimation of vectors with AKF.

The Extended Kalman Filter (EKF) is used to estimate satellite state vector in a reconfigurable version of fault-tolerant localization. The errors are detected by EKFs normalized innovation values, and the architecture of satellite state estimation model is eliminated by GPS satellite measurement, which includes errors by reconfigured satellite localization architecture.

The adaptive Kalman filter (AKF) is used for the adaptive version of the fault-tolerant estimation of the satellite state vector (Hajiyev, 2007). The errors are detected by EKF and eliminated by AKF. Comparison of these two fault-tolerant localization approaches are focused in this study.

The new approach in this study can be used not only for cluster satellite architectures but also for the detection of location vectors for terrestrial and offshore autonomous cluster architecture. In addition, using the proposed approach, it is possible to eliminate GPS measurement errors as well.

2 Methodology

In this study, the target and follower satellites state vectors are fault-tolerant estimated in two stages using GPS measurements.

First Stage
Using the distance measurement model (Pseudo-ranging model) for target and follower satellites state vector estimation via EKF.

Second Stage
Fault-tolerant Kalman Filter (FTKF) architectures are designed as Reconfigure EKF (REKF) and AKF versions in the presence of GPS measurement faults during relative localization within cluster satellites architecture using Hill—Clohessy–Wiltshire equation as the dynamic model of satellite formation geometry (Hill, 1878, New published 2020; Clohesy & Wiltshire, 1960, Published Online: 30 Aug 2012).

All equations mentioned in the study were applied for both target and follower satellites.

2.1 GPS-Based Pseudo-ranging Satellite Localization

Architecture created for locating the satellite is as shown in Fig. 1. The target satellites' positions obtained with the Keplerian orbital approach and measurement information from four GPS satellites are processed with the Extended Kalman Filter (EKF) method to obtain precise estimates of the satellite position vectors. Thus, satellites' location information based on "GPS based Pseudo-ranging model" is obtained (Bagcı & Hajiyev, 2019). All equations are valid for target and follower satellite separately.

The orbital equations with J2 perturbation of target satellites are

$$\ddot{\boldsymbol{r}} = \begin{bmatrix} \ddot{r}_x \\ \ddot{r}_y \\ \ddot{r}_z \end{bmatrix} = \begin{bmatrix} -\frac{\mu_\oplus}{r^3}x - \left\{\frac{3}{2}J_2\frac{\mu_\oplus}{r^2}\left(\frac{R_\oplus}{r}\right)^2\left(1-5\left(\frac{z}{r}\right)^2\right)\frac{x}{r}\right\} \\ -\frac{\mu_\oplus}{r^3}y - \left\{\frac{3}{2}J_2\frac{\mu_\oplus}{r^2}\left(\frac{R_\oplus}{r}\right)^2\left(1-5\left(\frac{z}{r}\right)^2\right)\frac{y}{r}\right\} \\ -\frac{\mu_\oplus}{r^3}z - \left\{\frac{3}{2}J_2\frac{\mu_\oplus}{r^2}\left(\frac{R_\oplus}{r}\right)^2\left(1-5\left(\frac{z}{r}\right)^2\right)\frac{z}{r}\right\} \end{bmatrix} \tag{1}$$

where $r = [x\ y\ z]^T$ stands for position vectors of satellites, $r = \|r\| = \sqrt{x^2+y^2+z^2}$ for the normal of position vector, G for the universal gravitational constant, $M_\oplus$ for the mass of Earth, $\mu_\oplus = GM_\oplus$ for the standard gravity parameter of the Earth, and a_{J_2} for the decay acceleration vector due to J_2 perturbation. $R_\oplus$ is the average Earth Equatorial radius, and J_2 the first component of the geopotential model.

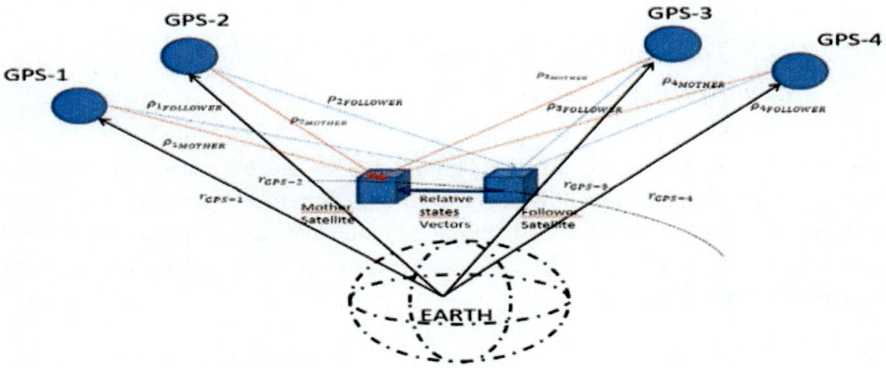

Fig. 1 Satellites relative localization with pseudo-ranging model

2.2 Statistical Evaluation in the Diagnostic Process

To determine whether the measurement information entering the Kalman filters within the satellite formation architecture contains errors or not, the normalized innovation values of the individual filters were used. The estimated error of estimation based on a finite number of measurements should be compatible with the theoretical statistical properties.

The sample mean of the normalized innovation is used as the monitoring statistics $S_c(k)$; here, $M$is the width of the moving window that is chosen 20, k is the iteration step. The GPS measurement errors are determined according to the threshold value (Berntorp & Di Cairano, 2018). The threshold value for the monitoring statistic Eq. (2) is determined as,

$$S_c(k) = \frac{1}{M}\left|\sum_{j=k-M+1}^{k} \tilde{\Delta}(j)\right| \tag{2}$$

Threshold = 0.288 (Degree of freedom: 6).

2.3 Reconfigurable EKF for Fault-Tolerant Satellite Localization

The measurements which come from GPS satellites are evaluated via statistical model and determined faulty GPS data. The GPS measurements which include faulty and bias distortions are ignored during satellite localization process via reconfigure EKF architecture. The GPS measurement sources are degraded due to faulty measurement via REKF.

2.4 Fault-Tolerant Satellite Localization Via Adaptive Kalman Filter

The measurements which come from GPS satellites are evaluated via statistical model and determine faulty GPS data. The GPS measurements which include faulty and bias distortions are corrected during satellite localization process via AKF architecture. The satellite localization state vectors are determined precisely even by fault GPS measurements.

3 Results and Discussion

Relative satellites' X- axis REKF fault-tolerant estimation results are shown in Figs. 2 and 3. Error variances and normalized innovation values increase due to faulty GPS measurements after the 50th second of simulation. However, relative position errors, error variances, and normalized innovation values converge to small values in REKF version after the 50th second because fault measurements are ignored by REKF version.

3.1 *Fault-Tolerant Cluster Satellite Relative Localization Via Adaptive Kalman Filter Results*

Target satellites' X-axis AKF fault-tolerant results are shown in Figs. 4 and 5. Position errors, error variances, and normalized innovation values increase due to

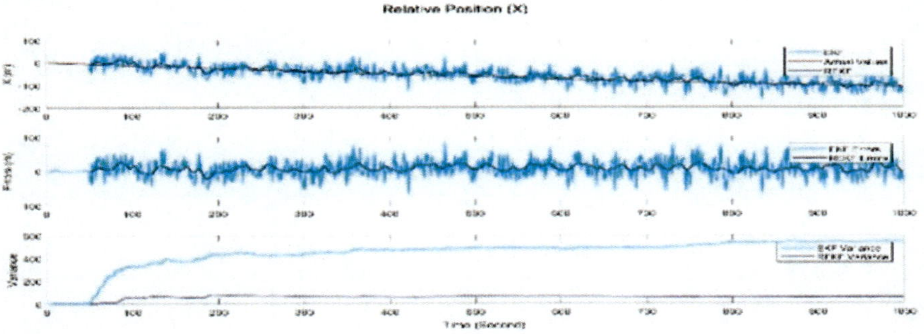

Fig. 2 Satellites relative X-axis estimation values, estimation errors, and error variance when REKF is used

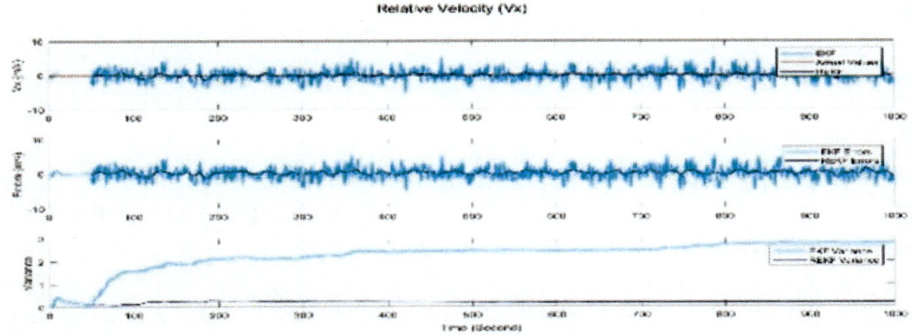

Fig. 3 Satellites relative Vx-axis estimation values, estimation errors, and error variance when REKF is used

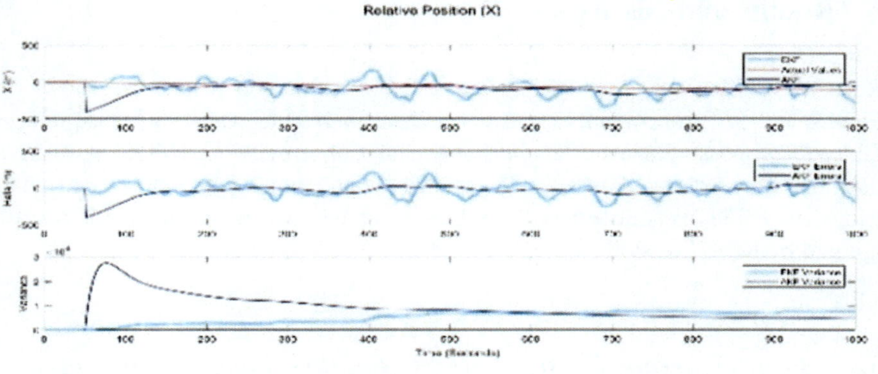

Fig. 4 Satellites relative X-axis estimation values, estimation errors, and error variance when AKF is used

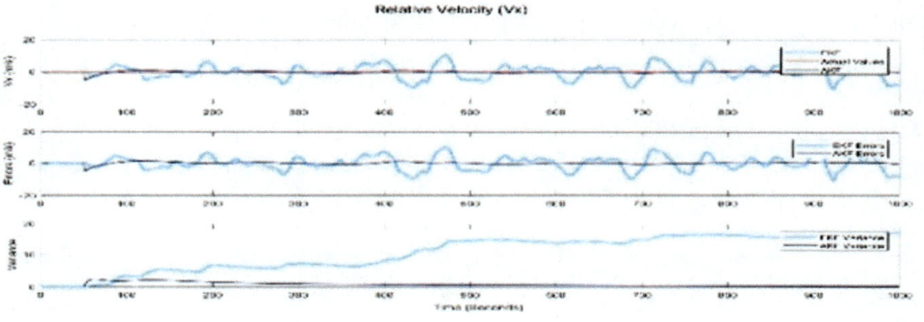

Fig. 5 Satellites relative Vx-axis estimation values, estimation errors, and error variance when AKF is used

faulty GPS measurements after 50th second of simulation. However, the relative position errors, error variances, and normalized innovation values converge to small values in fault-tolerant AKF version after 50th second because fault measurements are ignored by AKF version.

3.2 Comparison

Comparison of relative REKF and AKF version of cluster satellite architecture fault-tolerant localization algorithms for the target and follower satellites relative values are shown in Table 1. The algorithm step time is 1 s and the simulation observation time is 1000s. The GPS measurements which come from one of the constant and random biases were entered into one of the GPS measurements in both algorithms, 50 second after the start of the relative satellite positioning. Although the values in the tables are the average values obtained because of running the fault-tolerant

Table 1 Statistical error results of REKF and AKF algorithm

	Relative position vector estimation			Relative Velocity Vector estimation		
States	x(m)	y(m)	z(m)	v_x(m/s)	v_y(m/s)	v_z(m/s)
REKF μ	9.38936	2.94186	5.90800	0.36194	0.17413	0.34288
AKF μ	34.53058	8.35829	15.15488	0.49042	0.26863	0.31395

μ: Relative mean error values

relative satellite state estimation algorithms five times, they may vary due to random errors within the algorithms.

The statistical results show that REKF algorithm for the relative fault-tolerant localization is more accurate, because the GPS measurements which are faulty are ignored. On the other hand, AKF algorithm is estimated with target satellite position states even with the faulty measurements.

4 Conclusion

GPS-based cluster satellite localization is a traditional method of satellites' states estimation, on the other hand, it is more accurate and proved itself on all space missions. However, its accuracy is sensitive to GPS faulty data. In this study, fault-tolerant REKF and AKF approaches for estimation of the satellite states are proposed. The state estimation distortions due to GPS are corrected by both methods which are ignoring the fault GPS data (REKF) and AKF fault-tolerant versions.

The simulations show that the REKF version is more accurate than the AKF version within this scenario because it does not include any faulty data when the quality of the measurement sources is degraded. In this study, we assumed that only one GPS has faulty measurements. If all GPS have faulty measurements, or the GPS measurement number is increased, the AKF version of relative fault-tolerant satellite localization will be better for estimation of satellite states.

References

Bagcı, M., & Hajiyev, C. (2019) Measurement conversion based RKF for satellite localization via GPS. In *8th international conference on Recent Advances in Space Technologies (RAST)* (pp. 861–868).

Berntorp, K., & Di Cairano, S. (2018). Approximate noise-adaptive filtering using student-t distributions. *In 2018 annual American Control Conference (ACC)* (pp. 2745–2750).

Clohesy, W., & Wiltshire, R. (1960). Terminal guidance systems for satellite rendezvous. *Journal of Aerospace Sciences, 27*(9), 653–658. Published Online: 30 Aug 2012.

Erkec, T. Y., & Hajiyev, C. (2019). Traditional methods on relative navigation of small satellites. In *2019 9th international conference on Recent Advances in Space Technologies (RAST)* (pp. 869–874).

Erkec, T. Y., & Hajiyev, C. (2021). The methods of relative navigation of satellites formation flight. *International Journal of Sustainable Aviation, 6*(4), 260–279.

Hajiyev, C. (2007). Adaptive filtration algorithm with the filter-gain correction applied to integrated INS/Radar altimeter. *Proceedings of the Institution of Mechanical Engineers, Part G, 221*(5), 847–885.

Hill, G. W. (1878). Researches in the lunar theory. *American Journal of Mathematics, 1*(1), 5–26. New published 2020.

Sever, M., & Hajiyev, C. (2020). Satellite localization correction with extended Kalman filter. In *International symposium on electric aviation and autonomous systems*, 22–24 September, 2020.

Trajectory Tracking Control of an Unmanned Ground Vehicle Based on Fractional Order Terminal Sliding Mode Controller

Hayriye Tuğba Sekban and Abdullah Başçi

1 Introduction

Unmanned ground vehicles (UGV) are a typical kind of nonholonomic mechanical systems and they are often preferred because of their many advantages such as high mobility, good stability, fast motion and low energy consumption. In recent years, academic studies in the field of UGV have focused on the trajectory tracking control of the vehicle (Tzafestas, 2018). It has been seen that using controller designs based on both kinematics and dynamics of the vehicle for successful trajectory tracking control gives more realistic results (Koubaa et al., 2015). A large number of controllers have been used to perform trajectory tracking control of the UGV (Başçi et al., 2015; Orman et al., 2018). The sliding mode control (SMC) method has been frequently preferred to perform the trajectory tracking control of the UGV (Sharma & Panwar, 2016; Sekban et al., 2021). However, the chattering in the systems and the asymptotic stability in infinite time are the biggest disadvantages of this controller. The chattering effect can be reduced using a variety of methods. The problem of asymptotic convergence of system states in infinite time is solved by using a terminal sliding mode controller (TSMC) (Yu et al., 2020). With the development of calculus systems in recent years, fractional calculus has also been used in the area of control engineering. Various controllers have been developed one after the other, such as the fractional order PID controller, the fractional SMC, and the fractional adaptive controller. Studies have shown that fractional order controllers more flexible and more efficient than integer controllers (Sekban et al., 2016).

H. T. Sekban
Bayburt University, Bayburt, Turkey
e-mail: tugbasekban@bayburt.edu.tr

A. Başçi (✉)
Ataturk University, Erzurum, Turkey
e-mail: abasci@atauni.edu.tr

T. H. Karakoc et al. (eds.), *New Achievements in Unmanned Systems*, Sustainable Aviation, https://doi.org/10.1007/978-3-031-29933-9_24

In this study, FOTSMC, TSMC, and SMC controllers are used for trajectory tracking control of UGV under different linear and angular velocity references. Simulation results indicate that the FOTSMC controller performs better than TSMC and SMC.

2 Material and Method

UGVs are mobile robots that can move on land without the need for external guidance. Nonholonomic UGVs have one caster wheel at the front and two differential driving wheels at the rear, and their general representation is as in Fig. 1.

The rotation matrix ($R(\theta)$) is used to determine the vehicle position between the global axis (q^I) and the local axis (q^h).

$$q^I = \begin{bmatrix} x^I \\ y^I \\ \theta^I \end{bmatrix}, q^h = \begin{bmatrix} x^h \\ y^h \\ \theta^h \end{bmatrix}, R(\theta) = \begin{bmatrix} \cos\theta & -\sin\theta & 0 \\ \sin\theta & \cos\theta & 0 \\ 0 & 0 & 1 \end{bmatrix} \tag{1}$$

2.1 *Kinematic Modeling of the UGV*

The kinematic model of the UGV with the assumption that there is no lateral slip and pure rolling constraint is as follows (Azzabi & Nouri, 2021).

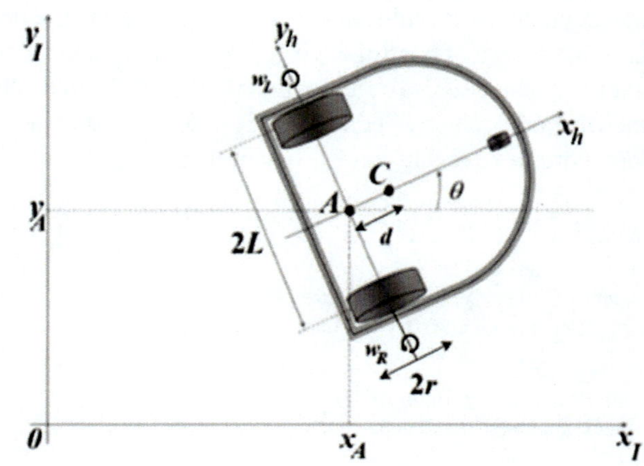

Fig. 1 Schematics of UGV

$$\dot{q}^I = \begin{bmatrix} \dot{x}_A^I \\ \dot{y}_A^I \\ \dot{\theta}_A^I \end{bmatrix} = \begin{bmatrix} \cos\theta & 0 \\ \sin\theta & 0 \\ 0 & 1 \end{bmatrix} \begin{bmatrix} v \\ \omega \end{bmatrix} = S(q)\eta \tag{2}$$

The change of the linear velocity and angular velocity of the vehicle is expressed mathematically in Eq. (3) and (4), respectively.

$$v = (v_R + v_L)/2 = r(\dot{\phi}_R + \dot{\phi}_L)/2 \tag{3}$$

$$\omega = (v_R - v_L)/2L = r(\dot{\phi}_R - \dot{\phi}_L)/2L \tag{4}$$

The kinematic model of point A according to the global coordinate axis is as follows.

$$\dot{q}^I = \begin{bmatrix} \dot{x}_A^I \\ \dot{y}_A^I \\ \dot{\theta}_A^I \end{bmatrix} = \begin{bmatrix} \frac{r}{2}\cos\theta & \frac{r}{2}\cos\theta \\ \frac{r}{2}\sin\theta & \frac{r}{2}\sin\theta \\ \frac{r}{2L} & -\frac{r}{2L} \end{bmatrix} \begin{bmatrix} \omega_R \\ \omega_L \end{bmatrix} \tag{5}$$

2.2 *Dynamic Modeling of the UGV*

The dynamic equation of the UGV is expressed as follows (Swadi et al., 2016):

$$M(q)\,\ddot{q} + V(q,\dot{q})\,\dot{q} + F(\dot{q}) + G(q) + \tau_d = B(q)\,\tau - A^T(q)\,\lambda \tag{6}$$

$M(q)$, $V(q,\dot{q})$, and $B(q)$ matrices are obtained by using the Lagrange dynamic approach assuming that the vehicle moves in a horizontal plane, and there is no disturbance, surface friction, and gravity (Azzabi & Nouri, 2021). It would be more accurate to express the dynamic equation of the vehicle in terms of linear and angular velocity. If Eq. (2) is replaced in Eq. (6) and arranged by multiplying both sides of the equation by $S(q)^T$, the dynamic equation of UGV is obtained as in Eq. (7) (Chen et al., 2009).

$$\widetilde{M}(q)\dot{\eta} + \widetilde{V}(q,\dot{q})\eta = \widetilde{B}(q)\tau \tag{7}$$

Using Eq. (7), the linearized dynamic model of the vehicle is obtained as follows.

$$\begin{bmatrix} \left(m+\frac{2I_w}{r^2}\right) & 0 \\ 0 & \left(I+\frac{2L^2}{r^2}I_w\right) \end{bmatrix}\begin{bmatrix} \dot{v} \\ \dot{w} \end{bmatrix}=\begin{bmatrix} \frac{1}{r} & 0 \\ 0 & \frac{L}{r} \end{bmatrix}\begin{bmatrix} u_1 \\ u_2 \end{bmatrix} \tag{8}$$

3 Controller Design

3.1 Kinematic Controller Design

The kinematic controller is used to estimate the linear and angular velocities that will stabilize the UGV asymptotically (Mevo et al., 2018). The kinematic controller based on the backstepping technique proposed by Kanayama is expressed as follows (Kanayama et al., 1990).

$$\begin{bmatrix} v_c \\ \omega_c \end{bmatrix}=\begin{bmatrix} v_r\cos\theta_e+k_x x_e \\ \omega_r+v_r\left(k_y y_e+k_\theta\sin\theta_e\right) \end{bmatrix} \tag{9}$$

where k_x, k_y, k_θ, are positive constants.

3.2 Dynamic Controller Design

Fractional calculus allows us to calculate the noninteger values of derivatives and integrals. ${}_aD_t^\mu$ is general fundamental operator. a and t are the limits of operations, μ is the fractional order and $\mu \in R$. The continuous integro-differential representation of ${}_aD_t^\mu$ is as follows.

$${}_aD_t^\mu=\begin{cases} \frac{d^\mu}{dt^\mu} & \mu \rangle\ 0 \\ 1 & \mu=0 \\ \int_a^t (dt)^{-\mu} & \mu \langle\ 0 \end{cases} \tag{10}$$

Since fractional expressions do not have exact solution methods, Laplace transforms are performed first in order to be able to solve them. Then, solutions are realized by using approaches such as Carlson, Matsuda, Tustin, Simpson, and Crone (Valerio & Costa, 2005). In this study, the Crone approach developed by Oustaloup and frequently used in the literature is discussed.

$$L\{{}_aD_t^\mu f(t)\} = s^\mu F(s) \tag{11}$$

$$s^\mu = j\prod_{n=1}^{N}\frac{1+\frac{s}{w_{zn}}}{1+\frac{s}{w_{pm}}} \tag{12}$$

where $\mu>0$, N is the degree of approximation, j is constant, w_{zn} and w_{pm} are the lower and upper cutoff frequencies, respectively. SMC is a nonlinear control method that can be easily designed and applied to nonlinear systems. The classical terminal sliding surface is as follows (Feng et al., 2013).

$$s = \dot{e}_c + \beta {e_c}^{p/q} \tag{13}$$

For the FOTSMC design, the sliding surface is first determined and a control signal is produced from its name. The fractional order terminal sliding surface is obtained as follows.

$$s = D^\mu e_c + \beta {e_c}^{p/q} \tag{14}$$

β is the positive sliding constant, p and q are both positive odd integers, e_c is the trajectory error, and $\dot{e}_c$ is the derivative of the trajectory error. In order to obtain the control signal (u), first the equivalent control signal (u_{eq}) and then the switching control signal (u_d) are generated. These two control signals are then summed.

u_{eq} is obtained by equating the derivative of the sliding surface to zero that is given below.

$$\dot{s} = D^\mu \dot{e}_c + \beta\frac{p}{q}{e_c}^{(p/q)-1}\dot{e}_c \tag{15}$$

The finite time t_s taken from $e_c \neq 0$ to $e_c(t_s) = 0$ is expressed as follows.

$$t_s = \frac{1}{\beta(p-q)}{e_c}^{1-(q/p)}(t_r) \tag{16}$$

Here t_r is the time when sliding surface reaches zero from an initial condition. In order to obtain $D^\mu(D^{-\mu}f(t))=f(t)$, if the derivative of Eq. (15) is taken from order ($-\mu$), the following equation is obtained (Podlubny, 1999).

$$0 = \dot{e}_c + D^{-\mu}\beta\frac{p}{q}{e_c}^{(p/q)-1}\dot{e}_c \tag{17}$$

Using Eq. (17), the equivalent control signals of the first and second controllers are obtained as follows.

$$\left.\begin{aligned} u_{eq1} &= r\left(m + \frac{2I_w}{r^2}\right)\left[\dot{v}_c + D^{-\mu}\beta\frac{p}{q}e_v{}^{(p/q)-1}\dot{e}_v\right] \\ u_{eq2} &= \frac{r}{L}\left(I + \frac{2L^2}{r^2}I_w\right)\left[\dot{w}_c + D^{-\mu}\beta\frac{p}{q}e_w{}^{(p/q)-1}\dot{e}_w\right] \end{aligned}\right\} \tag{18}$$

The switching control signal is as in Eq. (19).

$$u_d = k_i \operatorname{sgn}(s) \tag{19}$$

where k_i is a positive constant. The FOTSMC signals of the first and second controllers, respectively, are obtained as in Eq. (20).

$$\left.\begin{aligned} u_1 &= r\left(m + \frac{2I_w}{r^2}\right)\left[\dot{v}_c + D^{-\mu}\beta\frac{p}{q}e_v{}^{(p/q)-1}\dot{e}_v\right] + k_1 \operatorname{sgn}(s_1) \\ u_2 &= \frac{r}{L}\left(I + \frac{2L^2}{r^2}I_w\right)\left[\dot{w}_c + D^{-\mu}\beta\frac{p}{q}e_w{}^{(p/q)-1}\dot{e}_w\right] + k_2 \operatorname{sgn}(s_2) \end{aligned}\right\} \tag{20}$$

4 Simulation Results

In this section, the simulation results are presented. In order to demonstrate the performance of the proposed controller, trapezoidal and sinusoidal signals are used as reference for linear velocity and angular velocity, respectively. The simulation results are given in Figs. 2, 3, 4 and 5.

5 Conclusion

In this chapter, trajectory tracking control of an UGV was carried out using FOTSMC, TSMC, and SMC methods. The simulation results showed that FOTSMC have succeeded in reducing the errors occurring in the system by providing flexibility that TSMC and SMC could not show. In addition, the control signals produced by the proposed controller to follow the reference velocities is more smooth and contain lower amplitude chattering compared to other controllers. As a result, trajectory tracking performance of FOTSMC method was found to be better than TSMC and SMC.

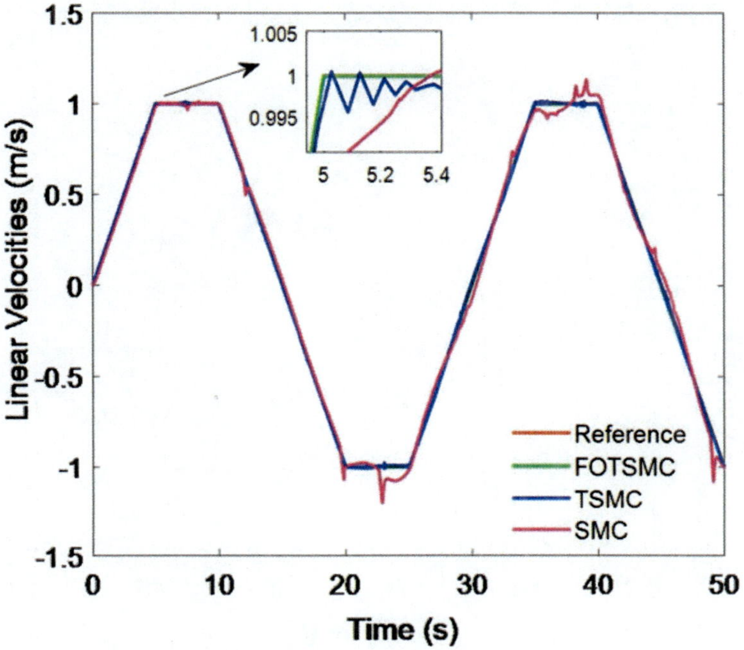

Fig. 2 The linear velocity tracking performances of both controllers for trapezoidal reference

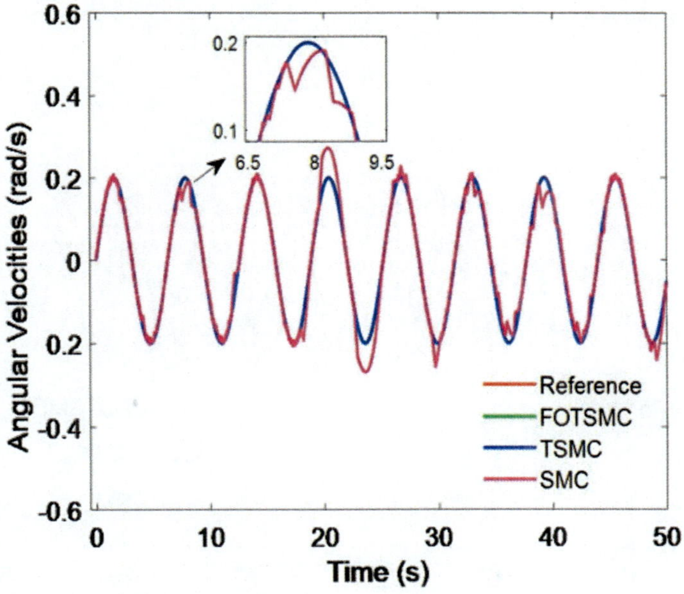

Fig. 3 The angular velocity tracking performances of both controllers for sinusoidal reference

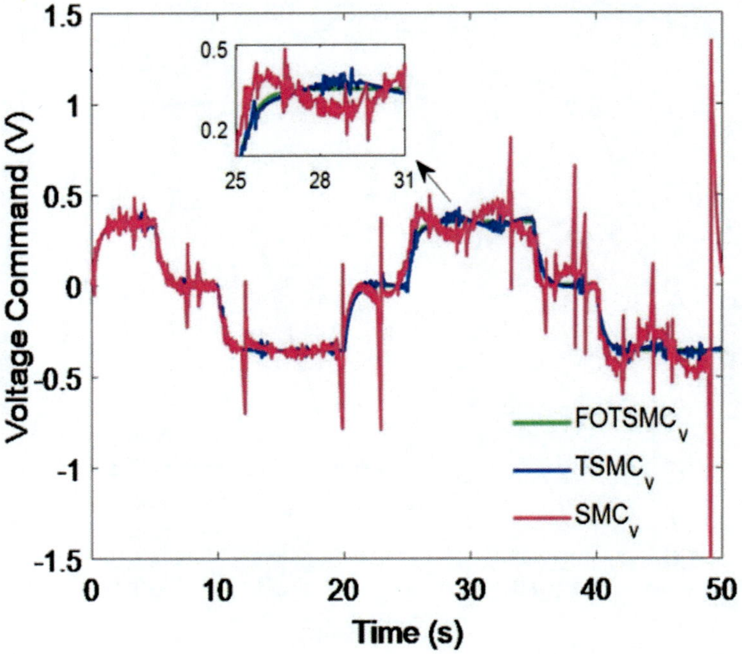

Fig. 4 The control signals generated by controllers for linear velocity tracking

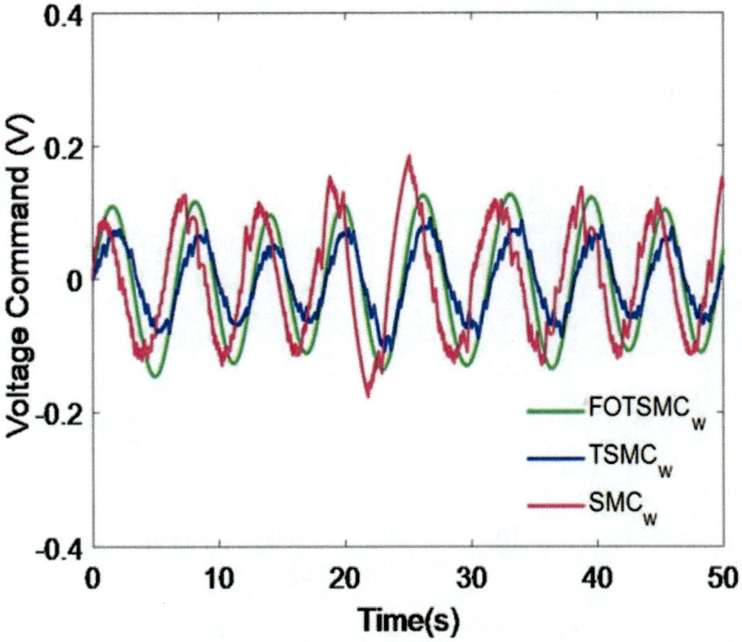

Fig. 5 The control signals generated by controllers for angular velocity tracking

References

Azzabi, A., & Nouri, K. (2021). *Design of a robust tracking controller for a nonholonomic mobile robot based on sliding mode with adaptive gain*. International Journal of Advanced Robotic Systems (Vol. 18, p. 172988142098708).

Başçi, A., Derdiyok, A., & Mercan, E. (2015). Otomatik Yönlendirmeli Bir Aracın Gövde Hızı Ve Gövde Açısının Bulanık Mantık İle Gerçek Zamanlı Kontrolü. *Pamukkale Univ Muh Bilim Derg, 21*.

Chen, C. Y., Li, T. H. S., Yeh, Y.-C., & Chang, C. C. (2009). Design and implementation of an adaptive sliding mode dynamic controller for wheeled mobile robots. *Mechatronics, 19*.

Feng, Y., Yub, X., & Hanc, F. (2013). On nonsingular terminal sliding-mode control of nonlinear systems. *Elsevier Automatica, 49*.

Kanayama, Y., Kimura, Y., Miyazaki, F. & Noguchi, T. (1990). A stable tracking control method for an autonomous mobile robot. In *1990 IEEE international conference*.

Koubaa, Y., Boukattaya, M., & Dammak, T. (2015). Adaptive sliding-mode dynamic control for path tracking of nonholonomic wheeled mobile robot. *Journal of Automation & Systems Engineering, 9*.

Mevo, B. B., Saad, M. R., & Fareh, R. (2018). Adaptive sliding mode control of wheeled Mobile robot with nonlinear model and uncertainties. In *2018 IEEE Canadian conference on electrical & computer engineering*.

Orman, K., Can, K., Başçi, A., & Derdiyok, A. (2018). An adaptive-fuzzy fractional-order sliding mode controller design for an unmanned vehicle. *Elektronika Ir Elektrotechnika, 24*.

Podlubny, I. (1999). Fractional order systems and PID controllers. *IEEE Transactions on Automatic Control, 44*, 208.

Sekban, H. T., Can, K., Orman, K. & Başçi, A. (2016). *Dörtlü Tank Sıvı-Seviye Sisteminin Kesir Dereceli PI Kontrolcü ile Kontrolü*. Otomatik Kontrol Ulusal Toplantısı.

Sekban, H. T., Can, K., & Başçi, A. (2021). Integral terminal sliding mode controller for trajectory tracking control of unmanned ground vehicle. In *ISASE 2021 International Symposium on Applied Sciences and Engineering*.

Sharma, A., & Panwar, V. (2016). Control of mobile robot for trajectory tracking by sliding mode control technique. In *International conference on electrical, electronics and optimization techniques*.

Swadi, S. M., Tawfik, M. A., Abdulwahab, E. N., & Kadhim, H. A. (2016). Fuzzy-Backstepping controller based on optimization method for trajectory tracking of wheeled Mobile robot. In *2016 UKSimAMSS 18th international conference on computer modelling and simulation*.

Tzafestas, G. S. (2018). Mobile robot control and navigation: A global overview. *Journal of Intelligent and Robotic Systems, 91*, 35.

Valerio, D., & Costa, J. S. (2005). Time-domain implementation of fractional order controllers. In *Control theory and applications, IEEE proceedings* (p. 152).

Yu, X., Feng, Y., & Man, Z. (2020). Terminal sliding mode control – An overview. *IEEE Open Journal of the Industrial Electronics Society, 2*, 36.

Nonlinear Control of Multi-quadrotor Flight Formations

Diogo Santos Ferreira, Afzal Suleman, and Paulo Oliveira

1 Introduction

Unmanned air vehicles (UAVs), which were originally developed for military purposes, have now been devoted to a myriad other uses, ranging from aerial photography, goods delivery, agriculture, and mapping and surveillance to pollution monitoring or infrastructure inspections. When appropriately synchronized, a swarm of UAVs can perform much more complex tasks with gains in efficiency and robustness. As an example, (Bacelar et al., 2019) describes how it is possible to deploy two UAVs to cooperatively carry heavy loads. (Rosalie et al., 2017) presents a strategy for area exploration and mapping carried out by a swarm of autonomous UAVs. For policing and surveillance missions in areas where the communication range is limited, (Scherer & Rinner, 2020) discusses how efficient a network of UAVs can be in covering the area. For agriculture applications, Ju and Son (2018) delve deeply into the advantages of using multiple UAVs with distributed control for better performance.

The most relevant formation control concepts are the leader-follower, the virtual leader and the behavior-based. In the leader-follower case, a formation is achieved when each follower drives into the desired position with respect to the leader, which has some known trajectory. In the second case, the virtual leader describes a reference trajectory and the formation is achieved when all the vehicles in the swarm follow the leader in a rigid structure. The behavior-based formation control

D. S. Ferreira (✉) · P. Oliveira
Department of Mechanical Engineering, IDMEC-Instituto Superior Técnico, Lisbon, Portugal
e-mail: diogo.s.ferreira@tecnico.ulisboa.pt; paulo.oliveira@tecnico.ulisboa.pt

A. Suleman
Department of Mechanical Engineering, University of Victoria, Victoria, BC, Canada
e-mail: suleman@uvic.ca

T. H. Karakoc et al. (eds.), *New Achievements in Unmanned Systems*, Sustainable Aviation, https://doi.org/10.1007/978-3-031-29933-9_25

approach defines different control behaviors for different situations of interest, as explained by Balch and Arkin (1998).

Contrary to many strategies found in the literature, which define the displacement in the inertial frame, the approach followed in this chapter for the 2D motion needs the displacement to be computed in the follower's body frame and measured by decentralized on-board sensors. For the 3D case, a centralized approach has been chosen for computational efficiency.

2 Quadrotor Model

Let {I} be an orthonormal reference frame according to the North-East-Down (NED) coordinate system. Let {B} be another orthonormal reference frame centered at point p. The orientation of {B} with respect to {I} is given by the Euler angles $\lambda = (\phi, \theta, \psi)$ that represent the rotation about their respective axes. The rotation matrix from {B} to {I} is given by an orthogonal matrix $R \in SO(3)$. From the definition of R, its derivative is $\dot{R} = RS(\omega)$, with ω the angular velocity of {B} expressed in {B}. Let p be the quadrotor's position, v its velocity, m its mass and J its inertia tensor. Let also f be the sum of external forces applied on the quadrotor expressed in {I} and n the sum of external moments expressed in {B}. The complete dynamics of the quadrotor is given by

$$\begin{cases} \dot{p} = v \\ m\dot{v} = f \\ J\dot{\omega} = -S(\omega)J\omega + n \\ \dot{R} = RS(\omega) \end{cases} \tag{1}$$

where $S(\omega)$ is a skew-symmetric matrix such that $S(x)y = x \times y$ for any $x, y \in \mathbb{R}^3$.

3 2D Formation Control

A trajectory tracking controller is implemented to make the follower track the leader while keeping a constant offset in its reference frame. Assuming the movement at constant height, Eq. (1) can be simplified for a purely kinematic model given by

$$\begin{cases} \dot{x} = u\cos\psi - v\sin\psi \\ \dot{y} = u\sin\psi + v\cos\psi \\ \dot{\psi} = r \end{cases} \tag{2}$$

If $\Delta = (\Delta_x, \Delta_y)^T$ is the desired displacement, R the 2D rotation matrix, $p, c \in \mathbb{R}^2$ the follower's and leader's positions, respectively, the position error can be written as

$$z_1 = R^T(c - p) - \Delta \tag{3}$$

and its derivative as

$$\dot{z}_1 = R^T\dot{c} - v - S(r)(z_1 + \Delta) \tag{4}$$

Consider a Lyapunov function $V_1 = 1/2\|z_1\|^2$. Its derivative is

$$\dot{V}_1 = z_1^T\left(R^T\dot{c} - v - S(r)\Delta\right) \tag{5}$$

Adding and subtracting a term $k_1\|z_1\|^2$ yields

$$\dot{V}_1 = k_1\|z_1\|^2 + z_1^T z_2 \tag{6}$$

where a new error $z_2 = k_1 z_1 + R^T\dot{c} - v - S(r)\Delta$ was introduced. Let now $V_2 = V_1 + 1/2\|z_2\|^2$ be an augmented Lyapunov function. Its derivative is

$$\dot{V}_2 = \dot{V}_1 + z_2^T\left[k_1\dot{z}_1 + R^T\ddot{c} - S(r)R^T\dot{c} - \begin{pmatrix} 1 & -\Delta_y \\ 0 & \Delta_x \end{pmatrix}\begin{pmatrix} \dot{v} \\ \dot{r} \end{pmatrix}\right] \tag{7}$$

If the accelerations $\dot{v}$, $\dot{r}$ are considered inputs of the system, the control law should be

$$\begin{pmatrix} \dot{v} \\ \dot{r} \end{pmatrix} = \begin{pmatrix} 1 & -\Delta_y \\ 0 & \Delta_x \end{pmatrix}^{-1}\left(k_1\dot{z}_1 + R^T\ddot{c} - S(r)R^T\dot{c} + z_1 + k_2 z_2\right) \tag{8}$$

which is well-defined for $\Delta_x \neq 0$. Under this control law, the error system can be written in strict feedback form (Khalil, 2014) as

$$\begin{cases} \dot{z}_1 = -(S(r) + k_1 I_2)z_1 + z_2 \\ \dot{z}_2 = -z_1 - k_2 z_2 \end{cases} \tag{9}$$

and the derivative of V_2 becomes

$$\dot{V}_2 = -k_1\|z_1\|^2 - k_2\|z_2\|^2 \tag{10}$$

which is negative for $(z_1, z_2) \neq 0$ if $k_1, k_2 > 0$. Thus, according to the Barbashin–Krasovskii theorem (Khalil, 2014), the error system is globally asymptotically stable around the origin.

Consider the existence of an unknown external disturbance such that

$$\dot{z}_2 = k_1 \dot{z}_1 + R^T \ddot{c} - S(r) R^T \dot{c} - \dot{v} - S(\dot{r})\Delta + Rd \tag{11}$$

Additionally, assume the controller has an estimator $\widehat{d}$ such that

$$\begin{pmatrix} \dot{v} \\ \dot{r} \end{pmatrix} = \begin{pmatrix} 1 & -\Delta_y \\ 0 & \Delta_x \end{pmatrix}^{-1} \left(k_1 \dot{z}_1 + R^T \ddot{c} - S(r) R^T \dot{c} + z_1 + k_2 z_2 + R\widehat{d} \right) \tag{12}$$

Considering the estimation error $\tilde{d} = d - \widehat{d}$ as an input to the error system with state $z = (z_1, z_2)$, one can design an adaptive controller for $\widehat{d}$. Now, consider the Lyapunov function $V_3 = V_2 + 1/(2k_d)\|\tilde{d}\|^2$. Following the same Lyapunov analysis, if the disturbance is constant, $\dot{\widehat{d}} = k_d(0_2 \;\; R)z$ stabilizes the error system.

3.1 Closed-Loop System

After deriving a control law, it is of interest to study the stability of the closed-loop system, i.e., the formation of one leader and one follower. When the position error is identically zero, $z_1 = \dot{z}_1 = 0$ and Eq. (4) becomes

$$R^T \dot{c} - v - S(r)\Delta = 0 \tag{13}$$

Assuming a general leader's trajectory $\dot{c} = C(\cos\psi_c, \sin\psi_c)^T$, the closed-loop equation can be expanded to isolate the control variables as

$$\begin{pmatrix} v \\ r \end{pmatrix} = \begin{pmatrix} C\cos(\psi - \psi_c) - C\Delta_y/\Delta_x \sin(\psi - \psi_c) \\ -C/\Delta_x \sin(\psi - \psi_c) \end{pmatrix} \tag{14}$$

These equations describe a nonlinear periodic system with a dynamics for ψ and output v. It is asymptotically stable around the points

$$\psi^* = \psi_c + 2k\pi, \quad \forall k \in \mathbb{Z} \tag{15}$$

within the region of convergence

$$\psi \in]\psi_c + (2k-1)\pi, \psi_c + (2k+1)\pi[, \quad \forall k \in \mathbb{Z} \tag{16}$$

In conclusion, the follower can have a heading difference relative to the leader of up to 180°. The bigger the difference, the slower is the convergence to the desired heading. In the limit, if a follower is set to track a leader describing a linear path, starting in opposite heading, it will not converge.

4 3D Formation Control

The motion of the quadrotor at constant height has been studied and a controller for the simplified model has been derived using the backstepping method applied to the position error. This method is now used to derive a similar nonlinear controller for the complete model. The dynamics from Eq. (1) can be written in a state-space form by introducing the state vector $X = \left(\phi, \dot{\phi}, \theta, \dot{\theta}, \psi, \dot{\psi}, z, \dot{z}, x, \dot{x}, y, \dot{y}\right)$ and the input vector $U = (T, n_x, n_y, n_z)$, according to (Bouabdallah & Siegwart, 2005). Defining the constants $a_\phi = (J_y - J_z)/J_x$, $a_\theta = (J_z - J_x)/J_y$, $a_\psi = (J_x - J_y)/J_z$, $b_\phi = 1/J_x$, $b_\theta = 1/J_y$ and $b_\psi = 1/J_z$, the dynamics becomes

$$f(X,U) = \begin{cases} \dot{\phi} \\ a_\phi \dot{\theta}\dot{\psi} + b_\phi n_x \\ \dot{\theta} \\ a_\theta \dot{\phi}\dot{\psi} + b_\theta n_y \\ \dot{\psi} \\ a_\psi \dot{\phi}\dot{\theta} + b_\psi n_z \\ \dot{z} \\ g - T/m \cos\phi\cos\theta \\ \dot{x} \\ T/m\, u_x \\ \dot{y} \\ T/m\, u_y \end{cases} \tag{17}$$

with $u_x = \cos\phi\sin\theta\cos\psi + \sin\phi\sin\psi$ and $u_y = \cos\phi\sin\theta\sin\psi - \sin\phi\cos\psi$. The system as it is posed highlights an important relation between the position and attitude of the quadrotor: the position components depend on the angles; however, the opposite is not true. The overall system can be thought of as the result of two semi-decoupled subsystems: the translation and the rotation, for which two controllers are designed separately.

4.1 Attitude Control

Let $z_\phi = \phi_{ref} - \phi$ be the roll angle error. Let $V_\phi = 1/2 z_\phi^2$ be a Lyapunov function with derivative

$$\dot{V}_\phi = z_\phi\left(\dot{\phi}_{ref} - \dot{\phi}\right) \tag{18}$$

If $\dot{\phi}$ is controlled to be

$$\dot{\phi} = \dot{\phi}_{\text{ref}} + k_\phi z_\phi \tag{19}$$

then $\dot{V}_\phi = -k_\phi z_\phi^2$. Let now $z_{\dot{\phi}}$ be the roll rate error given by

$$z_{\dot{\phi}} = \dot{\phi} - \dot{\phi}_{\text{ref}} - k_\phi z_\phi \tag{20}$$

and the augmented Lyapunov function

$$V_{\dot{\phi}} = V_\phi + 1/2 z_{\dot{\phi}}^2 \tag{21}$$

with its derivative given by

$$\dot{V}_{\dot{\phi}} = -k_\phi z_\phi^2 + z_{\dot{\phi}}\left(a_\phi \dot{\theta}\dot{\psi} + b_\phi n_x - \ddot{\phi}_{ref} - k_\phi \dot{z}_\phi\right) \tag{22}$$

If the control law for n_x is chosen to be

$$n_x = 1/b_\phi\left(\ddot{\phi}_{ref} + k_\phi \dot{z}_\phi - a_\phi \dot{\theta}\dot{\psi} - k_{\dot{\phi}} z_{\dot{\phi}}\right) \tag{23}$$

then

$$\dot{V}_{\dot{\phi}} = -k_\phi z_\phi^2 - k_{\dot{\phi}} z_{\dot{\phi}}^2 \tag{24}$$

which is negative for $\left(z_\phi, z_{\dot{\phi}}\right) \neq 0$ if $k_\phi, k_{\dot{\phi}} > 0$. Thus, according to the Barbashin–Krasovskii theorem (Khalil, 2014), the roll error system is globally asymptotically stable around the origin. Following the same backstepping procedure for the remaining angular variables, an attitude controller is derived as

$$\begin{cases} n_x = 1/b_\phi\left(\ddot{\phi}_{ref} + k_\phi \dot{z}_\phi - a_\phi \dot{\theta}\dot{\psi} - k_{\dot{\phi}} z_{\dot{\phi}}\right) \\ n_y = 1/b_\theta\left(\ddot{\theta}_{ref} + k_\theta \dot{z}_\theta - a_\theta \dot{\phi}\dot{\psi} - k_{\dot{\theta}} z_{\dot{\theta}}\right) \\ n_z = 1/b_\psi\left(\ddot{\psi}_{ref} + k_\psi \dot{z}_\psi - a_\psi \dot{\phi}\dot{\theta} - k_{\dot{\psi}} z_{\dot{\psi}}\right) \end{cases} \tag{25}$$

with gains $k_\phi, k_{\dot{\phi}}, k_\theta, k_{\dot{\theta}}, k_\psi, k_{\dot{\psi}} > 0$.

4.2 Position Control

Let $z_z = z_{\text{ref}} - z$ be the altitude error. Let $V_\phi = 1/2 z_z^2$ be a Lyapunov function with derivative

$$\dot{V}_z = z_z(\dot{z}_{\text{ref}} - \dot{z}) \tag{26}$$

If $\dot{z}$ is controlled to be

$$\dot{z} = \dot{z}_{ref} + k_z z_z \tag{27}$$

then $\dot{V}_z = -k_z z_z^2$. Let now $z_{\dot{z}}$ be the vertical speed error given by

$$z_{\dot{z}} = \dot{z} - \dot{z}_{\text{ref}} - k_z z_z \tag{28}$$

and the augmented Lyapunov function

$$V_{\dot{z}} = V_z + 1/2 z_{\dot{z}}^2 \tag{29}$$

with its derivative given by

$$\dot{V}_{\dot{z}} = -k_z z_z^2 + z_{\dot{z}}(g - T/m \cos\phi\cos\theta - \ddot{z}_{\text{ref}} - k_z \dot{z}_z) \tag{30}$$

If the control law for T is chosen to be

$$T = m/(\cos\phi\cos\theta)(g - \ddot{z}_{ref} - k_z \dot{z}_z - k_{\dot{z}} z_{\dot{z}}) \tag{31}$$

then

$$\dot{V}_{\dot{z}} = -k_z z_z^2 - k_{\dot{z}} z_{\dot{z}}^2 \tag{32}$$

which is negative for $(z_z, z_{\dot{z}}) \neq 0$ if $k_z, k_{\dot{z}} > 0$. Thus, according to the Barbashin–Krasovskii theorem (Khalil, 2014), the altitude error system is globally asymptotically stable around the origin. Following the same backstepping procedure for the remaining position variables, a position controller is derived as

$$\begin{cases} T = m/(\cos\phi\cos\theta)(g - \ddot{z}_{ref} - k_z \dot{z}_z - k_{\dot{z}} z_{\dot{z}}) \\ u_x = m/T(\ddot{x}_{\text{ref}} + k_x \dot{z}_x - k_{\dot{x}} z_{\dot{x}} u_y = m/T(\ddot{y}_{\text{ref}} + k_y \dot{z}_y - k_{\dot{y}} z_{\dot{y}} \end{cases} \tag{33}$$

with gains $k_x, k_{\dot{x}}, k_y, k_{\dot{y}}, k_z, k_{\dot{z}} > 0$.

To feed the attitude controller with the reference values of roll and pitch, it is assumed the quadrotor does not perform complex maneuvers, thereby keeping ϕ and θ small enough. From the definitions of u_x and u_y, it can be written as

$$\begin{pmatrix} \theta_{\text{ref}} \\ \phi_{\text{ref}} \end{pmatrix} = (\cos\psi \sin\psi \sin\psi - \cos\psi) \begin{pmatrix} u_x \\ u_y \end{pmatrix} \tag{34}$$

The controller scheme is shown in Fig. 1.

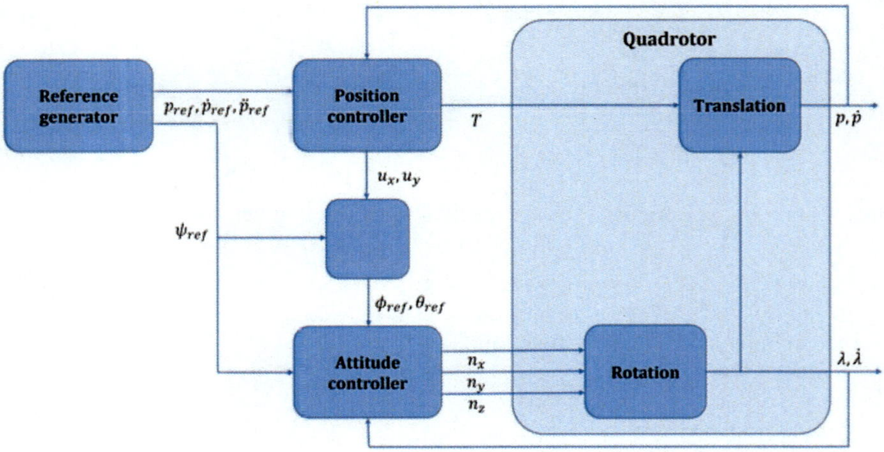

Fig. 1 3D controller scheme

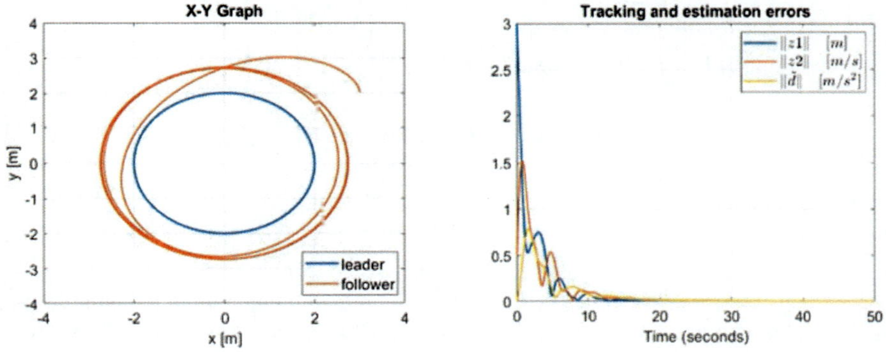

Fig. 2 2D simulation

5 Simulation Results

Several simulations have been carried out for the developed architectures. This section presents an illustrative simulation for both the 2D and 3D controllers, where the physical parameters are shown in SI units.

The simulation consists of a follower tracking a leader in a circular path. The displacement is $\Delta = (1, 1)$ and the gains are $k_1 = k_2 = k_d = 0.5$. The disturbance intensity is $d = (1, 1)$. Figure 2 shows the simulation position and the follower converges to a trajectory where it sees the leader at position $(1, 1)$ in its reference frame.

For the 3D case, the complete model simulation consists of a follower tracking a leader in a circular path at a certain height. The displacement is $\Delta = (1, 1, 0)$. The heavy quadrotor model developed by Pounds et al. (2010) was used.

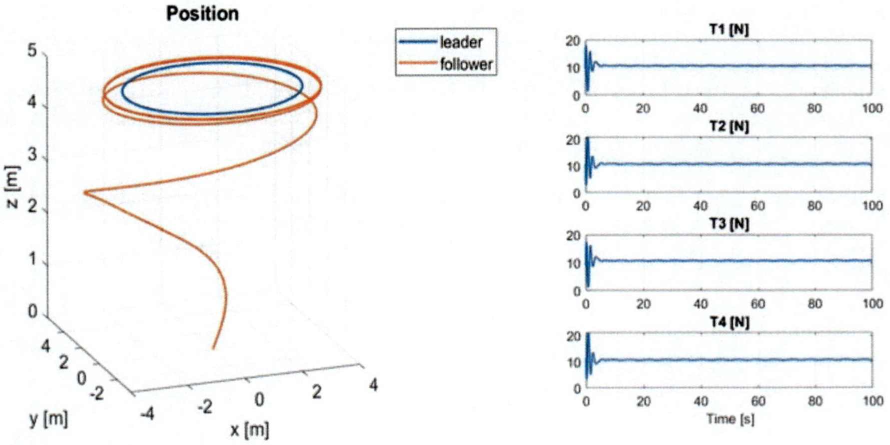

Fig. 3 3D simulation

The attitude gains are equal to 1 and the position gains are $k_x = k_y = 2.2, k_{\dot{x}} = k_{\dot{y}} = 0.18, k_z = 0.5, k_{\dot{z}} = 0.2$. Figure 3 displays the position of the vehicles and the actuation, respectively, for a 100-s simulation.

6 Concluding Remarks

This chapter proposes an approach to the formation control of quadrotors by solving the trajectory tracking problem using nonlinear control. The controller is applied to a leader-follower formation based on the backstepping method. The 2D controller is kinematic but robust to constant acceleration disturbances and the formation stability is guaranteed. As far as the 3D motion is concerned, a controller has been developed for a kinematic and dynamic model and the control laws have been proved stable. Future work aims to include disturbances in the simulation work and prove the overall formation stability.

References

Bacelar, T., Cardeira, C., & Oliveira, P. (2019). Cooperative load transportation with quadrotors. In *IEEE international conference on autonomous robot systems and competitions*.

Balch, T., & Arkin, R. (1998). Behavior-based formation control for multirobot teams. *IEEE Transactions on Robotics and Automation, 14*(6), 926–939.

Bouabdallah, S., & Siegwart, R. (2005). Backstepping and sliding-mode techniques applied to an indoor micro quadrotor. In *International conference on robotics and automation*. IEEE.

Ju, C., & Son, H. (2018). Multiple UAV systems for agricultural applications: Control, implementation, and evaluation. *Electronics, 7*(9), 162.

Khalil, H. (2014). *Nonlinear systems*. Pearson.

Pounds, P., Mahony, R., & Corke, P. (2010). Modelling and control of a large quadrotor robot. *Control Engineering Practice, 18*(7), 691–699.

Rosalie, M., Dentler, J., Danoy, G., Bouvry, P., Kannan, S., Mendez, M., & Voos, H. (2017). Area exploration with a swarm of UAVs combining deterministic chaotic ant colony mobility with position MPC. In *International conference on unmanned aircraft systems*. https://doi.org/10.1109/icuas.2017.7991418

Scherer, J., & Rinner, B. (2020). Multi-UAV surveillance with minimum information idleness and latency constraints. *IEEE Robotics and Automation Letters, 5*, 4812–4819.

Dynamic Modeling of Main Landing Gear of a High-Altitude Long Endurance UAV

Ali Dinc and Yousef Gharbia

Nomenclature

b	Shock absorber damping coefficient, Ns/m
EASA	European Union Aviation Safety Agency
F_1	Net force on m_1, N
F_2	Ground reaction force, N
g	Gravity, m/s^2
k_1	Shock absorber spring constant, N/m
k_2	Tire spring constant, N/m
L	Lift force acting on aircraft, N
m_1	Sprung mass (aircraft mass per LG), kg
m_2	Unsprung mass (mass of wheel & tire assy), kg
LG	Landing gear
V	Aircraft vertical speed (descent velocity), m/s
W_1	Aircraft weight minus lift force, N
x	Displacement of shock absorber, m
x_1	Displacement of aircraft, m
x_2	Displacement of tire, m

A. Dinc (✉) · Y. Gharbia
College of Engineering and Technology, American University of the Middle East, Egaila, Kuwait
e-mail: Ali.Dinc@aum.edu.kw; Yousef.Gharbia@aum.edu.kw

T. H. Karakoc et al. (eds.), *New Achievements in Unmanned Systems*, Sustainable Aviation, https://doi.org/10.1007/978-3-031-29933-9_26

1 Introduction

Similar to manned aircraft, the landing gear arrangement is a crucial part of unmanned aerial vehicle (UAV) design. The landing gear system acts as a suspension system that absorbs and dissipates kinetic energy during taxi, take-off, and especially, during landing of the air vehicle. Landing gears contain shock absorber component which is frequently used in a variety of industries. The primary role of a shock absorber is to absorb the impact energy of moving objects, and they are essentially made up of two main components: a spring and a damper. Shock absorbers assist an airplane or UAV land more gently by reducing the impact forces or travel smoothly over rugged terrain/ runway during taxi and takeoff. In today's modern landing gears, oleo-pneumatic shock struts are very common, especially above certain weight of aircraft.

Wahi (1976) conducted a complex analysis and real-time simulation of oleo-pneumatic shock struts. For aircraft design and research, Furnish and Anders (1971) performed an analytical simulation of landing gear dynamics. McBrearty (1948) studied various cases of landing-gear structural failure. Yadav and Ramamoorthy (1991) investigated nonlinear landing gear behavior during landing impact. Recent advances in the computational simulation of landing gear systems were presented by Krüger and Morandini (2011). Dinc and Gharbia (2020) investigated the impact of spring and damper elements by simulations on dynamics of aircraft landing gear. Dinc (2021) performed an optimization on the metering pin diameter of an aircraft landing gear shock absorber. Morrison et al. (1997) performed an aircraft landing gear simulation and analysis. Ross and Edson (1983) performed an application of active control technology to the A-10 airplane landing gear.

The originality of this chapter comes from the fact that the study focuses on calculations and performance estimation of a landing gear for a high- altitude long endurance UAV similar to "Global Hawk," manufactured by Northrop-Grumman company, as shown in Fig. 1 (Northrop-Grumman, n.d.), which has been a focus of several studies (Dinc et al., 2020; Dinc & Moayyedian, 2020; Drezner & Leonard, 2002; Tsach et al., 1996).

Fig. 1 Global Hawk UAV with landing gear

2 Method

In this analysis, a landing gear was treated as a mass-spring-damper system, as shown in Fig. 2, where m_1 represents the total mass of aircraft per landing gear, and m_2 represents the whole mass of the wheel and tire group. The k_1 shock absorber spring and the k_2 tire spring elements are the two spring elements in this model. The model's final component is the shock absorber damping "b." The purpose of this model is to calculate the F_1 force acting on the aircraft mass as well as the displacement of the aircraft and shock absorber. The following are the related equations of motion:

$$F_1 = m_1\ddot{x}_1 = W_1 - F_{spring,s.absorb} - F_{damping} \tag{1}$$

$$F_1 = m_1\ddot{x}_1 = W_1 - kx - b\dot{x} \tag{2}$$

$$x = x_1 - x_2 \tag{3}$$

$$m_1\ddot{x}_1 + k_1\,(x_1 - x_2) + \mathrm{b}(\dot{x}_1 - \dot{x}_2) = W_1 \tag{4}$$

$$W_1 = m_1 g - L \tag{5}$$

$$m_1\ddot{x}_1 + k_1\,(x_1 - x_2) + \mathrm{b}(\dot{x}_1 - \dot{x}_2) = m_1 g - L \tag{6}$$

Additionally, equations of motion for wheel & tire group mass (m_2) are written as follows:

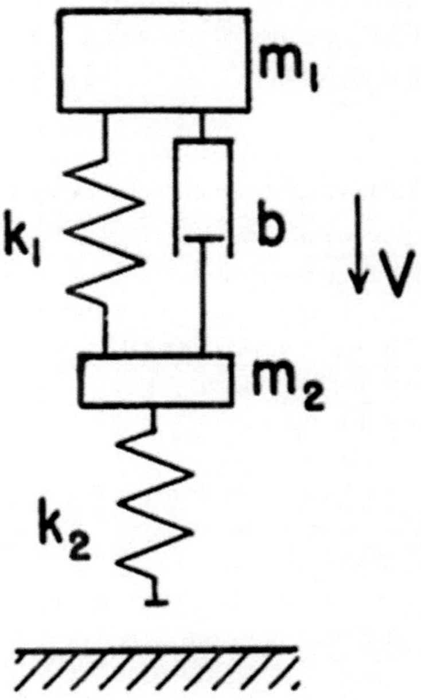

Fig. 2 Mass-spring-damper model for landing gear

$$F_2 = m_2\ddot{x}_2 = F_{spring,s.absorb} + F_{damping} - F_{spring,tire} \tag{7}$$

$$m_2\ddot{x}_2 = k_1 x + b\dot{x} - k_2\, x_2 \tag{8}$$

$$x = x_1 - x_2 \tag{9}$$

$$m_1\ddot{x}_1 + k_1\,(x_1 - x_2) + \mathrm{b}\,(\dot{x}_1 - \dot{x}_2) = W_1 \tag{10}$$

$$m_2\ddot{x}_2 - k_1\,(x_1 - x_2) - \mathrm{b}\,(\dot{x}_1 - \dot{x}_2) + k_2\,x_2 = 0 \tag{11}$$

3 Results and Discussion

In Matlab/Simulink, a computer model was built to solve the equation of motions simultaneously. The model iteratively calculates hydraulic, pneumatic, and tire forces from a set of inputs to determine displacements, velocities, and accelerations of the m_1 sprung (airplane) and m_2 unsprung (tire & wheel assembly) masses. Table 1 contains a list of input parameters gathered from literature.

A high-altitude long endurance UAV around the weight of "Global Hawk UAV" (which has maximum take-off weight of 12,133 kg) can be considered as "large aeroplane" as defined in CS-25 for civil certification requirements by EASA in European Union civil aviation system. However, if weight of airplane is below 5670 kg (12,500 lb), then it is certified under the CS-23 (EASA, 2012). Although civil certification is not compulsory for a UAV, from similarity, test cases can still be used for demonstration and validation purposes. Therefore, landing gear sections of CS-25 can be considered for simulation cases such as CS 25.723 (b) which mentions a case with 3.7 m/s (12 fps) descent velocity, also CS 25.473 (a) (2) which requires

Table 1 Input parameters for landing gear model

Design parameter	Assumed value
UAV gross weight (kg)	12,133
UAV mass per main landing gear m_1 (kg)	6066.5
Tire spring constant (kN/m)	2350
Discharge coefficient	0.65
Oil density (kg/m^3)	870
Orifice area (mm^2)	135
Shock absorber hydraulic area (m^2)	0.00645
Tire & wheel mass m_2 (kg)	80
Lift to weight ratio	1
Initial gas pressure in shock absorber (MPa)	2.65
Shock strut stroke (m)	0.33

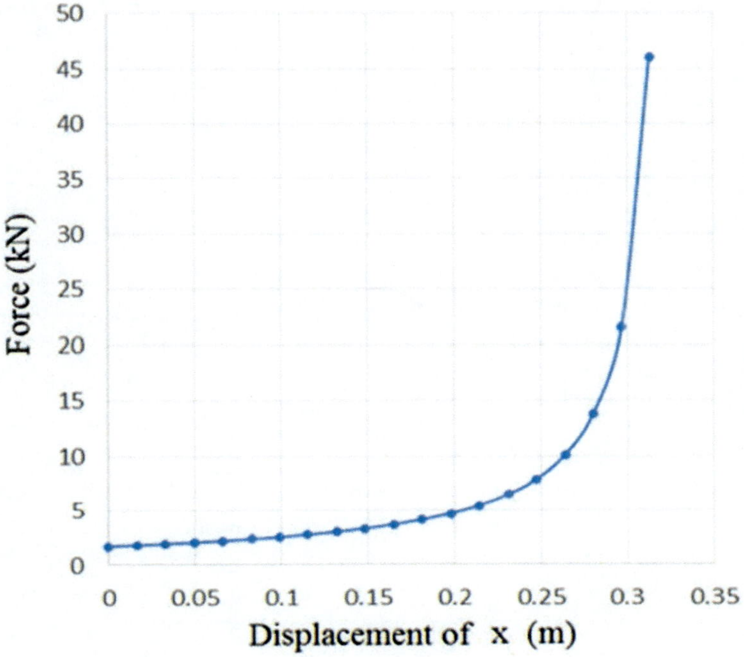

Fig. 3 Nitrogen gas behavior in shock absorber

3.05 m/sec (10 fps) descent velocity test and additionally CS 25.473 (a) (3) which cites 1.83 m/sec (6 fps) at the design take-off weight. During those test cases and related design calculations aeroplane lift, not exceeding aeroplane weight, (lift = weight) may be assumed per CS 25.473 (b). Therefore, simulation cases in this study were run at 1.83, 3.05, 3.7 m/s (6,10,12 fps) descent velocities and lift equals to weight condition.

Constraint was assumed to be 0.33 m for x (displacement of shock absorber) as a maximum value not to exceed the design value given in Table 1. For the vertical acceleration (g-force), 2 g and 3 g were assumed to be maximum acceptable values for 3.05, 3.7 m/s (10, 12 fps) descent velocity cases, respectively. These constraints should normally be set by the aircraft design company to withstand the loads.

Shock absorber contains nitrogen gas with an assumed initial pressure 2.65 MPa as indicated in Table 1. The behavior of nitrogen gas in shock absorber was calculated as a polytrophic process and depicted in Fig. 3.

Results obtained from the analysis include displacements, velocities, accelerations, and forces acting on m_1 sprung (aircraft mass per LG) and m_2 unsprung (mass of wheel & tire assembly) masses.

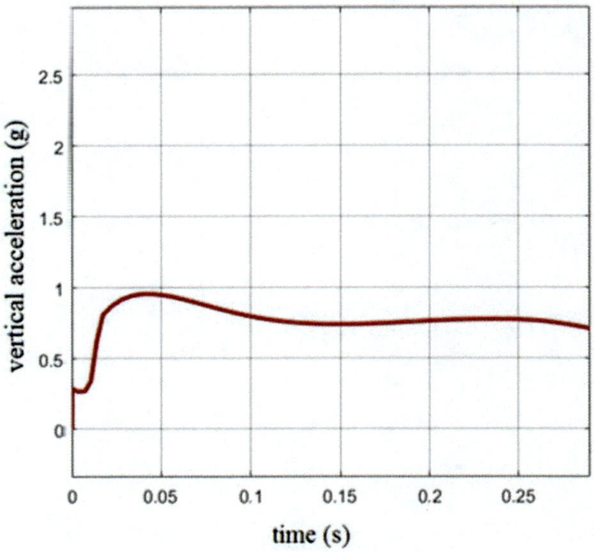

Fig. 4 Vertical acceleration (g) vs time at 1.83 m/s

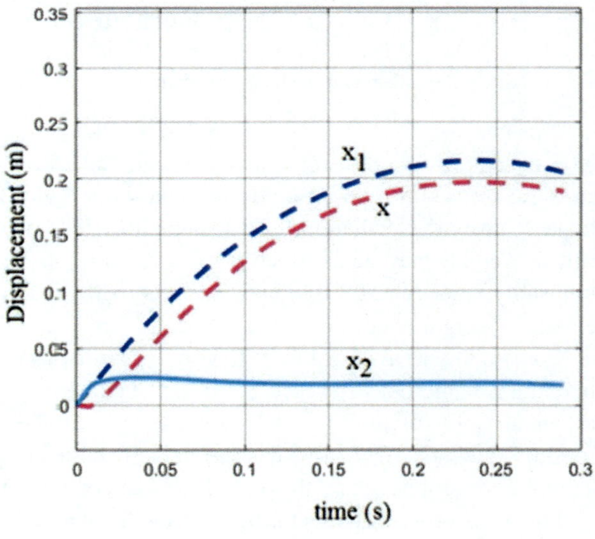

Fig. 5 Displacements vs time at 1.83 m/s

Figures 4 and 5 show the results for 1.83 m/s (6 fps) descent velocity. Maximum vertical acceleration is 0.957 g which is the lowest value in all cases and given in Table 2 together with displacement values.

Figures 6 and 7 show the results for 3.05 m/s (10 fps) descent velocity. Maximum vertical acceleration is 1.857 g, which is lower than the assumed constraint value of

Table 2 Summary of results for landing gear model

Parameter	At descent velocity (m/s)	Calculated value
Maximum x (m)	1.83	0.196
	3.05	0.268
	3.7	0.285
Maximum x_1 (m)	1.83	0.216
	3.05	0.314
	3.7	0.351
Maximum x_2 (m)	1.83	0.024
	3.05	0.047
	3.7	0.066
Maximum vertical acceleration (g)	1.83	0.957
	3.05	1.857
	3.7	2.638

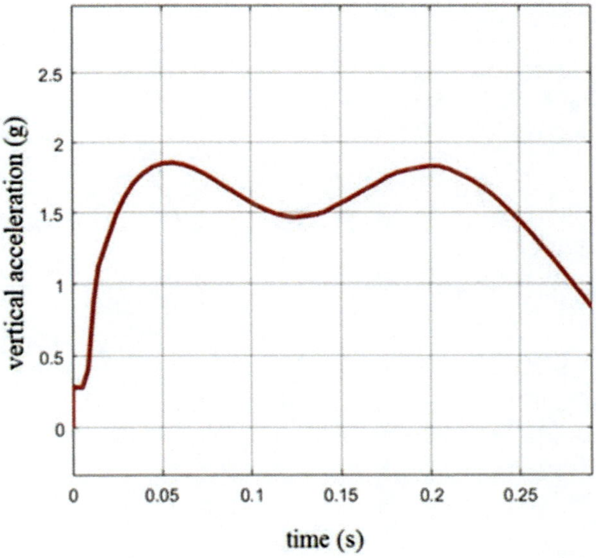

Fig. 6 Vertical acceleration (g) vs time at 3.05 m/s

2 g. Shock absorber displacement is 0.268 m, which is below 0.33 m design constraint, as given in Table 2.

Figures 8 and 9 show the results for 3.7 m/s (12 fps) descent velocity which is the most severe case. Maximum vertical acceleration is 2.638 g which is lower than the assumed constraint value of 3 g. Shock absorber displacement is 0.285 m which is below 0.33 m design constraint as given in Table 2.

Fig. 7 Displacements vs time at 3.05 m/s

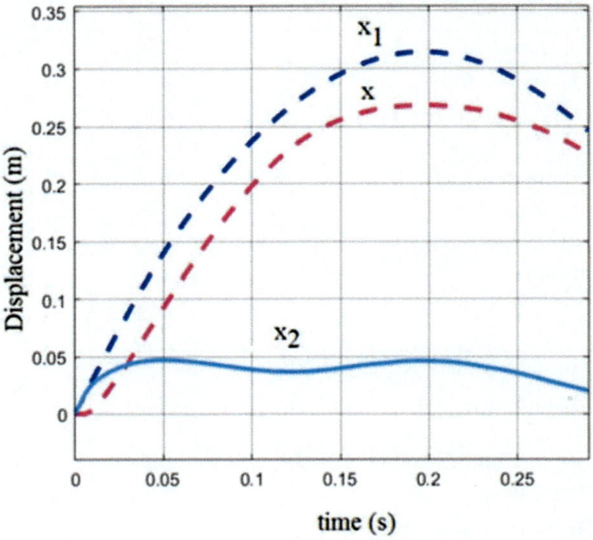

Fig. 8 Vertical acceleration (g) vs time at 3.7 m/s

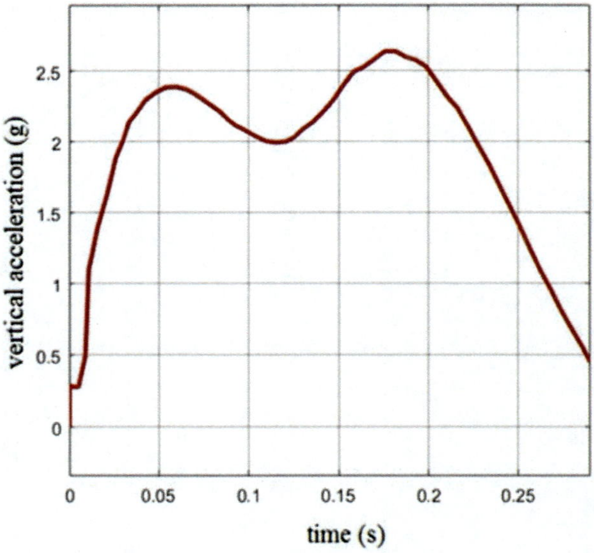

4 Conclusion

In this study, an analytical model of landing gear was constructed with a fixed orifice area in the shock absorber (spring-mass-damper dynamic model) for a high-altitude long endurance UAV which can be considered as in the class of large aeroplanes per

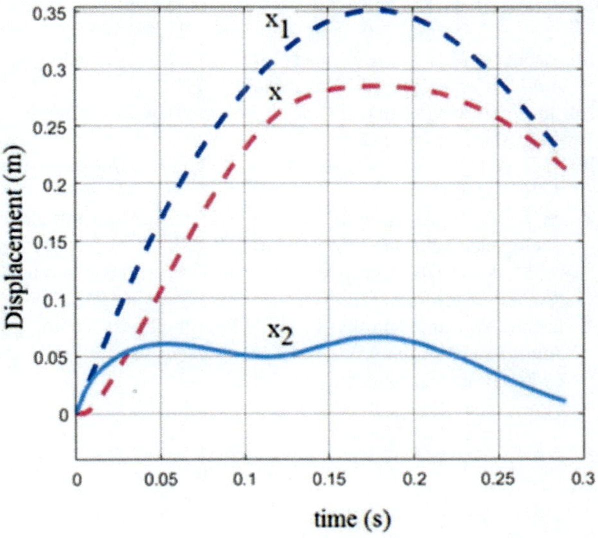

Fig. 9 Displacements vs time at 3.7 m/s

CS-25 EASA civil certification requirements. There different cases were run in line with certification requirements in CS-25. Results are satisfactory within the given constraints; however, they can be optimized in further studies.

References

Dinc, A. (2021). Metering pin diameter optimization of an aircraft landing gear shock absorber. *Eskişehir Technical University Journal of Science and Technology B - Theoretical Sciences, 9*, 37–46. https://doi.org/10.20290/estubtdb.900786

Dinc, A., & Gharbia, Y. (2020). Effects of spring and damper elements in aircraft landing gear dynamics. *International Journal of Recent Technology and Engineering, 8*, 4265–4269. https://doi.org/10.35940/ijrte.D9247.018520

Dinc, A., & Moayyedian, M. (2020). Predicting maximum endurance of a high altitude long endurance UAV with Taguchi method. *International Journal of Innovative Technology and Exploring, 7*, 15–21.

Dinc, A., Taher, R., Moayyedian, M., et al. (2020). A performance review of a high altitude long endurance drone. *International Journal of Progressive Sciences and Technologies, 24*, 32–39.

Drezner, J. A., & Leonard, R. S. (2002). *Innovative development: Global hawk and DarkStar - transitions within and out of the HAE UAV ACTD program.*

EASA. (2012). *CS-23 certification specifications.* https://www.easa.europa.eu/sites/default/files/dfu/agency-measures-docs-certification-specifications-CS-23-CS-23-Amdt-3.pdf. Accessed 23 Sept 2021.

Furnish, J. F., & Anders, D. E. (1971). Analytical simulation of landing gear dynamics for aircraft design and analysis. *SAE Technical Papers*, 0–8. https://doi.org/10.4271/710401

Krüger, W. R., & Morandini, M. (2011). Recent developments at the numerical simulation of landing gear dynamics. *CEAS Aeronautical Journal, 1*, 55–68. https://doi.org/10.1007/s13272-011-0003-y

McBrearty, J. F. (1948). A critical study of aircraft landing gears. *Journal of the Aeronautical Sciences, 15*, 263–280. https://doi.org/10.2514/8.11568

Morrison, D., Neff, G., & Zahraee, M. (1997). Aircraft landing gear simulation and analysis. *ASEE Annual Conference and Proceedings, 28*.

Northrop-Grumman. (n.d.) *Global Hawk*. https://www.northropgrumman.com/air/globalhawk/. Accessed 12 May 2020.

Ross, I., & Edson, R. (1983). *NASA-CR-166104 application of active control technology to the A-10*.

Tsach, S., Yaniv, A., Avni, H., & Penn, D. (1996). High altitude long endurance (HALE) UAV for intelligence missions. *ICAS, 1996*, 368–379.

Wahi, M. K. (1976). Oleopneumatic shock strut dynamic analysis and its real-time simulation. *Journal of Aircraft, 13*, 303–308. https://doi.org/10.2514/3.44526

Yadav, D., & Ramamoorthy, R. P. (1991). Nonlinear landing gear behavior at Touchdown. *The Journal of Dynamic Systems, Measurement, and Control, Transactions of the ASME, 113*, 677–683. https://doi.org/10.1115/1.2896474

GNSS-Aided Satellite Localization by Using Various Kalman Filters

Mert Sever and Chingiz Hajiyev

1 Introduction

The Global Satellite Navigation System (GNSS) is seen as an important system in terms of obtaining accurate positioning together with precise positioning. İn line with this importance, the use of GNSS receiver has started to be used widely in military and civilian areas. Various studies were created by using GNSS receiver simulations (Bagci & Hajiyev, 2016; Sever & Hajiyev, 2020; Sever & Hajiyev, 2021).

Considering the importance of strategic location determination, it is ensured that these systems make measurements with sufficient sensitivity bringing about the development of various methods such as Kalman Filters (Kalman, 1960).

2 Problem Statement

In this study, a low-earth orbit satellite's orbital position and velocity conditions were estimated by using pseudo-ranging method. By using simulation of four GNSS satellite data, orbital conditions were measured. Obtained model improved with application of traditional EKF, LKF, and NRM pre-processed EKF. With the use of Kepler's equations, orbital movement of the satellite was simulated. During the simulations, no orbital perturbations were considered.

M. Sever (✉)
Turkish National Defense University, Hezârfen Aeronautics and Space Technologies Institute, Istanbul, Turkey

C. Hajiyev
Faculty of Aeronautics and Astronautics, Istanbul Technical University, Istanbul, Turkey
e-mail: cingiz@itu.edu.tr

T. H. Karakoc et al. (eds.), *New Achievements in Unmanned Systems*, Sustainable Aviation, https://doi.org/10.1007/978-3-031-29933-9_27

$$\ddot{r} = \begin{bmatrix} \ddot{r}_x \\ \ddot{r}_y \\ \ddot{r}_z \end{bmatrix} = \begin{bmatrix} -\frac{\mu_\oplus}{r^3}x \\ -\frac{\mu_\oplus}{r^3}y \\ -\frac{\mu_\oplus}{r^3}z \end{bmatrix} \quad (1)$$

Here $r = [x \ y \ z]^T$ is the position of vector; x, y, and z are the position state components of the satellite. $\mu_\oplus = GM_\oplus$, where G is the universal gravitation constant and $M_\oplus$ is mass of the earth. MKS units were used.

Equations (2) and (3) are the formulas for orbital position and velocity vectors in discrete form:

$$\begin{aligned} x_{i+1} &= x_i + \Delta t V_{x_i} \\ y_{i+1} &= y_i + \Delta t V_{y_i} \\ z_{i+1} &= z_i + \Delta t V_{z_i} \end{aligned} \quad (2)$$

$$\begin{aligned} V_{y_{i+1}} &= V_{y_i} - \Delta t \mu_\oplus \frac{y_i}{r_i^3} \\ V_{z_{i+1}} &= V_{z_i} - \Delta t \mu_\oplus \frac{z_i}{r_i^3} \\ V_{x_{i+1}} &= V_{x_i} - \Delta t \mu_\oplus \frac{x_i}{r_i^3} \end{aligned} \quad (3)$$

Here V_x, V_y, *and* V_z are the velocity vector components of the satellite, Δt is the sample time. The estimated state vector of system (2) and (3) is presented below:

$$X(k) = [x(k) \quad y(k) \quad z(k) \quad V_x(k) \quad V_Y(k) \quad V_Z(k) \quad b(k)]^T \quad (4)$$

where $b(k)$ is the clock bias of the GNSS receiver.

By using the sphere equations, distance between each GNSS satellite and user were obtained. Communication between user and GNSS satellites are shown in Fig. 1.

In Fig. 1, M is the LEO satellite and S_i (i = 1,4) are the GNSS satellites. D_i expresses the distance between i'th GNSS satellite and user, and D expresses the distance between user and origin of the reference coordinate system. According to Fig. 1, distance measurement between GNSS satellite and user can be determined from sphere equation as

$$D_i = \sqrt{(x_i - x) + (y_i - y)^2 + (z_i - z)^2} + b_i + w_i \quad (5)$$

where x_i, y_i, and z_i ($i = 1,4$) are the position vector components of the GNSS satellite; x, y, and z are the position vector components of the LEO satellite; w_i is the random Gaussian measurement noise with zero mean.

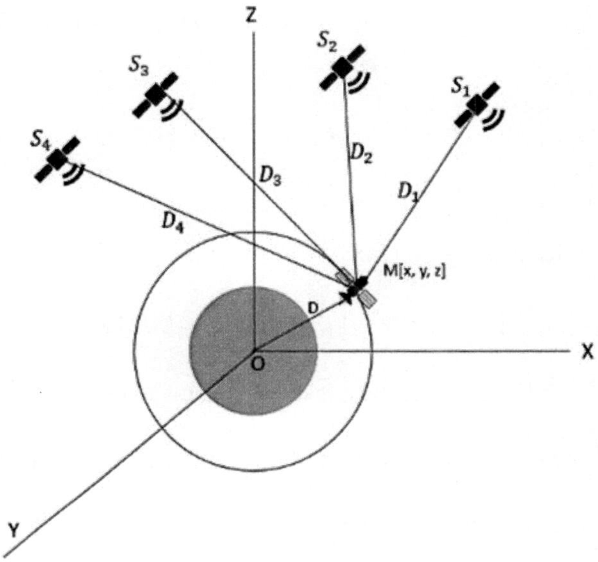

Fig. 1 Satellite localization with distance measurement method

3 Traditional Extended Kalman Filter

The Kalman Filter, which is applied by linearizing nonlinear models by considering the Taylor Series approach for the previous state estimation of the system function and the estimated position of the observation function in the current time, is called the Extended Kalman Filter (EKF). The EKF equations for the GNSS-aided satellite localization are given below:

Estimation equation:

$$\widehat{X}(k) = \widehat{X}(k/k-1) + K(k)\widehat{Z}(k) \tag{6}$$

Extrapolation equation:

$$\widehat{X}(k/k-1) = f\left(\widehat{X}(k-1/k-1)\right) \tag{7}$$

Innovation sequence equation:

$$\widehat{Z}(k/k-1) = D(k) - \widehat{D}(k) \tag{8}$$

Here

$$D(k) = [D_1(k) \quad D_2(k) \quad D_3(k) \quad D_4(k)]^T,$$
$$\widehat{D}(k) = \left[\overline{D}_1(k) \quad \overline{D}_2(k) \quad \overline{D}_3(k) \quad \overline{D}_4(k)\right]^T \tag{9}$$

$$\widehat{D}_i(k) = \sqrt{(x_i(k) - \widehat{x}(k-1))^2 + (y_i(k) - \widehat{y}(k-1))^2 + (z_i(k) - \widehat{z}(k-1))^2} + \widehat{b}(k-1) \tag{10}$$

Kalman gain matrix:

$$K(k) = P(k/k-1)\nabla h_x^T(k)\left[\nabla h_x(k)P(k/k-1)\nabla h_x^T(k) + R(k)\right]^{-1} \tag{11}$$

Predicted covariance matrix of estimation error:

$$P(k/k-1) = \phi(k, k-1)P(k-1/k-1)\phi^T(k, k-1) + G(k)Q(k-1)G(k)^T \tag{12}$$

System dynamic matrix is

$$\phi(k, k-1) = \left[\frac{\partial f}{\partial \mathrm{X}}\right]_{\widehat{X}(k-1)} \tag{13}$$

Estimated covariance matrix of estimation error:

$$P(k/k) = \left[I - K(k)\nabla h_x^T(k)\right]P(k/k-1) \tag{14}$$

Measurement matrix $\nabla h_x(k)$ is obtained by partially differentiating the distance measurement model between user and GNSS satellites and shown as below:

$$\nabla h_x(k) = \begin{bmatrix} \partial D_1/\partial x & \partial D_1/\partial y & \partial D_1/\partial z & \partial D_1/\partial b \\ \partial D_2/\partial x & \partial D_2/\partial y & \partial D_2/\partial z & \partial D_2/\partial b \\ \partial D_3/\partial x & \partial D_3/\partial y & \partial D_3/\partial z & \partial D_3/\partial b \\ \partial D_4/\partial x & \partial D_4/\partial y & \partial D_4/\partial z & \partial D_4/\partial b \end{bmatrix} \tag{15}$$

Measurement error covariance matrix:

$$R(k) = \sigma^2 I_{4x4} \tag{16}$$

Here, standard deviation of the measurement error is considered as 10 m for each measurement. ($\sigma = 10$ m).

4 Extended Kalman Filter Based on Linear Measurements

This approach was proposed by Hajiyev (2011). State vector of the user is given in Eq. (4). Estimation equation, extrapolation equation, and innovation equation are the same as Eqs. (6), (7), and (8), respectively.

The measurement vector $Y(k)$ is expressed as

$$Y(k) = \begin{bmatrix} \frac{1}{2}\left(L_1^2 - L_2^2 + D_2^2 - D_1^2\right) \\ \frac{1}{2}\left(L_1^2 - L_3^2 + D_3^2 - D_1^2\right) \\ \frac{1}{2}\left(L_1^2 - L_4^2 + D_4^2 - D_1^2\right) \end{bmatrix} \tag{17}$$

Here L_1, L_2, L_3, and L_4 are the distance between origin and i'th GNSS satellite ($i = 1,2,3,4$):

$$L_i = \sqrt{x_i^2 + y_i^2 + z_i^2} \tag{18}$$

Kalman gain matrix predicted covariance matrix of estimation error, and estimated covariance matrix of estimation errors are the same equations as Eqs. (11), (12), and (14), respectively.

Measurement matrix is presented as

$$H(k) = \begin{bmatrix} x_1 - x_2 & y_1 - y_2 & z_1 - z_2 & 0 & 0 & 0 & D_2 - D_1 \\ x_1 - x_3 & y_1 - y_3 & z_1 - z_3 & 0 & 0 & 0 & D_3 - D_1 \\ x_1 - x_4 & y_1 - y_4 & z_1 - z_4 & 0 & 0 & 0 & D_4 - D_1 \end{bmatrix} \tag{19}$$

Measurement error covariance matrix is

$$R(k) = \begin{bmatrix} (D_1-b)^2\sigma^2 + (D_2-b)^2\sigma^2 + \sigma^4 & (D_1-b)^2\sigma^2 + \frac{1}{2}\sigma^4 & (D_1-b)^2\sigma^2 + \frac{1}{2}\sigma^4 \\ (D_1-b)^2\sigma^2 + \frac{1}{2}\sigma^4 & (D_1-b)^2\sigma^2 + (D_3-b)^2\sigma^2 + \sigma^4 & (D_1-b)^2\sigma^2 + \frac{1}{2}\sigma^4 \\ (D_1-b)^2\sigma^2 + \frac{1}{2}\sigma^4 & (D_1-b)^2\sigma^2 + \frac{1}{2}\sigma^4 & (D_1-b)^2\sigma^2 + (D_4-b)^2\sigma^2 + \sigma^4 \end{bmatrix} \tag{20}$$

5 Newton–Raphson Method Aided Extended Kalman Filter

In this method, the position vectors are estimated via Newton–Raphson Method (NRM) then the obtained vectors become the measurement for EKF. Function of the model is shown in Eq. (21).

$$F(p_k) = \begin{bmatrix} f_1(x,y,z,b) \\ f_2(x,y,z,b) \\ f_3(x,y,z,b) \\ f_4(x,y,z,b) \end{bmatrix} \tag{21}$$

where

$$\begin{aligned} f_1(x,y,z,b) &= (x_1 - x)^2 + (y_1 - y)^2 + (z_1 - z)^2 - (D_1 - b)^2 \\ f_2(x,y,z,b) &= (x_2 - x)^2 + (y_2 - y)^2 + (z_2 - z)^2 - (D_2 - b)^2 \\ f_3(x,y,z,b) &= (x_3 - x)^2 + (y_3 - y)^2 + (z_3 - z)^2 - (D_3 - b)^2 \\ f_4(x,y,z,b) &= (x_4 - x)^2 + (y_4 - y)^2 + (z_4 - z)^2 - (D_4 - b)^2 \end{aligned} \tag{22}$$

By partially differentiating the model's function,

$$J(p_k) = \begin{bmatrix} \frac{\partial}{\partial x} f_1(x,y,z,b) & \frac{\partial}{\partial y} f_1(x,y,z,b) & \frac{\partial}{\partial z} f_1(x,y,z,b) & \frac{\partial}{\partial b} f_1(x,y,z,b) \\ \frac{\partial}{\partial x} f_2(x,y,z,b) & \frac{\partial}{\partial y} f_2(x,y,z,b) & \frac{\partial}{\partial z} f_2(x,y,z,b) & \frac{\partial}{\partial b} f_2(x,y,z,b) \\ \frac{\partial}{\partial x} f_3(x,y,z,b) & \frac{\partial}{\partial y} f_3(x,y,z,b) & \frac{\partial}{\partial z} f_3(x,y,z,b) & \frac{\partial}{\partial b} f_3(x,y,z,b) \\ \frac{\partial}{\partial x} f_4(x,y,z,b) & \frac{\partial}{\partial y} f_4(x,y,z,b) & \frac{\partial}{\partial z} f_4(x,y,z,b) & \frac{\partial}{\partial b} f_4(x,y,z,b) \end{bmatrix} \tag{23}$$

By using Eq. (27) ΔP is calculated and estimation of NRM starts. Estimation continues until the estimation errors become less than error tolerance.

$$\Delta P = -F.J^{-1}, P_{k+1} = P_k + \Delta P \tag{27}$$

$$NRM_x = P_{k+1}(1), NRM_y = P_{k+1}(2), NRM_z = P_{k+1}(3), NRM_b = P_{k+1}(4) \tag{24}$$

Estimated state vector via NRM is used as the input measurement for the EKF. Measurement vector is written in the following form:

$$Y(k) = [NRM_x \quad NRM_y \quad NRM_z \quad NRM_b] \tag{25}$$

Measurement matrix, $H(k)$ is

$$H(k) = \begin{bmatrix} 1 & 0 & 0 & 0 & 0 & 0 & 0 \\ 0 & 1 & 0 & 0 & 0 & 0 & 0 \\ 0 & 0 & 1 & 0 & 0 & 0 & 0 \\ 0 & 0 & 0 & 0 & 0 & 0 & 1 \end{bmatrix} \tag{26}$$

Measurement error covariance matrix:

$$R(k) = \text{diag}\left(\beta_x, \beta_y, \beta_z, \beta_b\right) \tag{27}$$

Here β_x, β_y, β_z, β_b are the variances of the errors of NRM estimations.

6 Results and Discussion

In this study, LEO satellite position was estimated by using traditional Extended Kalman Filter, Extended Kalman Filter based on linear measurements, and NRM pre-processed Extended Kalman Filter and the results were compared.

In Figs. 2 and 3, the X-axis position and velocity estimations of the LEO satellite with the help of traditional EKF are shown. GNSS receiver clock bias estimation results are given in Fig. 4.

In Figs. 5, 6, and 7, the X-axis position and velocity estimations of the LEO satellite and clock bias of the GNSS receiver obtained by the EKF based on linear measurements are shown.

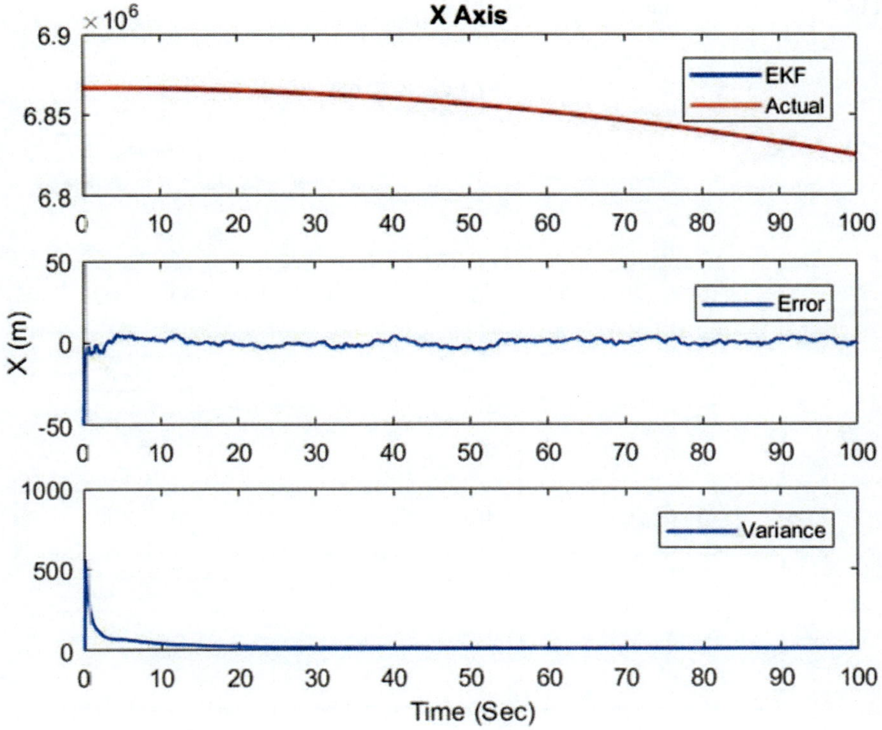

Fig. 2 X-Axis estimation with traditional EKF

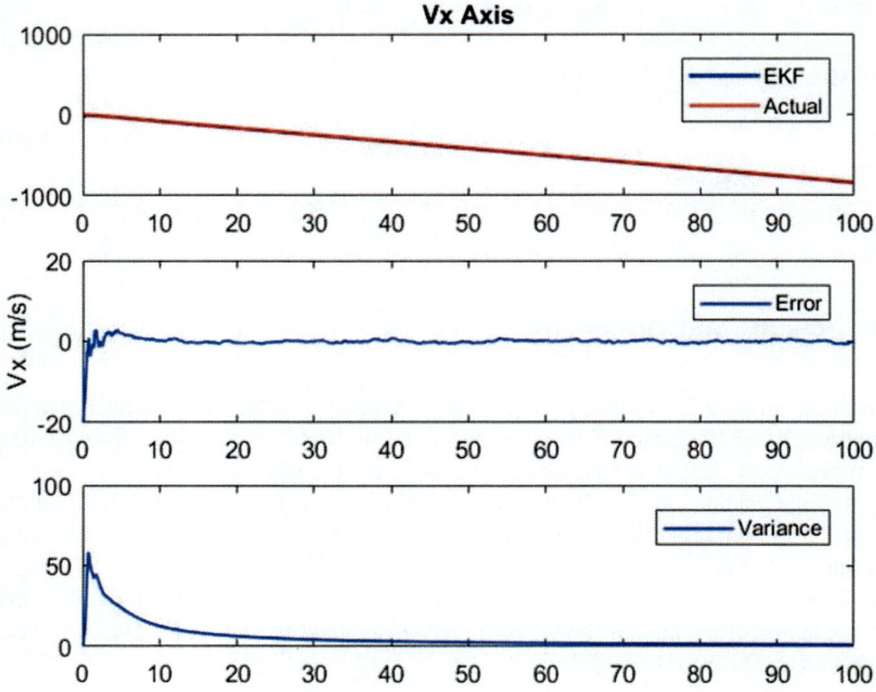

Fig. 3 Vx-Axis estimation with traditional EKF

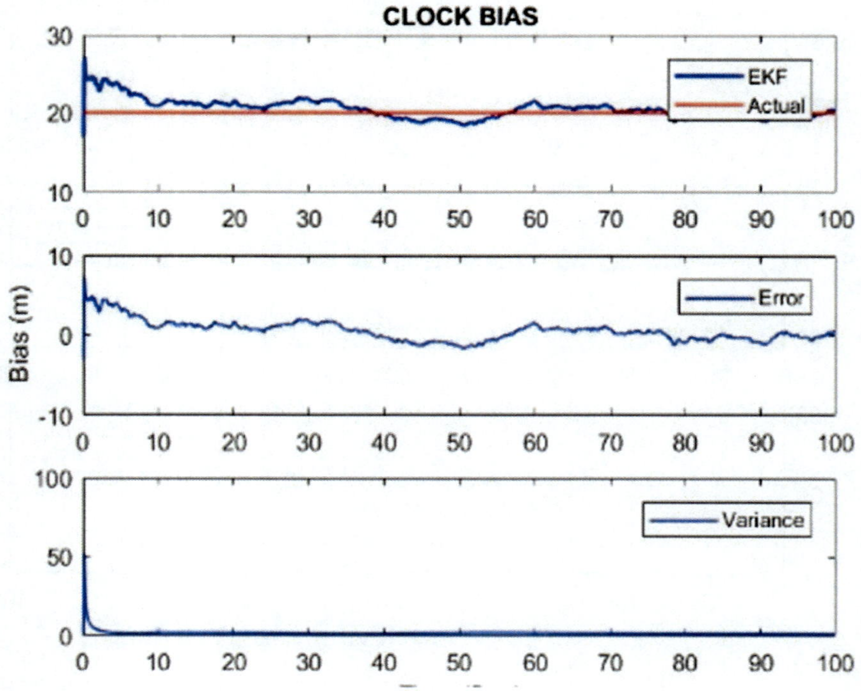

Fig. 4 Clock bias estimation with traditional EKF

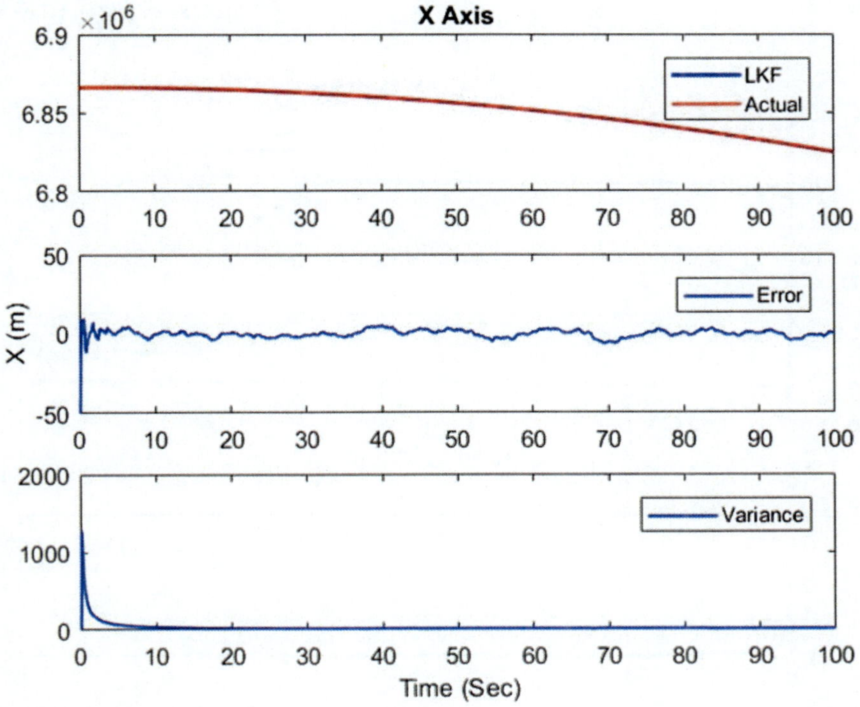

Fig. 5 X-Axis estimation with EKF based on linear measurements

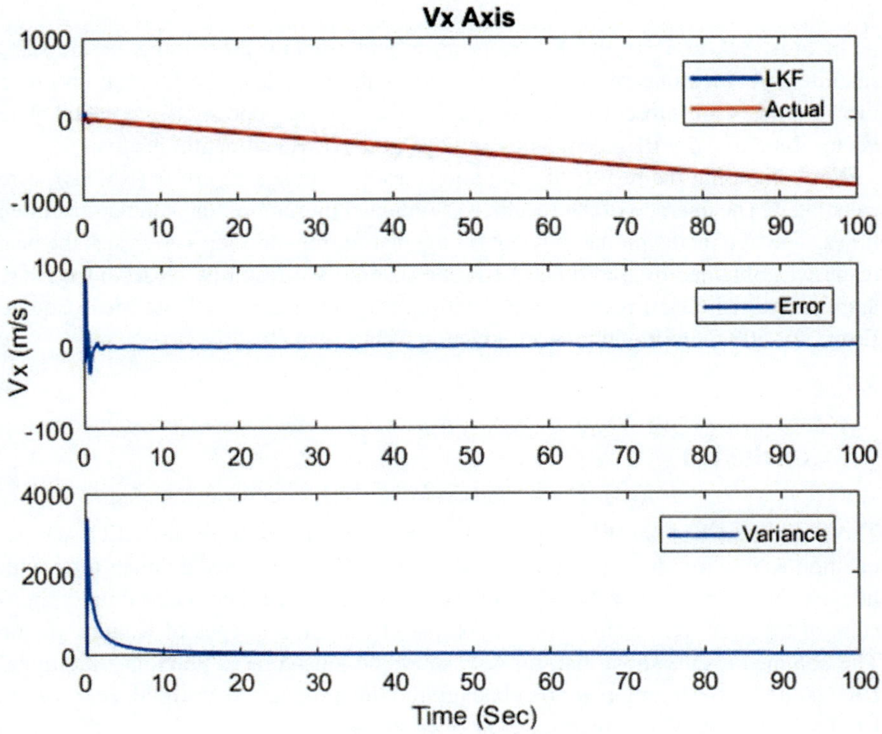

Fig. 6 Vx-Axis estimation with EKF based on linear measurements

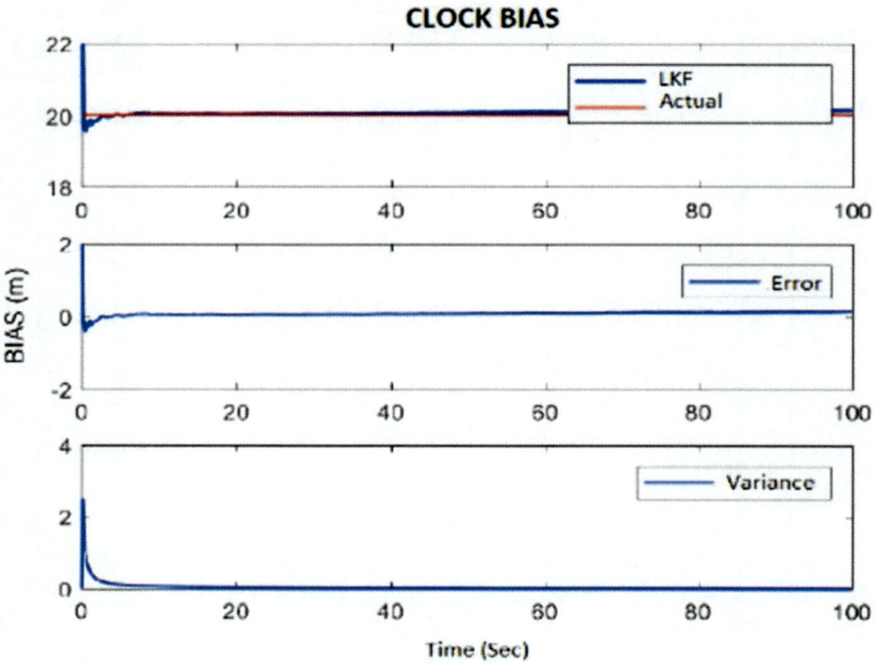

Fig. 7 Clock bias estimation with EKF based on linear measurements

In Figs. 8,9, and 10, the X-axis position and velocity estimations of the LEO satellite and clock bias of the GNSS receiver with the help of EKF based on linear measurements are presented. In the graphs, the red line represents the actual state vector, blue line the EKF estimates, and cyan the NRM estimation results.

For comparing the results of estimations, root mean square errors (RMSE) were calculated. The average of the results was obtained by running the simulations three times. The obtained simulation results show that for the long-term analyses, the best result was obtained by traditional EKF, the second-best result by NRM aided EKF, and finally the noisiest result by EKF with linear measurements. Root Mean Square Errors for 500 Sec Simulations are given in Table 1.

7 Conclusion

Three different Kalman filter approaches are used in this study for LEO satellite position estimation: traditional extended Kalman filter, extended Kalman filter with linear measurements, and NRM-aided extended Kalman Filter. The orbital parameters of the user were estimated and compared via various types of Kalman filters. The obtained results show that, for both short and long-time estimations, traditional EKF is the best estimation approach. Through this research, the orbital states of the LEO satellite can be obtained for their own problem.

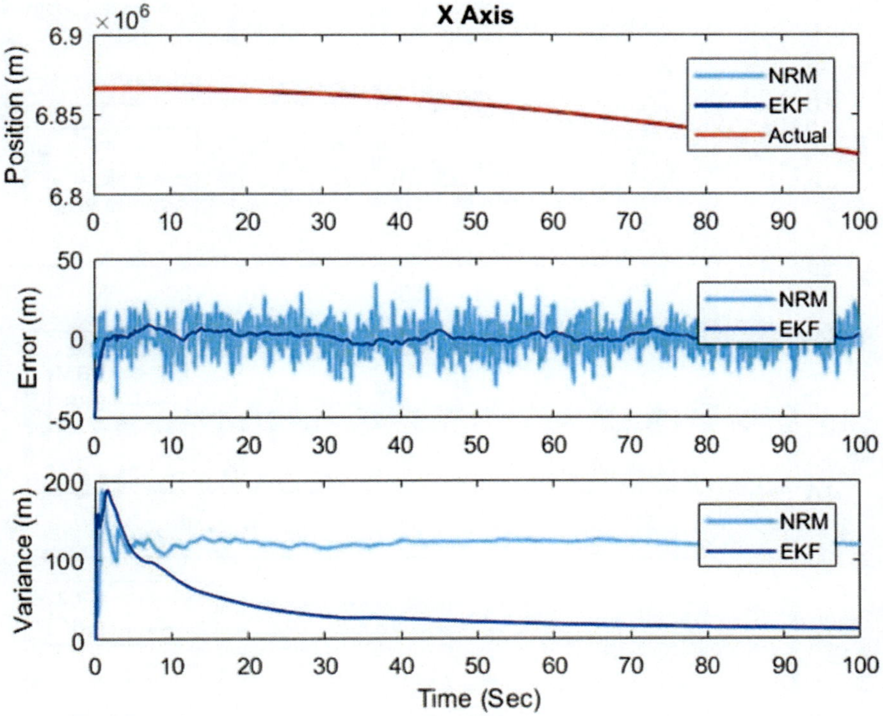

Fig. 8 X-Axis estimation with NRM pre-processed EKF

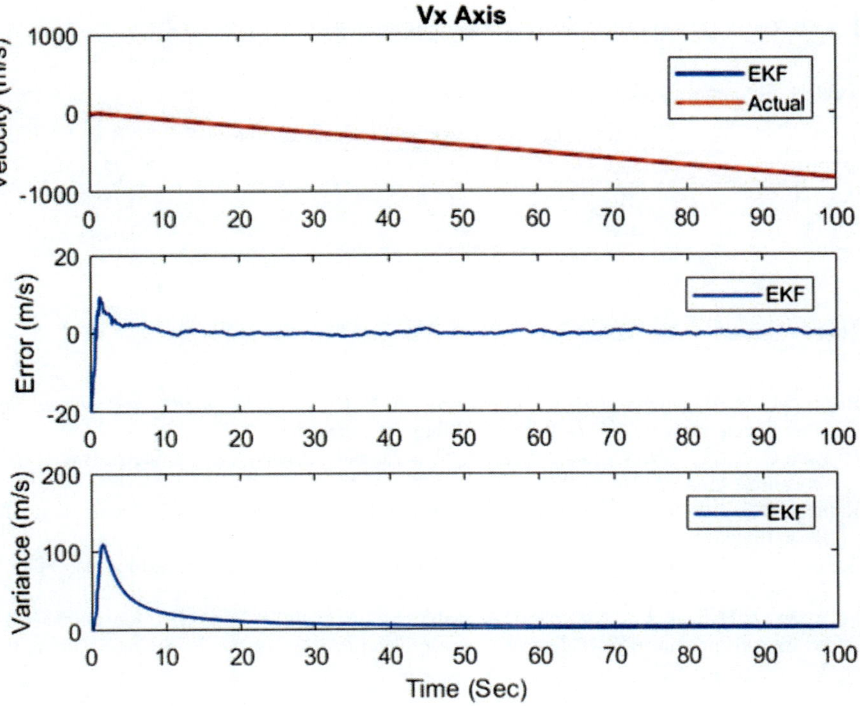

Fig. 9 Vx-Axis estimation with NRM pre-processed EKF

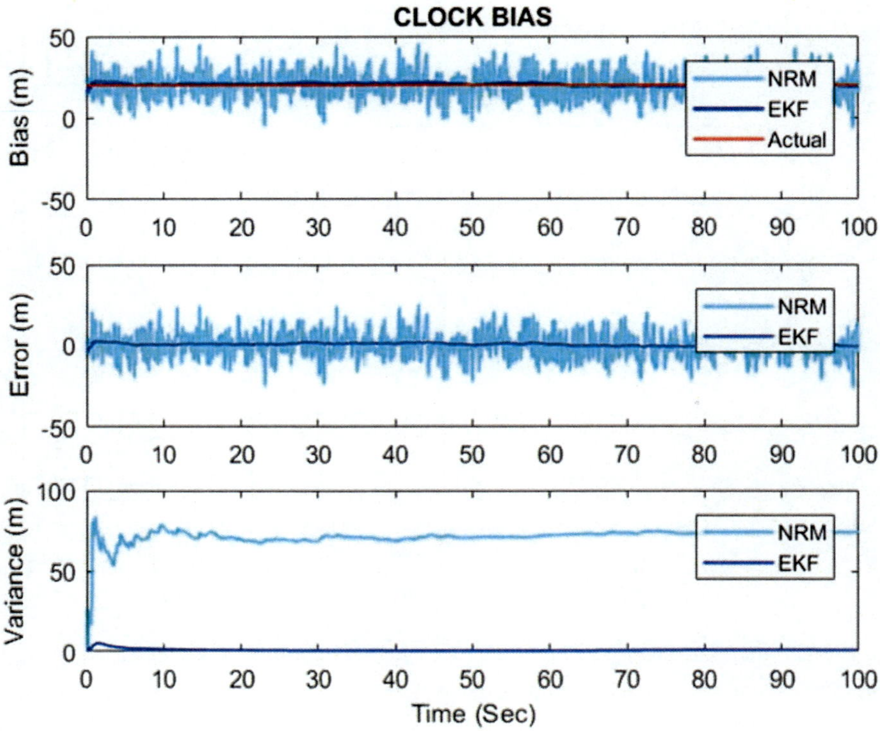

Fig. 10 Clock bias estimation with NRM pre-processed EKF

Table 1 Root mean square errors for 500 sec simulations

	Traditional EKF	EKF with linear measurements	NRM-aided EKF
X (m)	2.1333	2.2709	2.2223
Vx (m/s)	0.6846	1.6770	0.7588
Clock bias (m)	0.6406	2.2490	0.6495

References

Bagci, M., & Hajiyev, C. (2016). *Integrated NRM/EKF for LEO satellite GPS based orbit determination* (pp. 462–467). IEEE Metrology for Aerospace.

Hajiyev, C. (2011). GNSS signal processing via linear and extended Kalman filters. *Asian Journal of Control, 13*(2), 1–10.

Kalman, R.E. (1960). A New Approach to Linear Filtering and Prediction Problems. Journal of Basic Engineering

Sever, M., & Hajiyev, C. (2020). Satellite localization correction with extended Kalman Filter. In *International symposium on electric aviation and autonomous systems* (pp. 22–24).

Sever, M., & Hajiyev, C. (2021). GNSS signal processing with EKF and UKF for stationary user position estimation. *WSEAS Transactions on Signal Processing, 17*, 75–80. https://doi.org/10.37394/232014.2021.17.10

Thermal Study of Cylindrical Lithium-Ion Battery at Different Discharge Rates

Uğur Morali

Nomenclature

NTGK	Newman, Tiedemann, Gu, and Kim
ECM	Equivalent circuit model
P2D	Pseudo-two-dimensional
DoD	Depth-of-discharge

1 Introduction

Lithium-ion batteries hold great promise as energy storage materials. Lithium-ion batteries have been used in various energy-related applications owing to their high energy density and high power density (Liu et al., 2021; Zeng et al., 2021). Despite the high densities provided by lithium-ion batteries, thermally sensitive nature and temperature rise during discharge limit their use (Tarhan et al., 2021). In other words, temperature increase under abuse conditions continues to be the biggest challenge for practical applications. It is, therefore, necessary to perform battery thermal analysis when discharge conditions are imposed to enable thermally durable lithium-ion batteries.

Simulations have been used to determine thermal behaviour of batteries (Dattu et al., 2021; Mevawalla et al., 2021). Thermal battery models implemented by using simulation provide valuable information in analysing battery temperature rises. Moreover, thermal models provide a cost-effective assessment of battery thermal behaviour and a faster understanding of effective operating parameters without

U. Morali (✉)
Department of Chemical Engineering, Eskisehir Osmangazi University, Eskisehir, Turkey
e-mail: umorali@ogu.edu.tr

T. H. Karakoc et al. (eds.), *New Achievements in Unmanned Systems*, Sustainable Aviation, https://doi.org/10.1007/978-3-031-29933-9_28

extensive experimentation. In the literature, multi-scale, multi-dimensional models have been widely used to analyse battery thermal behaviour (Immonen & Hurri, 2021; Yang et al., 2021; Panchal et al., 2018). These models are Newman, Tiedemann, Gu, and Kim (NTGK), equivalent circuit model (ECM), and pseudo-two-dimensional (P2D) model.

In this study, the NTGK model was applied due to its simple computation and easy parameterization. The maximum battery temperature and average battery temperature of 26,650 cylindrical lithium-ion batteries were analysed under different discharge rates. The effect of discharge rate on the battery temperatures was interpreted in the light of simulation-based temperature results.

2 Method

A commercially available 26,650 cylindrical lithium-ion battery was used to simulate battery temperature rise. 26,650 means that the lithium-ion battery has dimensions of diameter 26 mm × length 65 mm. Maximum battery potential, minimum battery potential, and nominal capacity are 4.20 V, 2.75 V, and 4 Ah, respectively. Discharge rates applied in the simulation are 1.0C, 1.5C, 2.0C, 2.5C, and 3.0C. The battery was discharged from 4.2 V to 2.75 V.

ANSYS Fluent 2021R2 was used to apply the multi-scale, multi-dimensional NTGK model. Battery temperature expressions are presented in Eq. (1), (2), (3) and (4):

$$\frac{\partial \rho C_p T}{\partial t} = \sigma_+ |\nabla \phi_+| + \sigma_- |\nabla \phi_-| + Q_g \tag{1}$$

$$\nabla (\sigma_+ \nabla \phi_+) = -(j_{\mathrm{ECh}} - j_{\mathrm{short}}) \tag{2}$$

$$\nabla (\sigma_- \nabla \phi_-) = j_{\mathrm{ECh}} - j_{\mathrm{short}} \tag{3}$$

$$j_{\mathrm{ECh}} = cY \left[U - (\phi_+ - \phi_-) \right] \tag{4}$$

where electrical conductivity of positive and negative electrodes is indicated by σ_+ and σ_-, positive and negative electrodes' phase potentials, are indicated by ϕ_+ and ϕ_-, respectively. Q_g, and j_{ECh} are battery heat produced and volumetric current transfer rate, respectively. j_{short} denotes current transfer rate. U and Y are expressed as follows (Kwon et al., 2006; Kim et al., 2009; Gu, 1983):

$$U = a_0 + a_1(\mathrm{DoD}) + a_2(\mathrm{DoD})^2 + a_3(\mathrm{DoD})^3 \tag{5}$$

$$Y = a_4 + a_5(\mathrm{DoD}) + a_6(\mathrm{DoD})^2 \tag{6}$$

where a_0–a_6 are the fitting parameters. In this study, U and Y parameters were calculated using the fitting parameters obtained in (Kim et al., 2009).

3 Results and Discussion

Discharge curves obtained at different C rates are provided in Fig. 1. The flow time was the longest for the 1.0C rate, as expected. On the other hand, the discharge time at 3.0 C rate (flow time) exhibited the shortest flow time. When the applied discharged current was 4 A (1.0C-rate), the battery potential showed a sharp decrease from 4.2 V to 3.8 V at the beginning of the discharge procedure. This decrease observed in potential increased with increasing the C rate. For example, for the 3.0C rate, the battery potential decreased from 4.2 V to 3.30 V at the beginning of the discharge procedure. This could be attributed to the effect of discharge current that may induce an intense lithium-ion movement from the negative electrode to the positive electrode. Battery surface average temperature increased with increasing C rate, as shown in Fig. 2. Battery average temperature and maximum battery temperature are presented in Table 1. Maximum battery temperature with average battery temperature is depicted in Fig. 3. The highest average battery temperature was 345.164 K obtained at 3C rate. The lowest average battery temperature was 311.627 K obtained at 1C rate. The highest maximum battery temperature and the lowest maximum battery temperature were 346.110 K (obtained at the 3C rate) and 311.934 K (obtained at the 1C rate), respectively.

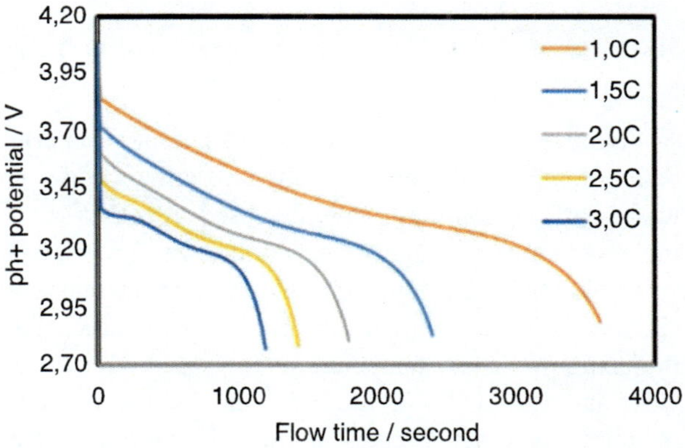

Fig. 1 Positive electrode potential during discharge at different C rates

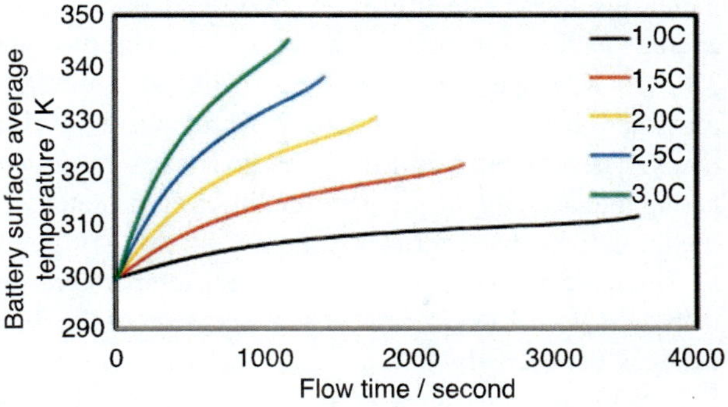

Fig. 2 Average battery temperature during discharge at distinct C rates

Table 1 Average battery temperature and maximum battery temperature at the end of discharge procedure

C rate	$T_{average}$	T_{max}
1.0	311.627	311.934
1.5	321.399	321.895
2.0	330.183	330.847
2.5	338.054	338.866
3.0	345.164	346.110

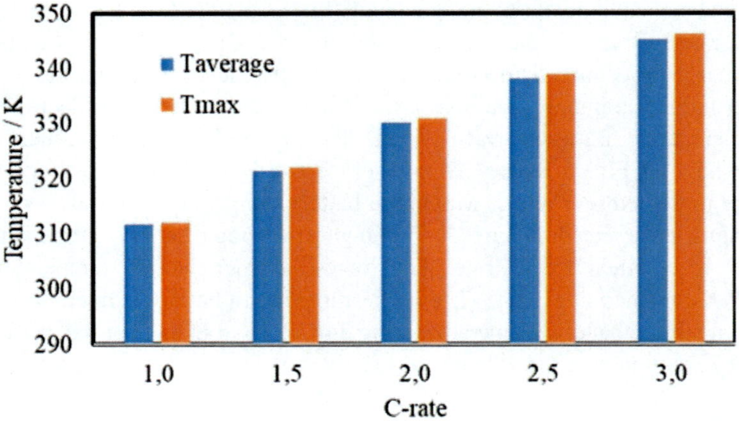

Fig. 3 Average battery temperature and maximum battery temperature versus C rate

Figure 3 showed that the rise in average battery temperature and maximum battery temperature trends are aligned. The increase in average battery temperature was 3.040% when the C rate increased from 1.0C to 1.5C. On the other hand, the increase in average battery temperature was 2.660% when the C rate increased from 1.5C to 2.0C. The temperature rise with increasing C rate from 2.0C to 2.5C (2.328%) was lower than that from 2.5C to 3.0C (2.060%). The highest increase in maximum battery temperature was observed between the discharges of 1.0C and 1.5C and was 3.094%. Similar trend for the maximum battery temperature was obtained for other C rates applied. Furthermore, the percentage of temperature change for the maximum battery temperature between the C ratios decreased with increasing the C rate from 1.0C to 3.0C. Concerning the 1.5C rate discharge, the average battery temperature and maximum battery temperature were 321.399 K and 321.895 K, respectively. On the other hand, the average battery temperature and maximum battery temperature at 2.0C rate were 330,183 K and 330,847 K, respectively. The suitability of the discharge C rates was interpreted in the section below, taking into account the temperature limits of the battery safety data sheet provided by the lithium-ion battery manufacturers.

The largest change in average battery temperature was observed when the C rate increased from 1.0C to 1.5C. The percentage of the difference between temperature changes between discharges at distinct C rates decreased with increasing C rate. Similar trends were observed for the maximum battery temperature. In other words, one-unit change does not alter both average battery temperature and maximum battery temperature at same order of magnitude. Therefore, one should be careful when C rate increased from one level to another level. The results also demonstrated that the temperature rise was more pronounced for the maximum battery temperature for all implemented C rates. This result showed that the maximum battery temperature dominated the battery temperature at the end of the discharge procedure. Therefore, the C rate should be carefully increased by monitoring the maximum battery temperature. The average battery temperature and maximum battery temperature results showed that the discharge rate of 2.0C was a key discharge rate for the 26,650 lithium-ion battery since the value of these temperatures was out of the battery application temperature limits generally provided by battery manufacturers.

4 Conclusion

This study showed the effect of C rate on temperature rise of a cylindrical lithium-ion battery by applying the multi-scale, multi-dimensional NTGK model. The maximum discharge rate for the 26,650 lithium-ion battery was the 2.0 C rate to discharge the battery in its temperature limits safely without any cooling system if the limiting maximum battery temperature is approximately 330 K. The battery thermal models can be successfully implemented using simulation software to determine and interpret the discharge conditions effects on battery temperatures. The battery thermal model can be used to improve battery thermal management systems in real-life applications.

References

Dattu, C. S., Chaithanya, A., Jaidi, J., Panchal, S., Fowler, M., & Fraser, R. (2021). Comparison of lumped and 1D electrochemical models for prismatic 20 Ah lifepo4 battery sandwiched between minichannel cold-plates. *Applied Thermal Engineering, 117*, 586.

Gu, H. (1983). Mathematical analysis of a Zn/Niooh cell. *Journal of the Electrochemical Society, 130*, 1459.

Immonen, E., & Hurri, J. (2021). Incremental thermo-electric CFD modeling of a high-energy lithium-titanate oxide battery cell in different temperatures: A comparative study. *Applied Thermal Engineering, 197*(117), 260.

Kim, U. S., Shin, C. B., & Kim, C.-S. (2009). Modeling for the scale-up of a lithium-ion polymer battery. *Journal of Power Sources, 189*, 841–846.

Kwon, K. H., Shin, C. B., Kang, T. H., & Kim, C.-S. (2006). A two-dimensional modeling of a lithium-polymer battery. *Journal of Power Sources, 163*, 151–157.

Liu, Z., Wang, C., Guo, X., Cheng, S., Gao, Y., Wang, R., Sun, Y., & Yan, P. (2021). Thermal characteristics of ultrahigh power density lithium-ion battery. *Journal of Power Sources, 506*(230), 205.

Mevawalla, A., Panchal, S., Tran, M.-K., Fowler, M., & Fraser, R. (2021). One dimensional fast computational partial differential model for heat transfer in lithium-ion batteries. *Journal of Energy Storage, 37*(102), 471.

Panchal, S., Mathew, M., Fraser, R., & Fowler, M. (2018). Electrochemical thermal modeling and experimental measurements of 18,650 cylindrical lithium-ion battery during discharge cycle for an EV. *Applied Thermal Engineering, 135*, 123–132.

Tarhan, B., Yetik, O., & Karakoc, T. H. (2021). Hybrid battery management system design for electric aircraft. *Energy, 121*, 227.

Yang, W., Zhou, F., Liu, Y., Xu, S., & Chen, X. (2021). Thermal performance of honeycomb-like battery thermal management system with bionic liquid mini-channel and phase change materials for cylindrical lithium-ion battery. *Applied Thermal Engineering, 188*(116), 649.

Zeng, Y., Chalise, D., Lubner, S. D., Kaur, S., & Prasher, R. S. (2021). A review of thermal physics and management inside lithium-ion batteries for high energy density and fast charging. *Energy Storage Materials*.

The Effect of Control Cylinder Placed at Different Angles in Front of a Heated Cylinder on Heat Transfer

Dogan Burak Saydam, Coskun Ozalp, and Ertaç Hürdoğan

Nomenclature

L/D	Length-to-diameter ratio
Re	Reynolds number
Nu	Nusselt number
D	Diameter
L	Length

1 Introduction

External flows on circular cylinders are of great importance in fluid mechanics due to wide engineering applications. External flows are practically used in heat exchangers, boilers, nuclear reactors, cooling towers, cooling of electronic devices, etc. encountered in different applications (Dehkordi et al., 2018). It is essential to eliminate or suppress vortex shedding, both to protect the structure from possible damage due to flow effects and to avoid high energy consumption (Sarioglu, 2017). For this purpose, flow control methods have been developed by researchers,

D. B. Saydam (✉) · E. Hürdoğan
Faculty of Engineering, Department of Energy Systems Engineering, Osmaniye Korkut Ata University, Osmaniye, Turkey

Energy Education-Etude Application and Research Center, Osmaniye Korkut Ata University, Osmaniye, Turkey
e-mail: doganburaksaydam@osmaniye.edu.tr; ehurdogan@osmaniye.edu.tr

C. Ozalp
Faculty of Engineering, Department of Energy Systems Engineering, Osmaniye Korkut Ata University, Osmaniye, Turkey
e-mail: coskunozalp@osmaniye.edu.tr

T. H. Karakoc et al. (eds.), *New Achievements in Unmanned Systems*, Sustainable Aviation, https://doi.org/10.1007/978-3-031-29933-9_29

including active and passive flow control methods (Chen et al., 2015). In the literature, there are different studies in which the flow structure around the objects is controlled by using active and passive flow control methods (Ozkan et al., 2017) experimentally investigated the effectiveness of permeable plates designed to control vortex shedding from a stationary cylinder. As a result, the authors demonstrated that the use of a permeable plate successfully suppressed vortex shedding by reducing velocity fluctuations in the cylinder wake region, further extending the vortex formation zone downstream, and reducing the vortex shedding frequency (Gupta & Saha, 2019) aimed to suppress the vortex shedding by using a passive method by placing a small control cylinder in the trailing area of the master cylinder. As a result, it was determined that the passive control method partially suppressed the vortex shedding, leading to significant reductions in the Strouhal number and lift/drag forces.

In this study, heat transfer by passive flow control method around the heated cylinder was investigated by placing a control cylinder with a smaller diameter than the main cylinder in the upstream region of a heated cylinder in different positions. In this respect, it is aimed that this study will make important contributions to the literature.

2 Method

The study was carried out in the Advanced Fluid Mechanics Laboratory of the Department of Energy Systems Engineering, Osmaniye Korkut Ata University. The water channel consists of two water tanks and a flow area section between these two tanks, the side and bottom walls of which are covered with acrylic sheets (Fig. 1). In the water channel, flow regulators in the form of honeycombs are placed in front of the narrowing areas of the channel to ensure a regular flow within the flow area (Ozalp et al., 2020, 2021b).

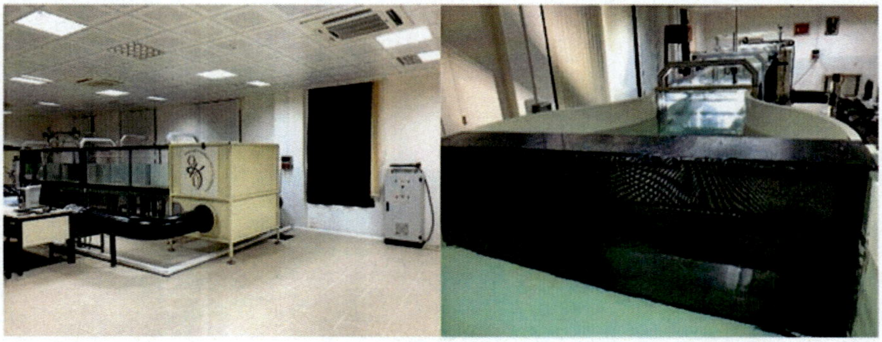

(a) (b)

Fig. 1 Image of the water channel

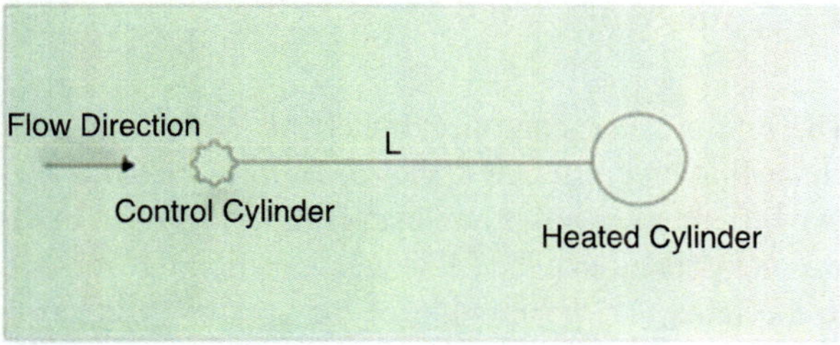

Fig. 2 Top-down view of heated cylinder and control cylinder

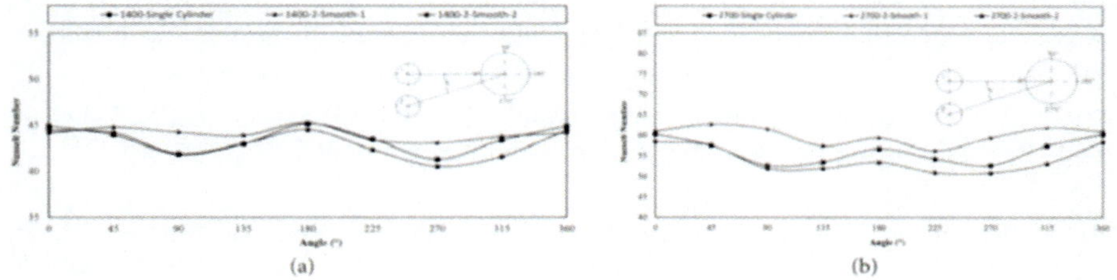

Fig. 3 Nu-angle change graph in case the heated cylinder is positioned at (**a**) Re = 1400 and (**b**) Re = 2700 at an angle of 21° to the smooth control cylinder at L/D = 2 distances

By directing the flow with the control cylinder (d diameter) connected to the servo system, temperature changes were investigated on the second cylinder (D diameter) at L distance (Fig. 2). To keep the cylinder surface at a constant temperature during the experiments, a variac was used to keep the electrical voltage constant in the system. With the voltage adjustment made on the variac, the temperature on the cylinder surface can be adjusted at the desired rate. Detailed information about the designed temperature measurement system is included in previous studies by the authors (Ozalp et al., 2021a; Saydam, 2020).

3 Results and Discussion

The effect of a control cylinder to a heated cylinder in a water channel placed in the upstream region of the flow control cylinder with d = 25 mm diameter at L/D = 2 position and at an angle of 21° was investigated at different Re (Re = 1400, Re = 2700) numbers. As a result, it is seen when the Nu-Angle graphs are examined that the heat transfer worsened by the 21° angled positioning in both Re numbers (Fig. 3). The shape format is Reynolds Number-L/D Ratio-Cylinder Position. The straight control cylinder in the L/D = 2 position produced better results than the cylinder in the 21° angled position. Especially in Re = 1400 and Re = 2700, heat transfer decreased as a result of the vortices breaking from the control cylinder hitting the measuring points located at 180–360° position of the heated cylinder. It is seen that this result also causes a decrease in Nu value at 180°–360° measuring points of the cylinder and especially at 270°.

4 Conclusion

The effect of the control cylinder placed in front of a heated cylinder in a water channel at two different angles (0 and 21°) at L/D = 2 distance was investigated in two different Reynolds numbers (1400 and 2700). In the investigated Re numbers, it was determined that the 21° angled positioning worsened the heat transfer compared to the fixed (0°) positioning of the control cylinder in the upstream region of the heated cylinder. The straight control cylinder in the L/D = 2 position produced better results than the cylinder in the 21° angled position. In Re = 1400 and Re = 2700, it was determined that the heat transfer decreased as a result of the vortices breaking off from the control cylinder hitting the measuring points located at 180–360° position of the heated cylinder. These results showed a decrease in Nu value at 180–360° measuring points of the cylinder and especially at 270°.

Acknowledgement This study was supported by the Scientific Research Projects Unit of Osmaniye Korkut Ata University (OKÜBAP) and the Scientific and Technological Research Council of Turkey (TUBITAK) within the scope of the project, named respectively, as OKÜBAP-2019-PT3-021 and TUBITAK-218 M357. Thanks to OKÜBAP and TUBITAK for their support.

References

Chen, W. L., Cao, Y., Li, H., & Hu, H. (2015). Numerical investigation of steady suction control of flow around a circular cylinder. *Journal of Fluids and Structures, 59*, 22–36.

Dehkordi, E. K., Goodarzi, M., & Nourbakhsh, S. H. (2018). Optimal active control of laminar flow over a circular cylinder using Taguchi and ANN. *European Journal of Mechanics – B/Fluids, 67*, 104–115.

Gupta, A., & Saha, A. K. (2019). Suppression of vortex shedding in flow around a square cylinder using control cylinder. *European Journal of Mechanics – B/Fluids, 76*, 276–291.

Özalp, C., Polat, C., Saydam, D. B., & Söyler, M. (2020). Dye injection flow visualization around a rotating circular cylinder. *European Mechanical Science, 4*, 185–189.

Ozalp, C., Saydam, D. B., Polat, C., Soyler, M., & Hürdoğan, E. (2021a). Heat transfer and flow structure around a heated cylinder by upstream installation of a grooved cylinder. *Experimental Thermal and Fluid Science, 128*, 110448.

Ozalp, C., Soyler, M., Polat, C., Saydam, D. B., & Yaniktepe, B. (2021b). An experimental investigation of a rotationally oscillating cylinder. *Journal of Wind Engineering and Industrial Aerodynamics, 214*, 104679.

Ozkan, G. M., Firat, E., & Akilli, H. (2017). Passive flow control in the near wake of a circular cylinder using attached permeable and inclined short plates. *Ocean Engineering, 134*, 35–49.

Sarioglu, M. (2017). Control of flow around a square cylinder at incidence by using a splitter plate. *Flow Measurement Instruments PB, 53*, 221–229.

Saydam, D. B. (2020). *Aktif Akiş Kontrol Tekniği Uygulanan bir Silindir Etrafinda Sicaklik ve Hiz Dağilimlarinin Deneysel Olarak Incelenmesi*. Osmaniye Korkut Ata University.

Aerodynamic Shape Optimization of the Morphing Leading Edge for the UAS-S45 Winglet

Ruxandra Mihaela Botez, Musavir Bashir, Simon Longtin-Martel, and Tony Wong

1 Introduction

The increasing demand for air transportation, the unpredictability of fuel price, and growing environmental concern lead to the need for more environmentally efficient aircraft. The global aviation industry was expected to reach 40.3 million commercial flights in 2020 before COVID-19 statistics, which signifies an increase of 50% from the last decade (ICAO, 2019). One way to achieve this desired efficiency by aircraft is to use new-generation morphing technology for its various lifting surfaces, which can be triggered and employed for all flight conditions by allowing a multi-point aircraft design, and improved aerodynamic performance (Sugar-Gabor et al., 2016). Researchers have proposed various techniques for achieving the necessary winglet adaptability which resulted in significant performance gains over the baseline wing design (Pecora, 2021; Communier et al., 2020; Sugar-Gabor, 2015; Barbarino et al., 2011; Gamboa et al., 2009; Eguea et al., 2018; Dimino et al., 2021).

All these challenges have led to the need for more innovative research ideas to design more efficient and environmentally friendly aircraft. Highly committed researchers at our Research Laboratory in Active Controls, Avionics, and AeroServoElasticity (LARCASE) are working on different methods to reduce aircraft fuel consumption (Botez, 2018; Hamy et al., 2016; Félix Patrón et al., 2014; Bashir et al., 2021; Khan et al., 2020; Dancila et al., 2012). Wind tunnel experiments are performed at the LARCASE's Price-Païdoussis subsonic wind tunnel. This wind tunnel has been used to validate other adaptive wing concepts (Communier et al. 2017, 2019a, b; Koreanschi et al., 2016; Kammegne et al., 2014; Hassig et al., 2013; Mosbah et al., 2013). Other research was done at the LARCASE on adaptive winglet

R. M. Botez (✉) · M. Bashir · S. Longtin-Martel · T. Wong
Research Laboratory in Active Controls, Avionics and Aeroservoelasticity (LARCASE), Université du Québec, École de Technolgie Supérieure, QC, Canada
e-mail: ruxandra.botez@etsmtl.ca

T. H. Karakoc et al. (eds.), *New Achievements in Unmanned Systems*, Sustainable Aviation, https://doi.org/10.1007/978-3-031-29933-9_30

for the Cessna Citation X (Segui et al., 2018; Segui & Botez, 2018), CRJ-700, and UAS-S4 and S45 (Aubeelack & Botez, 2019).

While most of winglets research has focused on their classical designs, few research has been conducted on morphing winglets, and therefore on the methods for optimizing the parameters needed for a high-performance winglet configuration. The winglet based on leading-edge morphing has not been studied using this technique, and therefore, we have focused this study to investigate the effect of leading-edge morphing on the winglet performance. Our design optimization of a leading-edge morphing winglet for the UAS-45 wing uses the modified CST parameterization technique and a hybrid optimization algorithm based on the Particle Swarm Optimization (PSO) algorithm coupled with Pattern Search and by employing the Vortex Lattice Method (VLM) in MATLAB to calculate the aerodynamic properties of the morphing winglet.

2 Methodology

The morphing winglet optimization study is based on the Reynolds number that is computed for the UAS-S45 cruise flight conditions. The design framework uses drag minimization as an optimization function to enhance the performance of the UAS-S45. The fundamental optimization framework consists of a standard optimization loop written and executed in MATLAB. The flowchart of the overall methodology is depicted in Fig. 1.

2.1 Airfoil Parameterization, VLM Solver, and Optimization Algorithm

The airfoil parameterization is based on a modified Class Shape Transformation (CST) method. The generalized CST equations are applied to the airfoil; therefore, the upper and lower surfaces are defined individually, using a class function $C\left(\frac{x}{c}\right)$ and a shape function $S\left(\frac{x}{c}\right)$ where x is the position on the chord c of a certain point where the two functions C and S are defined.

The PSO algorithm is based on a simplified social behavior closely related to the swarming theory, where the solutions are represented by a set of particles that heuristically navigate through a design space (Khan et al., 2020). The efficiency of PSO algorithms over genetic algorithms is due to their independence from parameters, such as crossovers and mutations; instead, the solution is updated by sharing its information among its population of particles.

In the PSO algorithm, each particle is a solution of a given optimization problem and is composed of two vectors: "position" and "velocity." A position vector x_i^n is

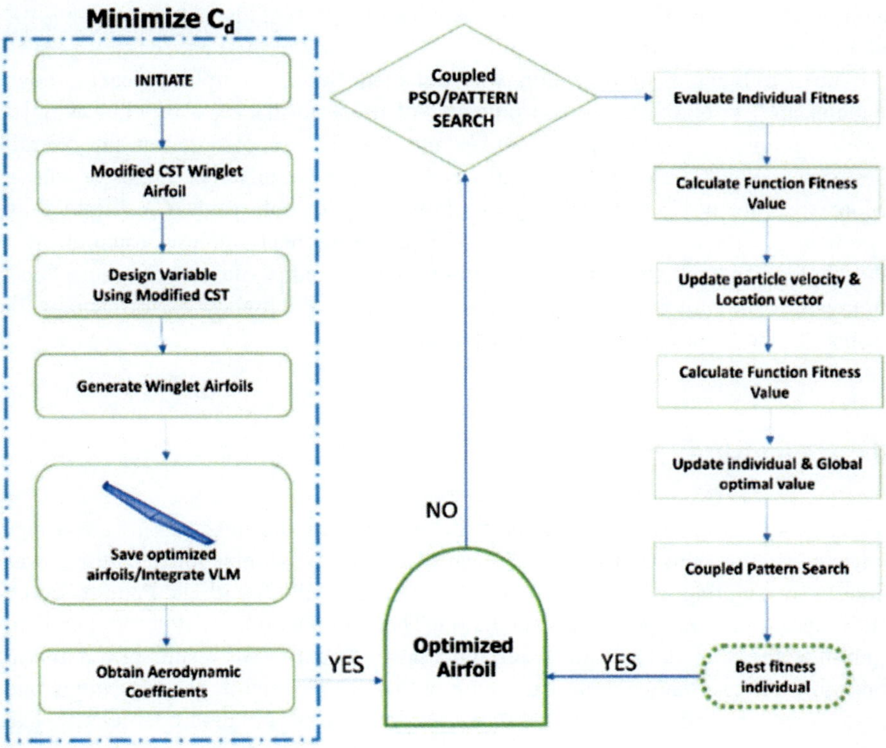

Fig. 1 Schematics of the optimization procedure

used to store the particles positioning (or location) in the given dimensional space. The velocity vector v_i^n is updated using the following equation for each iteration k:

$$v_i^n\,(k) = \left[v_i^n(k-1) + c_1 r_1\left(\text{pbest}_i^n - x_i^n(k-1)\right) + c_2 r_2\left(\text{pbestg}_i^n - x_i^n(k-1)\right)\right] \quad (1)$$

The optimization methods were built using the Vortex Lattice Method (VLM), while the UAS-S45 wing geometry was modelled in Tornado VLM. The objective fitness function at each design iteration was determined using the VLM, while Prandtl's lifting line theory supported the VLM theory.

3 Results and Discussion

The aerodynamic optimization considered in this study applied to morphing leading-edge winglet configurations. To use the full potential of the "morphing" concept, a baseline shape optimization was carried out followed by its morphing shape design at cruise flight condition. These cases were based on Reynolds number = 2.4×10^6 and Mach number of 0.1.

The aerodynamic optimization was obtained for the morphing leading-edge winglet and, to compare the full potential of its performance, a baseline winglet shape optimization was carried out. Figures 2a, b, and c represent the overall performance of three wing and winglet configurations at different angles of attack. More specifically, Figs. 2a, b, and c show the lift coefficients C_L, lift to drag coefficients ratios C_L/C_D, and the aerodynamic endurance values variations with the angles of attack for three wing configurations, and it can be seen from these figures that the morphing winglet configuration gave the highest performance with respect to the other two wing configurations.

4 Conclusion

This study was conducted to perform the aerodynamic optimization of a leading-edge morphing winglet for the UAS-S45. A hybrid optimization technique was chosen by coupling the Particle Swarm Optimization (PSO) with the Pattern Search (PS) algorithms and CST parameterization. The optimization function was aimed to minimize the drag and maximize aerodynamic endurance. The results of the hybrid optimization have shown that using a morphing winglet can improve aerodynamic performance by reducing drag and increasing overall lift and drag coefficients. The aerodynamic performances of the optimized winglet were found to be better than those of fixed geometry winglet.

Acknowledgments Special thanks are due to the Natural Sciences and Engineering Research Council of Canada (NSERC) for the Canada Research Chair Tier 1 in Aircraft Modelling and Simulation Technologies funding. We would also like to thank Odette Lacasse for her support at the ETS, as well as to Hydra Technologies' team members Carlos Ruiz, Eduardo Yakin, and Alvaro Gutierrez Prado in Mexico.

Fig. 2 Variations with the angle of attack of the (**a**) lift coefficient, (**b**) the CL/CD ratio, (**c**) the aerodynamic endurance performance for three wing configurations

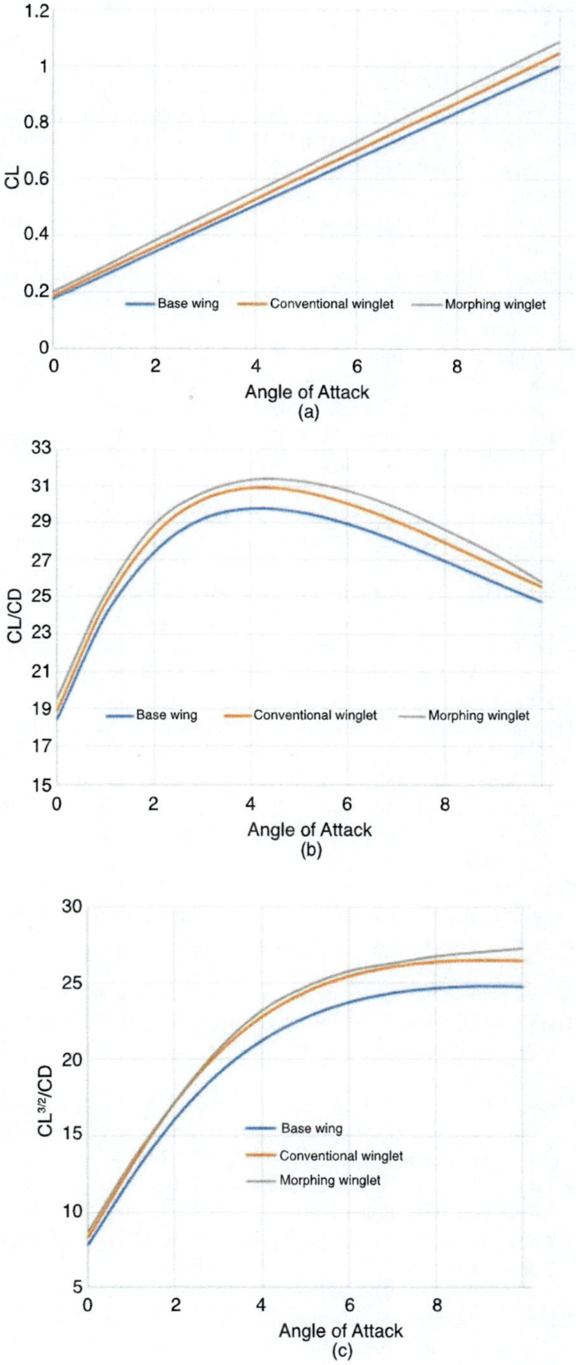

References

Aubeelack, H., & Botez, R. M. (2019). Simulation study of the aerodynamic force distributions on the UAS-S45 Baalam wing with an upswept blended winglet. *INCAS Bulletin, 11*, 21–38.

Barbarino, S., Bilgen, O., Ajaj, R. M., Friswell, M. I., & Inman, D. J. (2011). A review of morphing aircraft. *Journal of Intelligent Material Systems and structures, 22*, 823–877.

Bashir, M., Longtin-Martel, S., Botez, R. M., & Wong, T. (2021). Aerodynamic design optimization of a morphing leading edge and trailing edge airfoil – Application on the UAS-S45. *Applied Sciences, 11*, 1664.

Botez, R. (2018). *Morphing wing, UAV and aircraft multidisciplinary studies at the Laboratory of Applied Research in Active Controls, Avionics and AeroServoElasticity LARCASE* (pp. 1–11). Aerospace Lab.

Communier, D., Salinas, M. F., Moyao, O. C., & Botez, R. M. (2017). Aero structural modeling of a wing using CATIA V5 and XFLR5 software and experimental validation using the Price—Païdoussis wing tunnel. In *AIAA atmospheric flight mechanics conference*.

Communier, D., Botez, R., & Wong, T. (2019a). Experimental validation of a new morphing trailing edge system using Price–Païdoussis wind tunnel tests. *Chinese Journal of Aeronautics, 32*, 1353–1366.

Communier, D., Le Besnerais, F., Botez, R. M., & Wong, T. (2019b). Design, manufacturing, and testing of a new concept for a morphing leading edge using a subsonic blow down wind tunnel. *Biomimetics, 4*, 76.

Communier, D., Botez, R. M., & Wong, T. (2020). Design and validation of a new morphing camber system by testing in the price—Païdoussis subsonic wind tunnel. *Aerospace, 7*, 23.

Dancila, B., Botez, R., & Labour, D. (2012). Altitude optimization algorithm for cruise, constant speed and level flight segments. In *AIAA guidance, navigation, and control conference* (p. 4772).

Dimino, I., Andreutti, G., Moens, F., Fonte, F., & Pecora, R. (2021). Integrated design of a morphing winglet for active load control and alleviation of turboprop regional aircraft. *Applied Sciences, 11*, 2439.

Eguea, J. P., Catalano, F. M., Abdalla, A. M., De Santana, L. D., Venner, C. H., & Fontes Silva, A. L. (2018). Study on a camber adaptive winglet. In *2018 applied aerodynamics conference* (p. 3960).

Félix Patrón, R. S., Berrou, Y., & Botez, R. (2014). Climb, cruise and descent 3D trajectory optimization algorithm for a flight management system. In *AIAA/3AF aircraft noise and emissions reduction symposium* (p. 3018).

Gamboa, P., Vale, J., Lau, F., & Suleman, A. (2009). Optimization of a morphing wing based on coupled aerodynamic and structural constraints. *AIAA Journal, 47*, 2087–2104.

Hamy, A., Murrieta-Mendoza, A., & Botez, R. (2016). *Flight trajectory optimization to reduce fuel burn and polluting emissions using a performance database and ant colony optimization algorithm*.

Hassig, A., Brossard, J., & Botez, R. (2013). Calibration Issues in the Subsonic Pice–Païdoussis Wind Tunnel. In *Canadian Aeronautical Society Institute CASI AÉRO 2013 conference, 60th aeronautics conference and AGM*, April 30th–May 2nd.

ICAO. (2019). *Aviation's contribution to climate change* (Environmental report). International Civil Aviation Organisation.

Kammegne, M. J. T., Grigorie, T. L., Botez, R. M., & Koreanschi, A. (2014). Design and validation of a position controller in the Price—Païdoussis wind tunnel. In *IASTED modeling, simulation and control conference* (pp. 17–19).

Khan, S., Grigorie, T., Botez, R., Mamou, M., & Mébarki, Y. (2020). Novel morphing wing actuator control-based Particle Swarm Optimisation. *The Aeronautical Journal, 124*, 55–75.

Koreanschi, A., Sugar-Gabor, O., & Botez, R. M. (2016). Numerical and experimental validation of a morphed wing geometry using Price—Païdoussis wind-tunnel testing. *The Aeronautical Journal, 120*, 757–795.

Mosbah, A. B., Salinas, M. F., Botez, R., & Dao, T.-M. (2013). New methodology for wind tunnel calibration using neural networks-EGD approach. *SAE International Journal of Aerospace, 6*, 761–766.

Pecora, R. (2021). Morphing wing flaps for large civil aircraft: Evolution of a smart technology across the Clean Sky program. *Chinese Journal of Aeronautics, 34*, 13–28.

Segui, M., & Botez, R. M. (2018). Cessna citation X climb and cruise performance improvement using adaptive winglet. In *Advanced aircraft efficiency in a global air transport system.*

Segui, M., Bezin, S., & Botez, R. M. (2018). Cessna citation X performances improvement by an adaptive winglet during the cruise flight. *International Journal of Aerospace and Mechanical Engineering, 12*, 423–430.

Sugar-Gabor, O. (2015). *Validation of morphing wing methodologies on an unmanned aerial system and a wind tunnel technology demonstrator*. Ph. D. Thesis.

Sugar-Gabor, O., Koreanschi, A., Botez, R. M., Mamou, M. & Mebarki, Y. (2016). Analysis of the aerodynamic performance of a morphing wing-tip demonstrator using a novel nonlinear vortex lattice method. In *34th AIAA applied aerodynamics conference* (p. 4036).

Applications of Drones in the Field of Health and Future Perspectives

Kursat Alp Yigit, Alper Dalkiran, and T. Hikmet Karakoc

Nomenclature

UAV/UAS	Unmanned aerial vehicle/systems
OHCA	Out-of-hospital cardiac arrest
CPR	Cardio-pulmonary resuscitation
AED	Automated external defibrillator
EMS	Emergency medical services
GNSS	Global navigation satellite system

1 Introduction

Unmanned aerial systems/vehicles (UASs/UAVs) or drones are planes with neither a pilot nor passengers on board. Those planes only carry fit-for-purpose equipment such as a video camera, camera, global navigation satellite system (GNSS) receiver, laser scanner, or payload. UAVs can be used either remotely controlled or autonomously. Military, civil (hobby and commercial), and scientific uses of UAVs have rapidly increased worldwide. Thus, it is anticipated that this issue will be more

K. A. Yigit (✉)
Eskisehir Technical University, Eskisehir, Turkey

A. Dalkiran
School of Aviation, Suleyman Demirel University, Isparta, Turkey
e-mail: alperdalkiran@sdu.edu.tr

T. H. Karakoc
Eskisehir Technical University, Faculty of Aeronautics and Astronautics, Eskisehir, Turkiye

Information Technology Research and Application Center, Istanbul Ticaret University, Istanbul, Turkiye

T. H. Karakoc et al. (eds.), *New Achievements in Unmanned Systems*, Sustainable Aviation, https://doi.org/10.1007/978-3-031-29933-9_31

frequent in the world's plan shortly. High accuracy and the economy in time, cost, and personnel in professional uses of UAVs, such as surveying, can be considered primary reasons for its increasing demand.

There are many naming for UAVs. "Drones" or "UAV/UAS" can be found in the international literature, and they mean the same thing. There is not a distinct separation between those namings, except for a particular technical feature. When it comes to UAV, it should not be understood only as an airplane flying in the air. In this context, UAVs consist of three components:

- The aircraft itself.
- The payload on the aircraft.
- The ground control station.

In recent years, the requirement for UAVs in different areas of application, whether commercial, recreational, or public, has increased tenfold; currently, this demand is mainly consumed by military use (Townsend et al., 2020). Still, it is expected to shift to more recreational and public service exponentially.

The use of UAV is predicted to increase all over the world in the future. For example, the military UAV market volume will reach $26.8 billion in 2025 globally (Market, 2021a, b). The military and civilian UAV market in the world reached a total of $ 9.3 billion in 2019. This market is expected to increase by 16.4% annual compound growth rate and reach 58.4 billion dollars in 2026 (Market, 2021a, b). Therefore, it is clear that the use of UAVs will increase in all areas of the world in the next 5 years.

There are two types of UAVs: rotary-wing and fixed-wing. Rotary-wing UAVs consist of more than one rotor; those usually have four, six, or eight rotors. They do not need a runway or launcher as they can land and take off vertically. They stay in the air and are very agile, allowing them to maneuver more precisely. In addition, the disadvantages are that they carry smaller loads and have a limited working time due to the short battery life. Fixed-wing UAVs have a fixed-wing on the fuselage and are similar to an airplane with this structure. Fixed-wing UAVs have a more straight-forward structure than rotary-wing UAVs. Fixed-wing UAVs are easy to maintain and have low operating costs and longer flight times. Fixed-wing UAVs can carry larger loads over longer distances by reducing power consumption through their wings. Fixed-wing UAVs needing a runway or launching device for take-off and landing can be counted as a disadvantage.

2 Commercial Applications of Drone Technology

Severe developments in information technologies, which started in the 1980s, made it necessary for companies to reorganize their activities and brought radical changes in the modern economy. Today drone technology affects all sectors from infrastructure to agriculture on a large scale, creating effects similar to severe developments in information technologies (Mazur et al., 2016) In a short time, consumers will begin to see the effects of drones in all areas of business.

The fact that drones can travel their flight distances quickly and concisely has made their use very attractive in sectors requiring mobile and high data quality. The use of drones in daily activities in large-capital projects such as infrastructure and agriculture provides excellent benefits. The potential for drones to advance in the insurance and mining industries will drastically change the delivery process in the transportation industry. The increase in technological opportunities, legal regulations, and developments in investment support has enabled drones to be used in many new areas. The use of drone technology in these areas has led to new business and operation models. In some sectors, low cost and high quality, while in some sectors, delivery time and load capacity are decisive, enabling drones to be used according to these needs.

- The infrastructure industry uses drones maintenance and investment monitoring to reduce costs and accelerate the process.
- The agricultural industry uses drones for crop supervision, solid and field analyses, and health assessment. They prefer this device for an increase in yields and productivity.
- The mining industry uses drones efficiently for inspections. They prefer these devices to replace human interaction on dangerous tasks.
- The security industry uses drones for researching large and complicated areas very quickly. Therefore, drones are preferred in security applications day by day.
- The transport industry uses drones for short distances and light payloads. Because of the technological barriers, it has been minimal in recent years.
- Medical logistics use drones as two necessary facilities. One of them is flying defibrillators, and the second is drug or medical transportation.
- The telecommunication industry uses drones in various areas. Drones provide companies with reduced costs and higher speed advantages.
- The media and entertainment industry uses drones for aerial photography and filming. Drones use quality film and photos and lower costs compared to planes or helicopters.
- The insurance industry uses drones in three areas: risk monitoring, risk assessment, and claims management. They prefer this device for monitoring natural disasters.

3 Applications of Drones in the Field of Health

The primary use of drones, used in many areas in recent years, has been military activities. The inevitability of the integration of technological developments in the field of medicine has been realized by organizations. In recent years, the increase in civil and military drones (photography, racing, security, transportation, delivery) has led to drones in the healthcare system. Drone use in the medical field seems to be encountered anywhere in the healthcare system.

The most important sector where drones are being used is prehospital medical systems. The UAV-assisted health support area has emerged to provide audio and visual assistance to "bystanders" who can help the patient. Bystanders can efficiently perform "cardiopulmonary resuscitation" (CPR) maneuvers after delivering medical supplies to the scene. Although the healthcare industry will be excited to see drones flying in emergencies soon, it seems like the usage areas in the hospital are limited. However, the technological progress is that it can be used from the transport of the examinations taken from the patients to the transport of blood products.

The medical use of drones in the literature can be summarized as below.

- Facilitating AED access to the scene.
- Drug/Tissue, Blood products transportation.
- Trauma kit medical supplies transportation.
- Telemedicine.

One of the most significant gains from such widespread investigation of drones in prehospital systems is that cases will get the necessary medical attention at the scene more quickly. Factors such as faster defibrillation to shockable rhythms and faster control to catastrophic hemorrhages will impact all mortality rates we know when drones become widespread.

3.1 AED Access to Cases

The defibrillator device is essential for recovering the heart rhythm in the prehospital event. The UAVs can provide 32% faster access rates to deliver defibrillator devices to patients in urban areas (Claesson et al., 2016). This figure becomes 93% in rural areas where helicopters and ambulances cannot reach most cases.

3.2 Drug/Tissue, Blood Products Transportation

Studies have shown that drone use can be strikingly detailed on organ transfers. Transferred tissue can be affected by vibration during flight. Also, other physical effects like pressure and temperature may affect the transported tissue too. Transportation must be planned in detail overcome mentioned problems. Tissue transfers to complex regions can be effective compared to traditional logistic methods.

3.3 Telemedicine/Telesurgery

Beyond the initiation of resuscitation with the help of cameras and audio and the use of AEDs to people who do not have health education next to patients who need CPR, surgeries that can be performed via drones are no longer in the fantasy world.

The use of drones for medical purposes can deliver AED to the scene in developed or urban areas. Also, cameras and audio devices in drones made it possible to give live assistance to the bystanders. Bystanders can intervene with the patient who does not have health training at the scene. Furthermore, drugs, trauma kits can be delivered in rural areas too. It is used to facilitate equipment transportation or search and rescue efforts. Of course, these usage differences require the diversity of drones (flying distance/speed/carrying capacity). For this reason, we see dozens of different sizes and features of drones.

It is exciting to see studies increasing, as integrating technology into health always provides an exciting side and directs the studies on this subject. One of the best examples is that telesurgery studies that can be done remotely with UAVs and surgical robots appear in the literature.

The concept of using drones for emergency medical deliveries may prove valuable in several life-critical situations. There are examples of the transportation of low-weight autoinjectors carrying epinephrine for anaphylactic shock, nasal dispensers with naloxone for opioid overdoses, and single-use glucagon for hypoglycemia. However, one-use case, in particular, has gained worldwide interest in the past years; the possibility of transporting AED to the site of OHCA. OHCA affects some 275,000 individuals in Europe each year, and 30-day survival rates are generally low: 10% (Atwood et al., 2005). Despite tens of thousands of AEDs sold, the survival rate has not changed significantly: still, 90% of all people suffering from OHCA die. It has been proven that CPR and defibrillation are essential factors in improving survival rates. For each minute that passes from the time of collapse without treatment, the chance of survival decreases by 7–10%. Research shows that if CPR and early defibrillation are initiated within the first minutes, up to 50–70% of all patients may survive. The circumstances described above inevitably conclude that traditional EMS cannot reach this group of patients quickly enough, especially not in residential homes or suburban and rural areas. The method of dispatching AED-equipped drones to decrease the time from collapse to the first shock with a defibrillator has excellent potential. Research with theoretical models conducted by the Karolinska Institutet has shown that drones could have a slower response time than traditional EMS in 93% of rural OHCA cases with a mean time saving of 19 min. These models have further supported real-world test flights where drones have been dispatched in non-emergency situations to historical OHCA locations.

4 Medical Use Case Study

In recent years, studies have started in Europe regarding the use of drones to give AEDs to people who have had a heart attack. These studies have found that drones are faster than ambulances at delivering AEDs to people who have had a heart attack. Minutes or even seconds are vital for people who have had a heart attack. The survival rate for the patients in whom an AED has intervened increases dramatically. Drones with AED can be an excellent solution to provide the necessary devices for those patients.

In one of the studies, three drones, each with a range of 5 km, are placed in different regions around the settlement area. After receiving the emergency call, the drone and the ambulance are directed from the medical center to the incident area. The drone pilot remotely commands the drone and gets approval for the flight operations by the nearest air traffic control tower located in the same area. After approval, the drone takes off to reach the scene.

When the automated drone system arrives at the scene under surveillance by the drone pilot, the AED is slowly lowered from the drone at an altitude of 30 m. Bystanders received the AED at the scene and administered it from the emergency center to intervene in the patient who had a heart attack. In some cases caused by weather conditions (excessive rain, storm), it has been determined that drones cannot be used in areas with high buildings and no flight opportunity.

4.1 Drone Ambulance Project

People can face severe health problems at any time in their lives. The most important concept is to race against time to bring back to life a person who has had a heart attack. About 800,000 people in Europe have a heart attack every year, and unfortunately, only 8% survive (Khan et al., 2020). With the ambulance drone project, the survival rate can be increased up to 90% with interventions made for people who have had a heart attack. It is vital to use drone ambulances in order to minimize the loss caused by a heart attack.

The ambulance drone used to resuscitate people who had a heart attack includes one AED. The ambulance drone moves at a speed of 100 km, has a 10-min battery life, six propellers, and is painted yellow for easy identification. The drone in the ambulance has AED 50 shock delivery feature. A drone ambulance is integrated with GPS to locate the incoming emergency call.

When the emergency call comes to the command center, the drone ambulance is directed to where the call was made. The webcam on the ambulance drone is the most important assistant of the health personnel in the call center. Thanks to this camera, instructions can be given to the personnel who will intervene in the person who has a heart attack. The AED is removed from the hatch with the instructions given from the emergency center using the webcam. The responder applies the AED according to the instructions from the webcam.

5 Conclusions

Today the use of UAVs in both military and civilian areas has increased significantly. The health sector has started to use drones due to their sensitivity and the necessity to react quickly. The essential healthcare area where drones are used is prehospital medical systems. The medical use purpose can be summarized as follows:

- AED transportation to the scene, medicine, tissue, blood products transportation, trauma kit.
- Other materials transportation.
- Guiding for applications such as telemedicine.

The use of drones in the intervention of a heart attack, which is one of the most common conditions in the field of health, significantly increases the survival rate of the patient. AED transport with drones and the use of ambulance drones in the healthcare field has been a beacon of hope for patients who have had a heart attack. In the future, improvements in the technical features of drones (speed, battery life, enlargement of size) will make it inevitable to use them in many different applications in the healthcare field.

Drones play an essential role in non-hospital interventions in the health sector due to their technical features. This study showed how quickly drones were intervened in transporting AEDs to patients who had a heart attack. It is known how vital even seconds are in survival for a person who has had a heart attack. Drones provide the opportunity to intervene quickly with the AED they carry. This study will make it possible for us to see drones carrying blood tubes in the emergency departments of hospitals in a short time. Depending on the developments in drone technology, it will be possible to transport the patients who met with an accident to the emergency services of hospitals with drones after the first intervention.

References

Atwood, C., Eisenberg, M. S., Herlitz, J., & Rea, T. D. (2005). Incidence of EMS-treated out-of-hospital cardiac arrest in Europe. *Resuscitation, 67*(1), 75–80. https://doi.org/10.1016/j.resuscitation.2005.03.021

Claesson, A., Fredman, D., Svensson, L., Ringh, M., Hollenberg, J., Nordberg, P., Rosenqvist, M., Djarv, T., Österberg, S., Lennartsson, J., & Ban, Y. (2016). Unmanned aerial vehicles (drones) in out-of-hospital-cardiac-arrest. *Scandinavian Journal of Trauma, Resuscitation and Emergency Medicine, 24*(1), 1–9. https://doi.org/10.1186/s13049-016-0313-5

Khan, Z. H., Siddique, A., & Lee, C. W. (2020). Robotics utilization for healthcare digitization in global COVID-19 management. *International Journal of Environmental Research and Public Health, 17*(11), 3819. https://doi.org/10.3390/ijerph17113819

Market, M. (2021a). *Military drones market size, growth, trend and forecast to 2025 | MarketsandMarkets*. [online] Marketsandmarkets.com. Available at: https://www.marketsandmarkets.com/Market-Reports/military-drone-market-221577711.html. Accessed 11 Oct 2021.

Market, U. (2021b). *Unmanned Aerial Vehicle (UAV) market worth $58.4 billion by 2026*. [online] Marketsandmarkets.com. Available at: https://www.marketsandmarkets.com/PressReleases/unmanned-aerial-vehicles-uav.asp. Accessed 11 Oct 2021.

Mazur, M., Wisniewski, A., & McMillan, J. (2016). *PwC global report on the commercial applications of drone technology*. PricewaterhouseCoopers technical report.

Townsend, A., Jiya, I. N., Martinson, C., Bessarabov, D., & Gouws, R. (2020). A comprehensive review of energy sources for unmanned aerial vehicles, their shortfalls and opportunities for improvements. *Heliyon, 6*(11), e05285.

Comparison of 5th-Generation Fighters: Evaluation of Trends in Military Aviation

Murat Ayar and T. Hikmet Karakoc

Nomenclature

AHP	Analytic hierarchy process
MCDM	Multi-criteria decision-making

1 Introduction

Throughout their history, countries have considered protecting their sea, air, and land border integrity as their first aim in order to ensure their sovereignty. For this reason, while aiming to increase the strength of their armies beyond the desired level, they are also looking for ways to prefer national production of the equipment and systems to be used. Since the developments in the defense industry are not only related to the integrity of the country but also the technological developments in the defense industry are horizontally adaptable to the whole field, the defense industry is in a guiding position in technological developments.

Air forces have an important place in the defense and Attack plans of the countries. By its nature, the air force is seen as the only force that can cope with

M. Ayar (✉)
Department of Airframe and Powerplant Maintenance, Eskisehir Technical University, Eskisehir, Turkey
e-mail: muratayar@eskisehir.edu.tr

T. H. Karakoc
Eskisehir Technical University, Faculty of Aeronautics and Astronautics, Eskisehir, Turkiye

Information Technology Research and Application Center, Istanbul Ticaret University, Istanbul, Turkiye

T. H. Karakoc et al. (eds.), *New Achievements in Unmanned Systems*, Sustainable Aviation, https://doi.org/10.1007/978-3-031-29933-9_32

all other forces. The power of the air force is basically based on fighters. Even though the use of airplanes in wars dates back to the First World War, it is only in the 1940s that the 1st-generation fighters were up in the skies. Developed from the mid-1940s to the mid-1950s, these aircraft differed little in appearance from the piston models produced in the same years, but flew at subsonic speeds. The first jet-powered aircraft was the Messerschmitt Me 262, developed by the Germans during the Second World War and used in the last year of the war (Radinger & Schick, 1993).

In the 1960s, with the use of aluminum alloy in aircraft, delta wings and swept-back wings and afterburners, aircraft classified as 2nd-generation fighters began to be used (Johnson, 2004). In the 3rd-Generation Fighters, the maneuvering capabilities inherited from the previous generation have been increased and analog display systems have been adopted. In the 1990s, aircraft were developed to be used in multiple roles and also began to be equipped with very complex avionics systems and weapons. Thanks to the use of "fly-by-wire" method flight control systems in aircraft such as the F-16, which are designed to be somewhat unstable, a very effective maneuverability has been achieved (Collinson, 2011). Analog avionic systems have begun to be replaced by digital avionics systems. With the development of digital engine control systems, the performance of engines has increased considerably. When it comes to the 5th-generation fighters, this period has started with the F-22 aircraft entering service. The most basic feature of the 5th-generation aircraft is that they are designed to operate in a network-centered combat environment. Thanks to sensors that provide situational awareness, targets around the aircraft can be tracked continuously. In addition, with the opportunity to obtain and transmit effective data thanks to advanced avionics systems, situational awareness increases and therefore the workload of the pilot is reduced (Bongers & Torres, 2014). In this study, fighters were evaluated over the characteristics used in the definition of the 5th-generation fighter. By using the characteristics for evaluation, a selection was made among the fighters that are currently in production. Analytic hierarchy process (AHP) method, one of the multi-criteria decision-making methods (MCDM), was used for selection.

Multi-criteria decision-making is widely used in the literature in the fields of production, marketing, planning, human resources, facility location, group decision making, risk analysis, application evaluation, technology selection, capital investment, accounting and finance. Multi-criteria decision-making examines problems in the structure where the goal-means relationships are clearly stated, a certain number of alternatives can be clearly determined at the beginning, and the preference information initially obtained from the decision maker can be used to reach the results (Ho, 2019).

Considering the main criteria of mechanical performance, flight safety, control capability, avionics systems, weapon systems, and economy in the selection of multi-role fighter aircraft, 21 sub-criteria were evaluated with AHP and TOPSIS. Alternatives were compared with the sensitivity analysis, and as a result, the same ranking was obtained in the methods used in the selection of the optimum fighter aircraft. This situation has given the decision maker a more positive picture in terms of reliability in choosing the optimum multi-purpose fighter aircraft (Celikyay,

2002). In the study conducted in Taiwan, 16 different criteria, mostly technical, were evaluated for eight different alternative trainer aircraft. Fuzzy Set Theory was also used in this study (Wang & Chang, 2007). In the study conducted for the Spanish Air Force, a comprehensive study was conducted by evaluating 12 different criteria for five different alternatives. In the article, in which TOPSIS, AHP and Fuzzy Logic were used together, the scope of the criteria was kept wide and all requests were answered (Sánchez-Lozano et al., 2015). In their study in 2015, Dožić and Kalić selected passenger aircraft type for a hypothetical regional airline. AHP and Even Swaps Method were used as methods in the study (Dožić & Kalić, 2015). It has been observed that multi-criteria decision-making methods are usedz in aircraft selection problems as in different selection studies.

The fact that the AHP method allows qualitative decision making, especially in areas that require confidentiality such as military areas, has enabled it to be used in this decision problem. In addition, the decision to become more objective by blending the decisions taken from different decision makers is ensured by using this group decision-making method.

Analytic hierarchy process is a MCDM technique developed by Saaty in 1977 (Harker & Vargas, 1987). In the AHP method, it is a powerful and easy-to-understand mathematical method that evaluates all criteria together, taking into account the antecedents of individuals or groups. In this approach, the evaluations of decision-making experts or groups are taken as a basis. After determining the criteria affecting the decision, AHP finds the effects of these criteria on the decision by making pairwise comparisons. As a result, the qualitative and quantitative evaluations are transformed into the same format in a certain order and the decision problem is solved. In other words, in the AHP method, the knowledge, experience, and intuitions of the decision maker are combined mathematically (Saaty, 2013).

2 Method

In order to make a comparison of fighters, first, the essence of the study was brought into a decision problem structure. The decision problem was identified as "Which 5th Generation Fighters most innovative." In the next stage, alternatives and criteria

Decision Problem	*Which 5th Generation Fighters most innovative*

Criteria	Agility	Stealth	Battlespace Network	Airframe	Avionics	Sensor Fusion	Supercruise

Alternatives	Lockheed Martin F-22	Lockheed Martin F-35	Chengdu J-20	Sukhoi Su-5

Fig. 1 Structure of decision problem

were determined. Alternatives have been selected among manufactured aircraft that meet the definition of 5th-generation fighter. The criteria are compiled from the characteristic features that make up the definition of the 5th-generation fighter (Lockheed Martin, 2012; Bongers & Torres, 2014). By selecting all the elements of the decision problem, shown in Fig. 1.

After the structure of the decision problem was completed, pairwise comparisons from the stages of the method were made in consultation with the decision makers. At this stage, the criteria are compared with each other, the most important or effective is determined, and the decision matrix is started to be formed. After all pairwise comparisons are completed, operations are performed on the matrix obtained, and the matrix in which the results will be read is obtained.

3 Results and Discussion

As a result of solving the decision problem, we got a matrix. The weights of the alternatives and criteria were determined over this matrix. First, the weights of the criteria were taken from the matrix, which shows how much the criteria affect the decision problem. Figure 2 shows the ranking of the characteristics of the 5th-generation fighters in order of importance.

When the weights of the criteria are examined, it is seen that the stealth and agility features are the most dominant, taking the values of 20.61 and 19.69, respectively. At the same time, it is seen that features such as supersonic and airframes have the lowest effect on the decision.

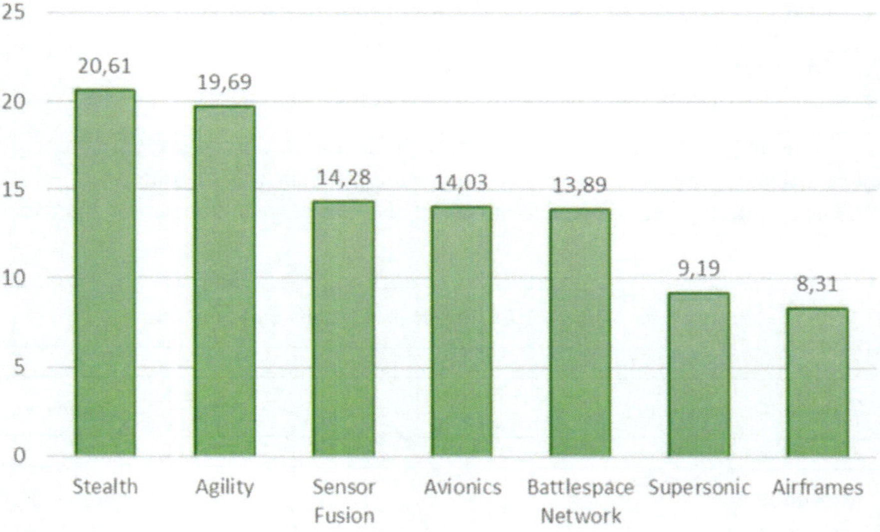

Fig. 2 Weights of the criteria

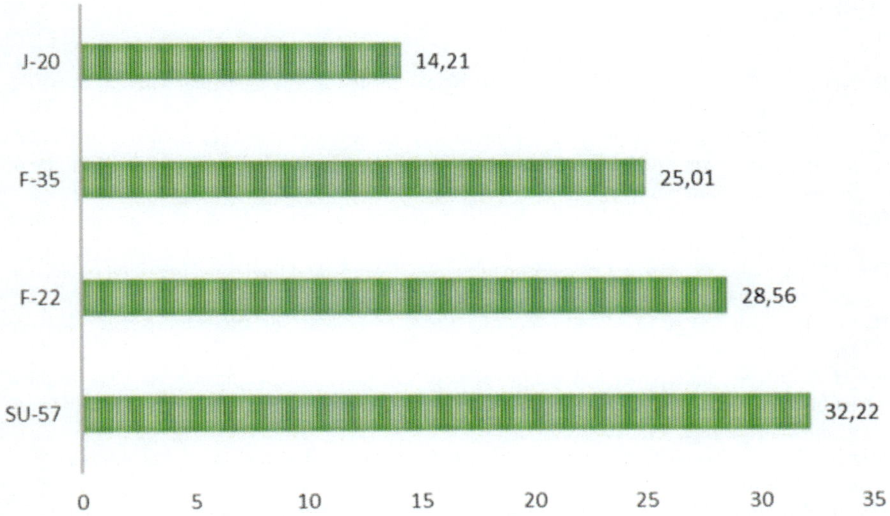

Fig. 3 Weights of the alternatives

When the comparison of the alternatives is examined, it is seen that the SU-57 is the best alternative with a value of 32.22. It is followed by the F-22 with a value of 28.56, the F-35 with a value of 25.01, and finally the J-20 fighter with a value of 14.21 (Fig. 3).

When the results are evaluated as a whole, the Sukhoi SU-57 fighter is seen as the best alternative in terms of its 5th-generation features. Although there are claims that stealth features are in the background among other alternatives, it has been seen that it stands out with its other features. It is obvious that the success of the SU-57 in agility, sensor fusion, and avionics criteria contributed to this result.

4 Conclusion

Today, competition emerges in fighters as it manifests itself in every field. Countries take part in this race as producers or users. In this study, the most recent 5th-generation fighters were examined and the best alternative and the weights of the criteria affecting this selection were found. It is thought that this study will help countries in a strategic decision problem such as fighter jet selection. It is also expected to guide aircraft manufacturers about the direction of the industry.

References

Bongers, A., & Torres, J. L. (2014). Technological change in US jet fighter aircraft. *Research Policy, 43*(9), 1570–1581.

Celikyay, S. (2002). *Application of multi-criteria decision-making methods in the selection of multi-purpose fighter aircraft* (Master Thesis), pp. 95–98. Istanbul Technical University, Institute of Science and Technology.

Collinson, R. P. G. (2011). Fly-by-wire flight control. In *Introduction to avionics systems* (pp. 179–253). Springer.

Dožić, S., & Kalić, M. (2015). Comparison of two MCDM methodologies in aircraft type selection problem. *Transportation Research Procedia, 10*, 910–919.

Harker, P. T., & Vargas, L. G. (1987). The theory of ratio scale estimation: Saaty's analytic hierarchy process. *Management Science, 33*(11), 1383–1403.

Ho, H. P. (2019). The supplier selection problem of a manufacturing company using the weighted multi-choice goal programming and MINMAX multi-choice goal programming. *Applied Mathematical Modelling, 75*, 819–836.

Johnson, E. J. (2004). *MiG! 6 o'clock high!: A history of the design bureau and an analysis of its aircrafts combat history*.

Lockheed Martin. (2012). *F-35 defining the future*. Lockheed Martin.

Radinger, W., & Schick, W. (1993). *Messerschmitt Me 262: Development, testing, production*. Schiffer Military/Aviation History.

Saaty, T. L. (2013). The modern science of multicriteria decision making and its practical applications: The AHP/ANP approach. *Operations Research, 61*(5), 1101–1118.

Sánchez-Lozano, J. M., Serna, J., & Dolón-Payán, A. (2015). Evaluating military training aircrafts through the combination of multi-criteria decision making processes with fuzzy logic. A case study in the Spanish Air Force Academy. *Aerospace Science and Technology, 42*, 58–65.

Wang, T. C., & Chang, T. H. (2007). Application of TOPSIS in evaluating initial training aircraft under a fuzzy environment. *Expert Systems with Applications, 33*(4), 870–880.

Index

T. H. Karakoc et al. (eds.), *New Achievements in Unmanned Systems*, Sustainable Aviation, https://doi.org/10.1007/978-3-031-29933-9